Niche Construction

MONOGRAPHS IN POPULATION BIOLOGY

EDITED BY SIMON A. LEVIN AND HENRY S. HORN

A complete series list follows the index.

Niche Construction

The Neglected Process in Evolution

F. JOHN ODLING-SMEE, KEVIN N. LALAND,
AND MARCUS W. FELDMAN

PRINCETON UNIVERSITY PRESS
Princeton and Oxford

Published by Princeton University Press, 41 William Street,
Princeton, New Jersey 08540
In the United Kingdom: Princeton University Press, 3 Market Place,
Woodstock, Oxfordshire OX20 1SY

Library of Congress Cataloging-in-Publication Data
Odling-Smee, F. John, 1935–
Niche construction : the neglected process in evolution / F. John Odling-Smee,
Kevin N. Laland, and Marcus W. Feldman.
p. cm. — (Monographs in population biology ; no. 37)
Includes bibliographical references (p.).
ISBN 0-691-04438-4 (alk. paper) — ISBN 0-691-04437-6 (pbk. : alk. paper)
1. Niche (Ecology) 2. Evolution (Biology) 3. Human ecology.
I. Laland, Kevin N. II. Feldman, Marcus W. III. Title. IV. Monographs in
population biology ; 37.

QH546.3 .O35 2003
577.8′8—dc21 2002031747

British Library Cataloging-in-Publication Data is available

This book has been composed in Times Roman

Printed on acid-free paper. ∞

www.pupress.princeton.edu

Printed in the United States of America

10 9 8 7 6 5 4 3 2 1

Contents

Figures

Tables

Preface

The seemingly innocent observation that the activities of organisms bring about changes in environments is so obvious, and has been familiar to science for so long, that it seems an unlikely focus for a new line of thinking about evolution. Yet *niche construction*, as this process of organism-driven environmental modification is now known, has hidden complexities. When we first began to explore the role of niche construction in evolution, in one case over twenty years ago, we had little idea of the scale of the problem. It rapidly became apparent that a serious investigation would take us into population and ecosystem ecology and lead us to reflect on the relationships between these disciplines. As our investigations proceeded, it appeared the ramifications of niche construction grew to frightening proportions. We now believe that understanding niche construction is a fundamental problem for science and the areas touched on in this book reflect its wide scope. Although we could have limited its sweep to ecology and evolution, we decided also to include a consideration of the consequences of a niche construction perspective for the human sciences, both because human niche construction is particularly powerful, and because all three of us have an interest in the relationship between human biology and cultural processes, especially in the context of evolution. We also believe that a focus on the niche construction of our own species has helped us to an understanding of niche construction in general.

We recognize that in a book of this breadth we cannot be experts on all the material covered. While we have done our utmost to produce a scholarly work, we accept that we will not have covered each topic as well as a specialist might. Nonetheless, we hope that we have done enough to bring into focus the problems that the process of niche construction introduces into evolutionary biology.

We would like to thank the following people for commenting on one or more chapters and for discussing the material in the book: Lauren Ancel, Robert Aunger, Culum Brown, Tom Burns, Gillian Brown, Claude Combes, John Crook, Rachel Day, Paul Ehrlich, Peter Godfrey-Smith, Barry Hewlett, Robert Hinde, Aaron Hirsh, Robert Holt, David Hull, Clive Jones, Jeremy Kendal, Ben Kerr, Adam Kuper, John Lawton, Robert Levin, Richard Lewontin, Lisa Lloyd, Lucy Odling-Smee, Eugene Odum, Sally Otto, Bernie Patten, Henry Plotkin, Vernon Reynolds, Sean Rice, Pete Richerson, Kate Robson Brown, Monty Slatkin, Stan Ulijaszek, and Steve Weinstein. We are particularly indebted to Robert Holt, David Hull, Ben Kerr, and an anonymous referee, who read the entire book and who made many extremely helpful suggestions that substantially improved the manuscript. We are also extremely grateful to Jean Doble and Larry Bond for their tireless efforts to solve technical and administrative problems, Anne Odling-Smee for producing the figures, and Sam Elworthy and Sarah Harrington of Princeton University Press for their assistance. The Leverhulme Trust, the Royal Society, and NIH provided financial support. Finally, we owe a particular debt to Richard Lewontin, for his enthusiastic support and encouragement, and for starting us off on this journey. This book is dedicated to our wives, Ros, Gillian, and Shirley.

JOS, KNL, MWF
April 2002

Niche Construction

CHAPTER 1

Introduction

To see what is in front of one's nose requires
a constant struggle.
(George Orwell)

Organisms play two roles in evolution. The first consists of carrying genes; organisms survive and reproduce according to chance and natural selection pressures in their environments. This role is the basis for most evolutionary theory, it has been subject to intense qualitative and quantitative investigation, and it is reasonably well understood. However, organisms also interact with environments, take energy and resources from environments, make micro- and macrohabitat choices with respect to environments, construct artifacts, emit detritus and die in environments, and by doing all these things, modify at least some of the natural selection pressures present in their own, and in each other's, local environments. This second role for phenotypes in evolution is not well described or well understood by evolutionary biologists and has not been subject to a great deal of investigation. We call it "niche construction" (Odling-Smee 1988) and it is the subject of this book.

All living creatures, through their metabolism, their activities, and their choices, partly create and partly destroy their own niches, on scales ranging from the extremely local to the global. Organisms choose habitats and resources, construct aspects of their environments such as nests, holes, burrows, webs, pupal cases, and a chemical milieu, and frequently choose, protect, and provision nursery environments for their offspring. Niche construction is a strongly intuitive concept. It is far more obvious than natural selection because it is far easier to observe individual organisms *doing* niche construction than to observe them *being* affected by natural selection. It is self-evident that all organisms must interact with their environments to stay alive, and equally obvious that, when they do, it is not just organisms that are likely to be affected by the consequences of these interactions, but also environments. That organisms actively contribute toward both the "construction" and "destruction" of their own and each other's niches is scarcely news. So why write a book about it?

The answer is that, when subject to close scrutiny, it becomes clear that niche construction has a number of important, but hitherto neglected implications for evolutionary biology and related disciplines. In fact, in this book we go so far as to argue that niche construction changes our conception of the evolutionary process. Niche construction should be regarded, after natural selection, as a second major participant in evolution. Rather than acting as an "enforcer" of natural selection through the standard physically static elements of, for example, temperature, humidity, or salinity, because of the actions of organisms, the environment will be viewed here as changing and coevolving with the organisms on which it acts selectively.

Using a combination of empirical data, comparative argument, and mathematical modeling we will try to convince the reader of the merits of this new way of thinking about evolution. We will illustrate how niche construction can change the direction, rate, and dynamics of the evolutionary process. Niche construction is a potent evolutionary agent because it introduces feedback into the evolutionary dynamic. Niche construction by organisms significantly modifies the selection pressures acting on them, on their descendants, and on unrelated populations. The later chapters of this book describe how niche construction can be incorporated into empirical and theoretical evolutionary analyses, and how it can be used to generate hypotheses. We will present methods for testing these hypotheses and point to the broad areas of biology and the social sciences to which they are applicable. Our hope is that the niche-construction perspective will prove fruitful by leading to the development of testable new theories and facilitating greater understanding of the evolutionary process.

In this first chapter we introduce the concept of niche construction, and spell out its major consequences with illustrative examples from natural history. We describe four major ramifications of niche construction. Niche construction may (1) in part, control the flow of

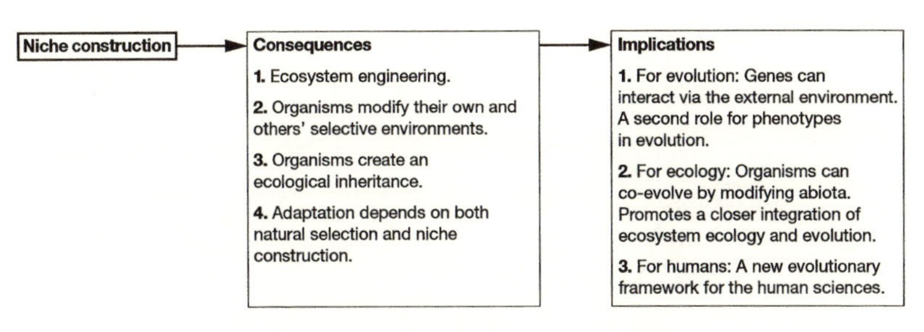

FIGURE 1.1. The consequences of niche construction and their implications for evolutionary theory, ecology, and the human sciences.

energy and matter through ecosystems (ecosystem engineering), (2) transform selective environments to generate a form of feedback that may have important evolutionary consequences, (3) create an ecological inheritance of modified selection pressures for descendant populations, and, finally (4) provide a second process capable of contributing to the dynamic adaptive match between organisms and environments (see fig. 1.1). We then consider some of the implications of these consequences for three different bodies of biological theory, namely, evolutionary theory itself, the relationship between evolutionary theory and ecosystem ecology, and the relationship between evolutionary theory and the human sciences.

1.1 THE CONSEQUENCES OF NICHE CONSTRUCTION

1.1.1 Ecosystem Engineering

We begin with an example of a potent niche constructor, the genus of leaf-cutter ants, *Atta*, as described by the myrmecologists Bert Hölldobler and Edward Wilson (1994). At present, 15 species of leaf-cutter ants are known to science. All of them live in the New World across a geographical range that stretches from the southern states of the United States of America to the south of Argentina. The most salient niche-constructing activity of this genus is "agriculture." Leaf-cutter ants grow fungi on substrates of fresh vegetation that they initially cut and collect from outside their nests and then carry into their nests to form the basis of fungal gardens (fig. 1.2). The fungal

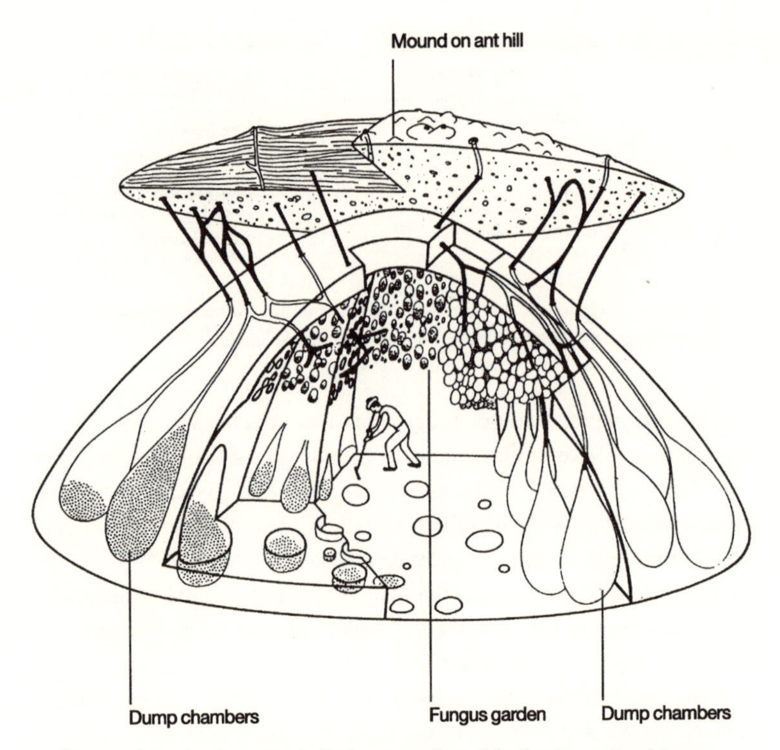

FIGURE 1.2. A giant nest built by a species of leaf-cutter ants, *Atta sexdens*, consisting of about a thousand chambers. In this particular example the nest is so huge that the loose soil brought out and piled on the ground by the ants while making it occupied more than 22 m^3 and weighed approximately 44 tons. The human figure shown inside the nest indicates the scale. (Reproduced from Hölldobler and Wilson 1994. From an illustration by J.C.M. Jonkman in *Insect-Fungus Symbiosis: Mutualism and Commensalism*, edited by L. A. Batra [Montclair, NJ: Allanheld and Osman, 1979]. Modified by N. A. Weber for Hölldobler and Wilson [1995], p. 116.)

crop that the ants grow consists of a fluffy white mold, resembling bread mold, made up of masses of thread-shaped hyphae. The ants' agriculture is so efficient that it not only provides them with an abundant supply of food, but enables individual colonies to reach staggeringly large sizes, with a single colony containing millions of workers. In one extreme case described by Hölldobler and Wilson, a nest of the species *Atta sexdens* consisted of about a thousand chambers, with the chambers varying in size from that of a closed fist to that of a soccer ball. Three hundred and ninety of its chambers were still in use when it was discovered, and they were filled with both fungal gardens and ants. This particular nest was so huge that the loose soil that had been brought out and piled on the ground by the ants in the course of making their nest occupied over 22 cubic meters and weighed approximately 40,000 kilograms, or 44 tons. Such an example makes it clear that the collective leaf-cutting activities of such large colonies of ants can have enormous impacts on the ants' surrounding environment.

Given such a prodigious capacity for niche construction, it is not surprising that several species of leaf-cutter ants, including *Atta cephalotes* and *Atta sexdens*, turn out to be among the worst pests of Central and South America. They destroy billions of dollars worth of agriculturally valuable crops each year. For instance, in Brazil, leaf-cutter ants are especially destructive in eucalyptus and citrus plantations. What is, perhaps, more surprising is that the same ants produce beneficial effects in ecosystems. For example, the ants turn over and aerate large quantities of soil in forests and grasslands, and they also circulate nutrients that are essential to the lives of many other species of organisms with whom they share their ecosystems. Moreover, it has recently been discovered that leaf-cutter ants can help the recovery of rainforests in areas where the primary forest has been destroyed by human farmers and loggers. Here, the ants' activities benefit newly established plants because the soil from their nests is much easier than the surrounding soil for young plant roots to penetrate. Also, the decomposition of the plant material that the ants store in their nests increases the soil's pH, thereby increasing its capacity to retain its nutrients, preventing them from being washed away out of reach of the plants.

Leaf-cutter ants are a good illustration of the first major consequence of niche construction. The activities of organisms can result in significant, consistent, and directed changes in their local environments. Simply by choosing or perturbing their habitats, for example, by repeatedly consuming the same resource, or repeatedly emitting the same detritus, organisms can substantially modify their worlds, and do so in a nonrandom or predictable manner. As a consequence, niche-constructing organisms frequently modify the environments of other organisms too, including organisms in other species. They also affect some of the properties of the ecosystems that they share with other species, in ways that may either harm or benefit other organisms. For instance, as the major herbivores of the neotropics, leaf-cutter ants clearly have an impact on the growth and density of those species of plants that they exploit, as well as on those plants that grow in the improved soil of their nests and those species that rely on the ants to disperse their seeds. Moreover, leaf-cutter ants have glands that secrete substances that kill virtually all bacteria and fungi, except for the single fungus that they cultivate.

While the leaf-cutter ants provide a particularly striking example, there is nothing remarkable about the fact that they have an impact on their local ecology. In chapter 2 of this book we will demonstrate that niche construction is extremely common. Population-community ecologists know a good deal about how organisms can affect each other's environments, both inter- and intraspecifically, and how, by doing so, they can influence such phenomena as the distribution and abundance of organisms, population and community structures, food webs, and trophic dynamics (Begon et al. 1996; DeAngelis 1992; Rosenzweig 1995). Similarly, ecosystem ecologists already have a good understanding of the many ways in which organisms can influence energy and matter flows through ecosystems when they take resources from them, or return detritus to them, and also how their influence can, in turn, affect the structure and function of ecosystems, the resistance and the resilience of ecosystems to perturbations, and the nature of various biogeochemical cycles (O'Neill et al. 1986; Odum 1989; Jones and Lawton 1995; Patten and Jorgensen 1995).

For our purposes, however, a recent insight from a team of ecosystem ecologists, Jones et al. (1994, 1997) and Jones and Lawton (1995), is particularly valuable. Jones et al. describe organisms that choose or perturb their own habitats as "ecosystem engineers," where "ecosystem engineering" is essentially the same as "niche construction." Jones et al. claim that when organisms invest in ecosystem engineering they not only contribute to energy and matter flows and trophic patterns in their ecosystems but in part also *control* them. They propose that organisms achieve their control via an extra web of connectance in ecosystems, which they call an "engineering web," and which is established by the interactions of diverse species of engineering organisms (Jones et al. 1997). This engineering web operates in conjunction with the familiar material (stoichiometric) and energy (thermodynamic) webs of connectance in ecosystems that are already studied by ecologists (Reiners 1986). Jones et al. also suggest that it is not always necessary for ecosystem engineers to contribute directly to a particular energy or material flow among a set of trophically connected organisms in an ecosystem for them to control the flow (Jones et al. 1997, p. 1952).

We can illustrate these ideas by using two of Jones et al.'s own examples, both taken from the Negev desert in Israel. The first is a case of engineering by microorganisms. In many deserts, including the Negev, the soil is extensively covered by dominant microphytic communities of blue-green algae, cyanobacteria, and fungi. Although these microorganisms are barely visible to the naked eye, they nevertheless have a powerful engineering effect because they secrete polysaccharides that bind the desert's soil and sand together to form a crust that not only protects their own colonies from heat, but also controls erosion, runoff, and site availability for the germination of higher plants in the desert (West 1990; Zaady and Shachak 1994; Jones et al. 1997). After rain, the asphaltlike patches that are created by these microorganisms reduce the absorption of water by about 30%, and this increases the runoff of water, allowing the water to form pools in pits previously dug, for example, by desert porcupines digging for geophytes. Windblown seeds then germinate in these moist pits and give rise to lush oases that may eventually harbor dozens of other species (Alper 1998). Yet all of this ultimately depends on the long reach of the engineering activities of microorganisms.

The second example is provided by three species of snail, *Euchondrus* spp., that eat endolithic lichens that grow under the surface of limestone rocks in the Negev desert. One consequence of this unusual form of herbivory is that the snails are major agents of rock weathering and also of soil formation in this desert. Their agency, however, is not due to the amount of lichens they consume, which is actually rather little. Instead, it is due to the unexpected fact that these snails have to physically disrupt and ingest the rock substrate in order to consume the lichens. They later excrete the rock material ingested as feces, which they deposit on the soil under the rocks. Shachak et al. (1987) estimated that the annual rate of biological weathering of these rocks by snails is 0.7 to 1.1 metric tons per hectare per year, which is sufficient to affect the whole desert ecosystem (Shachak et al. 1987; Shachak and Jones 1995). By converting rock to soil at this rate, the snails become major agents in soil formation.

So ecosystem control is one major new idea associated with the ecological effects of niche construction. It stems from the capacity of niche-constructing organisms to modify not only their own environments but also the environments of other organisms in the context of shared ecosystems.

1.1.2 The Modification of Selection Pressures

The second consequence of niche construction, and its first evolutionary consequence, derives from these ecological effects. If organisms modify their environments, and if in addition they affect, and possibly in part control, some of the energy and matter flows in their ecosystems, then they are likely to modify some of the natural selection pressures that are present in their own local selective environments, as well as in the selective environments of other organisms. In fact, it is difficult to see how organisms can avoid doing this. Environmental change modifies natural selection pressures (Endler 1986), while organisms are a known source of environmental change in ecology (Jones et al. 1997).

However, in order for niche construction to be a significant evolutionary process, it is not sufficient for niche-constructing organisms to modify one or more natural selection pressures in their local environments temporarily, because whatever selection pressures they do modify must also persist in their modified form for long enough, and with enough local consistency, to be able to have an evolutionary effect. Often this criterion will not be met. Moreover, independent agents in a population's environment may erase or overwhelm the effects of the population's niche construction, thereby ensuring that there is no persistent environmental change caused by the population's activities. For instance, other environmental agents may disperse a population's detritus by dissipating it over time, or, if the agents are detritivores, they may consume the population's detritus, or recycle it, instead of allowing the detritus to accumulate.

There are, however, at least two ways in which this persistence criterion can be satisfied. If, in each generation, each individual repeatedly changes its own ontogenetic environment in the same way, for instance, because each individual inherits genes that express the same niche-constructing phenotypes, then ancestral organisms may modify a source of natural selection by repetitive niche construction. The immediate environmental consequences of this kind of niche construction may be transitory, and may be restricted to single generations only, but if the same environmental change is reimposed sufficiently often and persists for a sufficient number of generations, it may modify the pressures of natural selection in local environments and therefore drive a new evolutionary episode.

For example, individual web spiders repeatedly build webs in their local environments, generation after generation, because they repeatedly inherit genes from their ancestors that are expressed in web construction. Even though spiders' webs are transitory objects, and are only too likely to be destroyed on a daily basis by other agents in the environment, such as other animals, or the weather, every time a spider's web is destroyed the spider's genes "instruct" the spider to make a new one. As a result there is almost always a web in the local environments of these spiders. The omnipresent web appears to have fed back, over many generations, in the form of modified natural selection. For instance, spiders on a web are exposed to the threat of avian predators, but they frequently engage in courtship and process prey on the web. Thus the web may have been a source of selection to favor further phenotypic changes in these species, including the marking of their webs to enhance crypsis, differential responses to the frequency of web vibration for prey and for a potential mate, or, as in the case of one genus of African orb-web spider, *Cyclosa*, the building of dummy spiders in the web probably to divert the attention of birds that prey on them (Edmunds 1974; Preston-Mafham and Preston-Mafham 1996).

The second way of satisfying the same persistence criterion occurs when all or a part of the consequences of one generation's niche-constructing activities persist in their modified form in the selective environments of a succeeding generation. By this means, ancestral organisms can bequeath legacies of modified natural selection pressures to their descendants via the external environment. This between-generational transmittal may be restricted to just two generations, as happens, for example, in maternal inheritance (Kirkpatrick and Lande 1989; Cowley and Atchley 1992; Schluter and Gustafsson 1993; Mousseau and Fox 1998a,b) when mothers modify the selection pressures in the local environments of their offspring. Alternatively, it may be a multiple-generation phenomenon in which the cumulative effects of generations of niche construction modify the selective environments of more distant descendants.

The common cuckoo, *Cuculus canorus*, provides a familiar two-generation example. In this species of brood parasite, cuckoo mothers repeatedly select a host belonging to some other bird species and lay their eggs in the chosen host's nests, subsequently relying entirely on this host to incubate the cuckoo eggs and raise the cuckoo young to independence. Cuckoo mothers have parasitized other birds in this way for generations and, as a result, have apparently bequeathed modified natural selection pressures to their offspring, in the form of these alien nurseries. The modified natural selection pressures have probably contributed to several novel adaptations in cuckoo chicks, including an extremely short incubation period, which ensures that the cuckoo chicks usually hatch before the host's chicks, and the ejection of the host's eggs from the nest or the killing of any of the host's chicks that have managed to hatch. These latter acts are themselves further examples of niche construction, this time via the agency of the cuckoo chicks rather than their mothers. The effect is that each cuckoo chick is raised on its own by its host and does not have to compete with any rival chicks when its foster parent arrives with food. However, having killed its rivals, the cuckoo chick must stimulate an adequate rate of feeding by its host. It appears to accomplish this task by behaving as if it were the equivalent of a whole brood of its host's chicks, instead of just a singleton. It does so by emitting a rapid begging call that mimics the begging sounds, as well as the calling rate, of a complete brood of its host's chicks (Davies et al. 1998). The initial choice of host's nests by cuckoo mothers may also have made possible some additional adaptations in their offspring when the latter become parents. For example, cuckoos that were raised in the nests of a particular host species subsequently tend to parasitize the same host species, possibly because as they developed they learned their hosts' characteristics (Krebs and Davies 1993). The mother's niche construction has modified the selection on her offspring, resulting in a cascade of evolutionary events, including the selection of further niche construction on the part of the chick. In recent years there has been increasing recognition that such maternal effects are both taxonomically widespread and evolutionarily significant (Wade 1998; Mousseau and Fox 1998a).

Earthworms provide an equally familiar multigenerational example, one which has the added distinction of having been described by Darwin (1881). Through their burrowing activities, their dragging organic material into the soil, their mixing it up with inorganic material, and their casting, which serves as the basis for microbial activity, earthworms dramatically change the structure and chemistry of the soils in which they live, often on a scale that exceeds even the soil-perturbing activities of leaf-cutter ants. For instance, in temperate grasslands earthworms can consume up to 90 tons of soil per hectare per year. Similarly, as a result of their industry, earthworms affect ecosystems by contributing to soil genesis, to the stability of soil aggregates, to soil porosity, to soil aeration, and to soil drainage. Also, because their casts contain more organic carbon, nitrogen, and polysaccharides than the parent soil, earthworms can also affect plant growth by ensuring the rapid recycling of many plant nutrients. In return, the earthworms probably benefit from the extra plant growth they induce by gaining an enhanced supply of plant litter (Kretzschmar 1983; Hayes 1983; Stout 1983; Lee 1985; Ellis and Mellor 1995). All of these effects typically depend on multiple generations of earthworm niche construction, leading only gradually to cumulative improvements in the soil. It follows that most contemporary earthworms inhabit local selective environments that have been radically altered, not just by their parent's generation, but by many generations of their niche-constructing ancestors. It is likely that some earthworm phenotypes, such as epidermis structure, or the amount of mucus secreted, coevolved with earthworm niche construction over many generations. Moreover, because these originally aquatic creatures are able to solve their water- and salt-balance problems through tunneling, exuding mucus, eliminating calcite, and dragging leaf litter below ground, that is, through their niche construction, earthworms have retained the ancestral freshwater kidneys (or nephridia) and have evolved few of the structural adaptations one would expect to see in an animal living on land (Turner 2000). For instance, earthworms produce the high volumes of urine characteristic of freshwater rather than terrestrial animals.

The production of oxygen by photosynthetic organisms is another multiple-generation example, which illustrates the extreme effects that niche construction can have on a global scale if its consequences happen to build up over long periods of time. When photosynthesis first evolved in bacteria, particularly in cyanobacteria, a novel form of oxygen production was created. The contribution of these ancestral organisms to the earth's 21% oxygen atmosphere must have occurred over billions of years, and it must have taken innumerable generations of photosynthesizing organisms to achieve. It is highly likely that modified natural selection pressures, stemming from the earth's changed atmosphere, played an enormous role in subsequent biological evolution. For example, many organisms have evolved a capacity for aerobic respiration, and they have also evolved other mechanisms, such as the enzyme superoxide dismutase, that protect cells against oxidation (Futuyma 1998).

In the next chapter we illustrate traits in many species that appear to have evolved as a consequence of selection generated by prior niche construction. However, if organisms evolve in response to selection pressures modified by their ancestors, there is feedback in the evolutionary dynamic, and it is well established that biological systems with feedback behave quite differently from those without it (Robertson 1991). This is a further point to which we shall repeatedly return in this book.

1.1.3 Ecological Inheritance

With the exception of the special cases of maternal and cultural inheritance (reviewed in chapter 3) standard evolutionary theory is typically concerned with only a single general inheritance system in evolution. It assumes that natural selection among individual organisms influences which individuals survive and reproduce to pass on their genes to the next generation (fig. 1.3a), and this genetic inheritance is generally regarded as the only inheritance system to play a major role in biological evolution. This assumption is not affected by niche construction as long as the physical consequences of the niche-construction process are erased in the selective environments of populations between each generation, and therefore last only a single generation. For instance, in the orb-web spider case, the repetitive construction of webs by spiders owes its capacity to influence the evolution of populations of spiders not to any between-generation persistence of the webs themselves (spiders' webs are far too transitory for that), but rather to the spider's genetic inheritance system. This ceases to be true, however, when the physical consequences of one generation's niche construction are not completely erased in the environments of its descendants but are instead bequeathed, either wholly or in part, from one generation to the next, in the form of legacies of modified natural selection pressures. This is what happens in the case of cuckoos over two generations, and in earthworms and in cyanobacteria over multiple generations. Here, then, is a third major consequence of niche construction. Where niche construction affects multiple generations, it introduces a second general inheritance system in evolution, one that works via environments. This second inheritance system has not yet been widely incorporated by evolutionary theory.

We call this second general inheritance system *ecological inheritance* (Odling-Smee 1988; Odling-Smee et al. 1996). It comprises whatever legacies of modified natural selection pressures are bequeathed by niche-constructing ancestral organisms to their descendants. Ecological inheritance differs from genetic inheritance in several important respects. First, genetic inheritance depends on the capacity of reproducing parent organisms to pass on replicas of their genes to their offspring. Ecological inheritance, however, does not depend on the presence of any environmental replicators, but merely on the persistence, between generations, of whatever physical changes are caused by ancestral organisms in the local selective environments of their descendants. Thus, ecological inheritance more closely resembles the inheritance of territory or property than it does the inheritance of genes. Although the inheritance of property is common enough among human beings, it is not restricted to humans. As

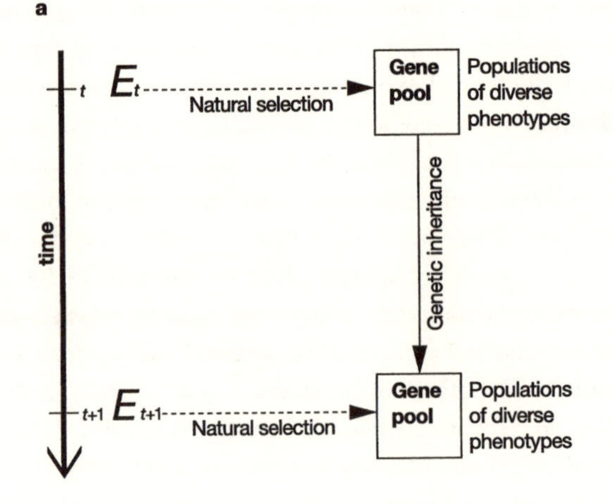

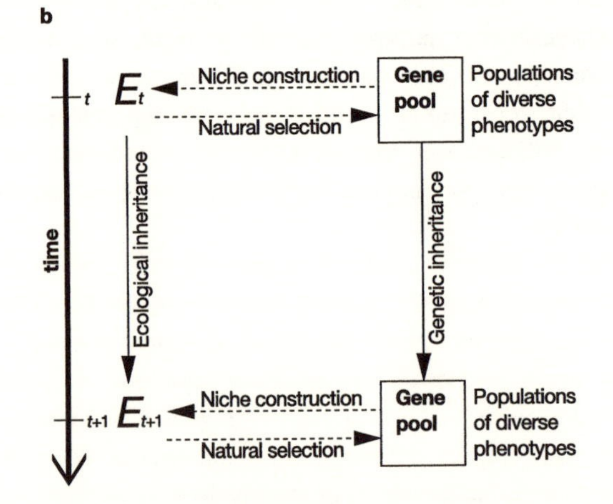

FIGURE 1.3. (a) Standard evolutionary perspective: Organisms transmit genes from generation t to generation $t + 1$ with natural selection acting on phenotypes. (b) With niche construction: Organisms also modify their local environments (**E**), as depicted by the arrow labeled "niche construction." Each generation inherits from ancestral organisms both genes and a legacy of modified selection pressures, described as "ecological inheritance." (From Laland et al. [2000b], fig. 1. Reprinted with the permission of Cambridge University Press.)

we have seen, cuckoos inherit an alien nest while earthworms inherit a modified soil environment. Ecological inheritance also has a lot in common with the more familiar concept of ecological succession, except that it has evolutionary as well as ecological consequences because it involves the inheritance by populations of modified natural selection pressures, via a succession of environmental states, which may then drive further evolutionary changes in those populations.

Second, when organisms inherit naturally selected genes, they are, in effect, inheriting information molecularly encoded in the nucleotide sequences of DNA. Genetic information is, of course, noncognitive (see chapter 4). Nevertheless, it is information that is used to inform the expression of phenotypes in ontogenetic environments, relative to their local selective environments (J. Holland 1992, 1995; Eigen 1992). In contrast, when organisms inherit legacies of modified natural selection pressures they typically do not inherit information. Instead they inherit some of the agents in their environments that select for their genes and that thereby determine which information the organisms express (J. Holland 1995).

Third, genes and biotically modified natural selection pressures are passed on from one generation to the next by completely different processes. Genetic inheritance depends on the between-generation processes of reproduction, including sexual reproduction, which means that genes can only be transmitted to new organisms once during their lives. It also means that genes can only be transmitted to organisms by parents, and in one direction only, from parents to offspring, rather than the other way round. However, an ecological inheritance, in the form of one or more biotically modified natural selection pressures, can potentially be bequeathed by any organism to any other organism, at any stage during an organism's lifetime, and therefore within as well as between generations. It is also possible for an ecological inheritance to travel backward in generational terms because offspring may sometimes modify their parents' selective environments, as well as their own and those of their descendants.

Finally, the selective environments of organisms can be modified either by their genetic relatives or by other unrelated organisms. In fact, any organism's selective environment is potentially modifiable by any other organism that happens to be a neighbor or that shares, or that has previously shared, some common physical aspect of a mutual environment or that is capable of exerting an indirect influence by affecting the flow of energy or materials through that environment. All such neighbors are ecologically related but they need not be genetically related. Ecological and genetic ancestors are not necessarily identical.

The way in which the two general inheritance systems operate in evolution, and how they interact with each other, is summarized in figure 1.3b. On the right of figure 1.3b genes are shown being transmitted by genetically related ancestral organisms at time t, to their genetic descendants at time $t + 1$, in the usual way. On the left, however, selected habitats, modified habitats, artifacts, or in general, ancestrally modified sources of natural selection persist or are actively or effectively transmitted by these same organisms to their descendants in their local environments (E). Thus, the selective environments encountered by the descendent organisms at time $t + 1$ do not just comprise independent sources of natural selection pressures as evolutionary theory currently implies. They stem partly from such independent environmental agents, for example, climate, weather, or physical or chemical events, but they also stem in part from sources of natural selection that have previously been modified by ancestral niche construction.

1.1.4 Adaptation

This capacity of organisms to modify some of their own selection pressures, whether between generations or within generations, also has a fourth consequence. It requires us to revise the concept of adaptation in evolution and to adjust its meaning along lines anticipated by Richard Lewontin (1983). Lewontin pointed out that contemporary evolutionary theory implicitly assumes that natural selection pressures in environments are decoupled from the adaptations of the organisms for which they select. Therefore, with some exceptions (reviewed in chapter 3), for example, those that involve frequency-dependent or habitat selection, standard theory treats sources of natural selection in environments and adaptations in organisms as independent of each other or, as Lewontin puts it: "The environment 'poses the problem'; the organisms 'posit solutions,' of which the best is finally 'chosen'" (1983, p. 276). What this classical approach overlooks, and what we are stressing here, is that the selective environments of organisms are themselves partly built by the niche-constructing activities of the organisms that they are selecting for. To quote Lewontin again: "Organisms do not adapt to their environments; they construct them out of the bits and pieces of the external world" (1983, p. 280). Therefore, some selection pressures cannot be decoupled from the adaptations of organisms. Instead they must be participants in a system of feedbacks between natural selection pressures in environments and adaptations in organisms.

We have already encountered several examples of this kind of feedback in action. For instance, the cuckoo chicks, having destroyed their host's brood, adapt to mimic the missing broods they have killed (Davies et al. 1998). Other equally simple examples are found in spiders. One discussed by Dawkins (1996) on the basis of work by Vollrath (1988, 1992) concerns how, when its prey crashes into its web, the prey neither breaks the web nor bounces off it, but sticks to it. Many web spiders evolved the ability to make the threads of their webs sticky enough to hang on to the prey. But how are the spiders to ensure that they themselves do not get stuck to their own webs yet are free to move around on them? Dawkins offers two answers. One involves the anointing of spiders' legs with a special oil that provides the spiders with some protection against the stickiness of their own webs, while the other involves spiders making some of the spokes of their own webs nonsticky, to allow themselves free movement along these spokes. Such examples nicely illustrate Lewontin's point.

Lewontin (1983, 2000) argued that the classical picture of evolution can be represented formally as a pair of differential equations in time:

$$\frac{dO}{dt} = f(O, E), \tag{1.1}$$

$$\frac{dE}{dt} = g(E). \tag{1.2}$$

Equation 1.1 states that evolution, or change in the organism over time, depends on both the current state of the organism and its environment, while equation 1.2 states that environmental change depends only on environmental variables. The crucial point is that these two equations are separable. Adapted organisms are not supposed to cause any of the environmental changes that subsequently select for adapted organisms. Hence, the evolution of organisms is generally assumed to be directed exclusively by independent natural selection pressures in environments, and not at all by the niche-constructing activities of organisms. Lewontin argued that what is actually happening in nature is better represented by a pair of *coupled* differential equations

$$\frac{dO}{dt} = f(O, E), \tag{1.3}$$

$$\frac{dE}{dt} = g(O, E), \tag{1.4}$$

in which the histories of both environment and organism are functions of both environment and organism. Equations 1.3 and 1.4 describe a situation in which niche-constructing organisms and their environments are, in effect, coevolving, because they are codetermining and codirecting changes in each other. Equations 1.3 and 1.4 describe the coevolution of organism and environment in which both are acting as both causes and effects.

Evolutionary biology has provided a compelling explanation for why organisms appear so extraordinarily well suited to the environments in which they live: namely, through the action of natural selection, species have come to exhibit those characteristics that enable survival and reproduction. However, there are in fact two logically distinct routes to the evolving match between organisms and their environments: either the organism changes to suit the environment, or the environment is changed to suit the organism. The first alternative is brought about through the process of natural selection, and the second is one possible outcome of the process of niche construction. Of course, in reality these two processes can seldom be separated.

Yet the standard view is that niche construction should not be regarded as a process in evolution because it is determined by prior natural selection. The unstated assumption is that the environmental source of the prior natural selection is independent of the organism (as formalized by eq. 1.2). However, in reality, the argument that niche construction can be disregarded because it is partly a product of natural selection makes no more sense than would the counter proposal that natural selection can be disregarded because it is partly a product of niche construction. One cannot assume that the ultimate cause of niche construction is the environments that selected for niche-constructing traits, if prior niche construction had partly caused the state of the selective environments (as formalized by eq. 1.4). Ultimately, such recursions would regress back to the beginning of life, and as niche construction is one of the defining features of life (see chapters 2 and 4), there is no stage at which we could say natural selection preceded niche construction or that selective environments preceded niche-constructing organisms. From the beginning of life, all organisms have, in part, modified their selective environments, and their ability to do so was, in part, a consequence of their naturally selected genes.

1.2 THE IMPLICATIONS

We can now start to consider some of the implications of adding niche construction to contemporary evolutionary theory. In doing so we introduce the three principal fields we shall be dealing with in later chapters: evolutionary theory itself, the relationship between evolutionary theory and ecosystem-level ecology, and the evolutionary basis of human cultural processes.

1.2.1 Implications for Evolutionary Theory

What difference does it make if the selection pressures acting on organisms stem from an independent environment or a niche-constructed environment? The principal difference is equivalent to the difference between Lewontin's coupled and uncoupled equations and can be encapsulated by one word, namely, "feedback." If organisms evolve in response to selection pressures modified by themselves and their ancestors, there is feedback in the system. In chapters 3 and 6 of this book we will describe and analyze theoretical models that illustrate some of the differences that this feedback makes to the evolutionary process. We show how traits whose fitness depends on alterable sources of selection (recipient traits) coevolve with traits that alter sources of selection (niche-constructing traits), resulting in very different evolutionary dynamics for both traits from what would occur if each had evolved in isolation. Our models demonstrate how feedback from a population's niche construction can cause either evolutionary inertia or momentum, lead to fixation of otherwise deleterious alleles, support stable polymorphisms where none are expected, eliminate what would otherwise be stable polymorphisms, and influence levels of linkage disequilibrium. There is no escaping the conclusion that niche construction is evolutionarily consequential.

A second difference is ecological inheritance. The niche-construction perspective stresses two legacies that organisms inherit from their ancestors, genes and a modified environment with its associated selection pressures. As we document in chapter 2, ecological inheritance is likely to be ubiquitous, particularly when the widespread evidence for maternal inheritance is taken into account (Mousseau and Dingle 1991; Roach and Wulf 1987; Bernado 1996; Mousseau and Fox 1998a). Consider, for instance, the observation that most species of insects are oviparous, with the female depositing eggs on or near the food required by the offspring upon hatching (Gullan and Cranston 1994). These offspring inherit from their mother the legacy of a readily available, nutritious larval food and a nursery environment. When one considers that careful selection of appropriate sites by ovipositing females is found in the vast majority of insects and that estimates of the number of insect species range from 5 to 80 million, the pervasiveness of ecological inheritance becomes clear.

The analyses that we will present in chapters 3 and 6 demonstrate that, because of the multigenerational properties of ecological inheritance, niche construction can generate unusual evolutionary dynamics. Theoretical population-genetic analyses have established that processes that carry over from past generations can change the evolutionary dynamic in a number of ways, generating time lags (in the response to selection of the recipient trait), momentum effects (populations continuing to evolve in the same direction after selection has stopped or reversed), inertia effects (no noticeable evolutionary response to selection for a number of generations), opposite responses to selection, and sudden catastrophic responses to selection (Feldman and Cavalli-Sforza 1976; Kirkpatrick and Lande 1989; Robertson 1991; Laland et al. 1996; Mousseau and Fox 1998a,b; Wolf et al. 2000). Wherever there is ecological inheritance, a product of niche construction, the evolutionary process may include some or all of these complications.

A third implication of niche construction is that it allows acquired characteristics to play a role in the evolutionary process, in a non-Lamarckian fashion, by their influence on selective environments through niche construction. When phenotypes construct niches, they become more than simply "vehicles" for their genes (Dawkins 1989), as they may now also be responsible for modifying some of the sources of natural selection in their environments that subsequently feed back to select their own genes. However, relative to this second role of phenotypes in evolution, there is no requirement for the niche-constructing activities of phenotypes to result directly from naturally selected genes before they can influence the selection of genes in populations. Animal niche construction may depend on learning and other experiential factors, and in humans it may depend on cultural processes.

The Galápagos woodpecker finch provides a specific example (Alcock 1972). These birds create a woodpecker-like niche by learning to use a cactus spine or similar implement to peck for insects under bark (Tebbich et al. 2001). While true woodpeckers' (Picidae) bills are adaptive traits fashioned by natural selection for grubbing, the finch's capacity to use spines to grub for insects is not an adaptation. Rather, the finch, like countless other species, exploits a more general and flexible adaptation, namely, the capacity to learn, to develop the skills necessary to grub in environments that reliably contain cactus spines and similar implements. The finch's use of spines develops reliably as a consequence of its ability to interact with the environment in a manner that allows it to benefit from its own experience (Tebbich et al. 2001). Moreover, the finch's learning certainly opens up resources in the bird's environment that would be unavailable otherwise and is therefore an example of niche construction. This behavior probably created a stable selection pressure favoring a bill able to manipulate tools rather than the sharp, pointed bill and long tongue characteristic of woodpeckers. Since tool manipulation can depend in part on learning, there is a further twist to this example. Niche-constructing skills influenced by learning could modify natural selection in favor of an enhanced learning ability, and it would certainly be interesting to know whether the learning capabilities and their neural substrates in this species differ from those in closely related non-tool-using species. While the information acquired by individuals through ontogenetic processes cannot be inherited because it is erased when they die, processes such as learning can nonetheless still be of considerable importance to subsequent generations because learned knowledge can guide niche construction.

Beyond individual learning, a few species, including most vertebrates, have also evolved a capacity to learn from other individuals, and to transmit some of their own learned knowledge to others. The resulting "protocultural" processes may also underlie niche construction. An example is the spread of milk-bottle-top opening in a variety of British birds (Fisher and Hinde 1949; Hinde and Fisher 1951). These birds learned to peck open the foil cap on milk bottles and to drink the cream, and this behavior spread throughout Britain and into several other countries in Europe. Hinde and Fisher found that this behavior probably spread by local enhancement, where the birds' attention was drawn to the milk bottles by a feeding conspecific, and after this initial tip-off, they subsequently learned on their own how to open the tops. However, further analysis by Sherry and Galef (1984) revealed that, in addition to social learning by local enhancement, milk-bottle-top opening could be acquired by other means, for example, it could also spread if the birds were merely exposed to opened milk bottles, even if there were no other birds present and performing the opening behavior. In this example, the birds' niche-constructing behavior is propagated by local enhancement. However, by creating opened milk bottles, this niche construction biases the probability that other birds will learn to open bottles. Moreover, any selection acting on genetic variation at loci affected by milk-bottle opening would be modified in essentially the same manner as if genes were directly responsible for the behavior. For example, the niche construction might influence selection acting on the birds' learning capacities, foraging behavior, or digestive enzymes.

Acquired niche-constructing traits have almost certainly played a significant role in the evolution of hominids among whom cultural transmission processes are ubiquitous. In chapter 6 we will describe theoretical models that reveal circumstances under which cultural transmission can overwhelm natural selection, accelerate the rate at which a favored gene spreads, initiate novel evolutionary events, and trigger hominid speciation.

1.2.2 Implications for Ecology

The niche-construction outlook may also shed light on problems traditionally considered within the domain of ecology. This is largely because of ecosystem engineering, which modulates and partly controls the flow of energy, matter, and *information* through ecosystems. Genes that interact via niche construction's effects on an external environment do not always have to be in the same population. In later chapters we will demonstrate how genes in different populations may interact with each other via biotic and even abiotic components in the environment to form environmentally mediated genotypic associations (EMGAs). Such associations may, of course, be present within a population as well (Wolf et al. 1998).

If, in a single population, genetic variation is expressed in a niche-constructing phenotype that affects natural selection acting on other genes in the same population, then the population will merely codirect its own evolution through niche construction. However, if the niche construction modifies natural selection acting on genes in a second population, then the first population will now codirect the evolution. Conceivably, the induced change in the second population could feed back to the first population in the form of another modified natural selection pressure. The two populations would therefore coevolve through niche construction.

This coevolution could also be indirect. For instance, the first population's niche construction could influence the evolution of the second population by changing an intermediate component of their shared environment. An example here could be two species that are competing for the same environmental resource or nutrient and that coevolve because of this competition (DeAngelis 1992).

It may be possible to model many cases of coevolution by standard coevolutionary models, in terms of standard evolutionary ecology or genetics, without making any reference to either niche construction or ecological inheritance (Futuyma and Slatkin 1983; Thompson 1994; Heesterbeek and Roberts 1995; Abrams 1996). This is either because niche construction is already implicit in some of these standard models or because in a lot of cases the explicit inclusion of niche construction would make no difference.

In some cases, however, for instance, where there is interspecific exploitative competition or where prey species share a common predator, niche construction cannot be omitted from formal analyses without distorting the processes involved, and in order to describe coevolution accurately it is necessary to treat niche construction as a process in its own right (Tilman 1982; Holt 1985; Abrams 1988; DeAngelis 1992; Holt et al. 1994). When the coevolution of populations is indirect and depends on the modification of an intervening environmental component by the niche-constructing phenotypes of either one or more coevolving populations, then the explicit inclusion of niche construction and ecological inheritance adds significantly to the models. This is especially likely to be true when the intermediate environmental component concerned is abiotic. For example, if niche construction resulting from a gene in a plant population causes the soil chemistry to change in such a way that the selection on genes in a second population of plants, or possibly of microorganisms, is also changed, then the first population's niche construction will drive the evolution of the second population simply by changing the physical state of the intervening abiotic environmental variable, in this case the soil. This kind of indirect coevolution via intermediate abiota is not well described by conventional population-genetic coevolutionary models for the simple reason that abiotic components are not alive, they do not carry genes, and they cannot evolve. While the demographics of such interspecific interactions, and some issues, such as the conditions for coexistence, are well captured by ecological models, the evolutionary ramifications are comparatively underexplored. Yet abiota are continuously subject to change by niche-constructing organisms (Jones et al. 1997), and any changes brought about through the activities of one population of organisms may easily serve as a legacy of modified natural selection for another. Thus adding niche construction and ecological inheritance to population-genetic coevolutionary models may make it possible to capture these interspecific interactions. As the dynamics of physical change in abiota are likely to be quite different from the dynamics of evolutionary change in populations, this kind of indirect feedback among coevolving species via intermediate abiota may generate some interesting and as yet underexplored behavior in coevolutionary systems.

Ecosystem engineering (Jones et al. 1994, 1997) further illustrates the utility of the niche-construction perspective. Jones et al. point to several ecosystem phenomena that cannot be understood in terms of energy and matter flows only. They stress the critical role played by the creation of physical structures and other modifications of their environments by organisms that partly control the distribution of resources for other species. Ecosystem engineering does not always conform to the principles of mass flow and the conservation of energy, nor to stoichiometry requirements, because ecosystem engineers are not necessarily part of these flows or cycles, but they can control them (Jones et al. 1997). We elaborate on this point in chapter 5. Gurney and Lawton (1996) have demonstrated theoretically how the efficacy with which niche construction acts to degrade a virgin habitat determines not only whether there will be no engineers, a stable population of engineers, or population cycles in the frequency of engineering, but also the extent of virgin and degraded habitat.

Evolutionary phenomena associated with niche construction complement and add to Jones et al.'s observations of the ecological repercussions of engineering. For example, when they engineer, niche-constructing organisms frequently influence their own evolution by modifying their own selective environments, perhaps by changing abiotic components or chains of such components. Second, niche-constructing organisms also influence the evolution of other populations, again often indirectly via intermediate abiotic components. Third, some organisms create new niches for themselves, for example, through technological innovation or relocation to a novel environment, which again can influence the dynamics of their ecosystems. Fourth, evolutionary and coevolutionary events can operate on ecological time scales, which means that the dynamics of abiotic components may reflect gene frequency changes in evolving engineering species. However, these complications do not necessarily mean that ecological analyses become intractable, and in chapter 8 we describe empirical methods and theory that can be used to investigate the ecological ramifications of niche construction.

A niche-construction perspective might also promote a much closer integration between ecosystem-level ecology and evolutionary theory. Hitherto, it has proved difficult to apply evolutionary theory to ecosystems, or even to much reduced ecosystem modules, because of the presence of nonevolving abiota in ecosystems. However, the proposed extension of evolutionary theory, illustrated in figure 1.3b, is indifferent to whether any source of natural selection that is modified by niche construction is biotic or abiotic. In chapters 5 and 8 we will show how extending evolutionary theory along these lines allows abiotic ecosystem variables to be included in both evolutionary and coevolution models.

With the omission of niche construction, standard evolutionary theory underplays the full set of interactions that occur between biotic and abiotic components in ecosystems and ignores diverse forms of feedback that contribute to coevolutionary scenarios and ecosystem dynamics. This is one reason why it has hitherto been difficult to integrate process-functional and population-community ecology with each other and with standard evolutionary theory (O'Neill et al. 1986). When niche construction is incorporated, information (in the sense spelled out in chapter 4) can be seen to flow through ecosystems, and evolutionary control webs begin to emerge.

1.2.3 Implications for the Social Sciences

We shall also address the relationship between human cultural processes and human genetic evolution. At present, contemporary evolutionary theory provides a restricted basis for understanding how human cultural processes relate to human genetic processes in evolution (Laland et al. 1999). Most theory includes only one evolutionary inheritance system, genetic inheritance. It can therefore assign only one role to phenotypes in evolution, that of contributing to genetic inheritance through their differential survival and reproduction. The theory does concede that human cultural activities may influence or may actually be human adaptations, or be the result of other human adaptations, and that cultural processes may also influence human fitness, but it does not concede anything more. In effect, the assumed exclusiveness of the genetic inheritance system, as espoused by classical sociobiology (Wilson 1975), renders all the other consequences of human cultural activities evolutionarily irrelevant.

Niche construction extends contemporary evolutionary theory by the introduction of two liberating innovations. First, as we have already seen, niche construction assigns a second role to phenotypes in evolution, while ecological inheritance provides a second inheritance system to which phenotypes can potentially contribute. In chapter 6 we will see that ecological inheritance is likely to have been of paramount importance in human evolution, where material culture has played a number of roles. Second, there is no requirement for niche construction to result directly from genetic variation before it can influence the selection of genetic variation. For example, niche construction may depend on learning, as in the case of the woodpecker finch and British birds discussed above, and in humans niche construction may also depend on cultural processes. To cite one well-known example, when our ancestors first domesticated cattle by agricultural niche construction, they apparently modified a natural selection pressure on a gene that enables the enzyme lactase, needed for the digestion of milk, to be synthesized by human adults (Feldman and Cavalli-Sforza 1989; Durham 1991; Holden and Mace 1997). This demonstrates how cultural processes are not just a product of human genetic evolution, but also a cause of human genetic evolution. Adding niche construction and ecological inheritance to contemporary evolutionary theory may therefore improve our understanding of the relationship between human genetic and cultural processes.

There have been two principal reasons why many human scientists have found it difficult to make use of evolutionary theory. One is that the theory appears to offer too little. Human scientists are predominantly interested in human behavior and cultural processes, rather than just genes, and as a consequence they see little useful point of contact with evolutionary theory. Our niche-construction framework may provide such a bridge because it emphasizes the active role that organisms play in the evolutionary process. Humans are not just passive vehicles for genes, they actively modify sources of natural selection in environments. They are the ultimate niche constructors. A second reason why human scientists have difficulty with evolution is the simplicity of adaptationist accounts. Adding niche construction inevitably makes evolutionary theory more complicated, and any extra complexity must prove worthwhile to those scholars for whom environmental effects and interactions between organisms and environments are the focus of study. The relevance of the niche-construction perspective to these issues is discussed in chapter 9, where we illustrate how our framework can apply in the human sciences, providing methods and making empirically testable predictions. Indeed, many social scientists have already started to use niche construction as a useful theoretical tool.

1.3 PREVIOUS APPROACHES

If niche construction has as many consequences and implications as those we have now listed, why has it not already been incorporated into contemporary evolutionary theory? There are some theoretical devices by which contemporary evolutionary theory deals with niche construction and we discuss these in chapter 3. Here it is more appropriate to introduce some of the early forerunners of the idea, both to indicate how long the concept of niche construction has been appearing in the margins of evolutionary theory and to show that, in spite of its frequent appearances, the concept itself has received surprisingly little attention from biologists.

Perhaps the first person to draw attention to the idea of niche construction in a clear way was not even a biologist, but a physicist, Schrödinger (1944, p. 108), who did so in a lecture "Mind and Matter" given at Cambridge in 1956, as a companion to his earlier and more famous "What is Life?" lecture. It may have been because he was not a biologist that Schrödinger was able to take the outsider's advantage of being able to discriminate between the forest and the trees more easily than those who are already in the forest.

The evolutionary biologist Ernst Mayr also made an early contribution with a much-cited quotation from his book *Animal Species and Evolution*:

> A shift into a new niche or adaptive zone, is almost without exception, initiated by a change in behavior. The other adaptations to the new niche, particularly the structural ones, are acquired secondarily. With habitat and food selection—behavioral phenomena—playing a major role in the shift into new adaptive zones, the importance of behavior in initiating new evolutionary events is self-evident (Mayr 1963, p. 604).

In this passage Mayr is clearly drawing attention not just to the importance of behavior in evolution but also to how organisms can, in part, actively determine their own selective environments by niche-constructing-type activities, which *then* select for different structural adaptations. However, as Plotkin (1988) pointed out, having made this emphatic claim, Mayr himself did not follow it up. The idea was left, floating and unexploited.

Conrad Waddington (1959, 1969), another biologist, thought about niche construction in the same decade, but primarily in the context of organismal development, rather than for evolving populations. Waddington was also an early advocate of bringing developmental biology and evolutionary biology closer together, and it may have been this concern that drew his attention to the many ways in which organisms modify their own selective environments throughout their lives, by choosing and changing their own environmental niches. He called this phenotype-dependent component of both development and evolution "the exploitive system," and he pointed out that, as far as evolutionary theory was concerned, the exploitive system had originally been left out of the modern synthesis (Huxley 1942) and that it was still being left out by contemporary evolutionary theory. Once again, possibly because Waddington was a developmental rather than an evolutionary biologist, his concept of the exploitive system was not taken up.

The next important figure in this story was the Harvard population geneticist Richard Lewontin. In the 1970s and 1980s Lewontin wrote a series of articles on adaptation. For example, Gould's and Lewontin's (1979) influential article "The Spandrels of San Marco and the Panglossian Paradigm: A Critique of the Adaptationist Programme" made many biologists think again about adaptation. However, that part of Lewontin's attack on adaptationism that was based on niche construction proved much less influential, and it has drawn little response (Futuyma 1998). Even for those biologists who accepted that Lewontin was correct, it was not immediately clear what to do about it.

Writing at roughly the same time as Lewontin, although from a very different point of view, Richard Dawkins came up with a pragmatic partial solution to this puzzle. In his book *The Extended Phenotype*, Dawkins (1982) proposed that genes not only express phenotypes, but that some of them also express "extended phenotypes" that, through the activities of organisms, reach beyond the bodies of the organisms themselves to change various components of their selective environments. To cite just one of his examples, Dawkins argued that the lodges, lakes, and dams that are built by beavers are extended phenotypes of beaver genes.

As far as this argument goes it is obviously right, but it is also too restricted. For instance, Dawkins recognized that any genes that are expressed in an extended phenotype should affect the probability of the survival and reproduction of the organism that is carrying them, and therefore their own representation in the next generation. However, Dawkins did not consider that the same gene might also affect the fitness of other genotypes, at other genetic loci, by changing their selective environment. A beaver's dam modifies many selection pressures in the beaver environment, some of which are likely to feed back to affect the fitness of genes that are expressed in quite different traits, such as their teeth, tails, feeding behavior, susceptibility to predation, diseases, life-history strategies, and social systems. Similar limitations constrain almost all the other approaches to niche construction in contemporary evolutionary theory, which are discussed in more detail in chapter 3.

Aside from the advocates of niche construction that we have mentioned, a number of other researchers have pursued and in some cases continue to pursue related ideas (Levins and Lewontin 1985; Wilson 1985; West-Eberhart 1987; West et al. 1988; Bateson 1988; Plotkin 1988; Wcislo 1989; Holt and Gaines 1992; Michel and Moore 1995; Brandon and Antonovics 1996; Moore et al. 1997; Wolf et al. 1998; Oyama et al. 2001; Sterelny 2001; Jablonka 2001; Griffiths and Gray 2001). There were, and still are, other scientists who have resisted the idea, for a variety of different reasons. For the moment, we will introduce only the two principal reasons. The first and probably the most straightforward reason for rejecting niche construction is a belief that it does not exist. For example, both George Gaylord Simpson (1949) and Theodore Dobzhansky (1955) maintained that, humans aside, organisms either do not construct or regulate their niches to any significant degree, or their impact on their environments is invariably too weak, too transient, or too capricious to have any substantial effect on selection pressures. They argued that there are always other more potent independent agents in environments that invariably override the effects of the organisms themselves, thereby preventing organisms from influencing either their own natural selection or the natural selection of their successors. Ultimately, this is an empirical issue, but there is already sufficient evidence to show that organisms can, and indeed do, modify at least some of their own natural selection pressures with sufficient consistency to render this older critical position implausible (see chapter 2).

Originally, this kind of criticism may have stemmed from an intuition that the environment is so vast, and organisms are so small, that the capacity of organisms to change their environments must be negligible. This intuition overlooks two points. One is that natural selection is local—indeed, it is famous for being "myopic." Niche construction becomes an effective codirecting agent in evolution through the modification of local selection pressures. The second point is that, in spite of its local ramifications, because niche construction may be influenced by inherited genes and the same genes may be inherited for many generations, niche construction may sometimes generate some truly large-scale changes in the wider world through the accumulation of effects over long spans of time. The production of oxygen by photosynthetic organisms is a clear example.

Resistance to the idea of niche construction usually takes a different form today. From many personal communications, we have found that most contemporary biologists are prepared to admit that niche construction occurs, and that when it occurs it is bound to have some ecological consequences, but they may still doubt whether it has anything other than trivial evolutionary consequences. Advocates of this position typically maintain that it does not matter much whether natural selection pressures originate from niche-constructing organisms or from other independent sources in environments, as the process of evolution will still be the same. Others accept that sometimes niche construction does affect the process but argue that the effect is not great enough to require anything more than some ad hoc adjustments to contemporary evolutionary theory. Such protagonists would suggest that niche construction is not sufficiently consequential to justify the kind of major revision of evolutionary theory that we are proposing here (fig. 1.3b). In the subsequent chapters we expand on the consequences of niche consequences that have been badly underestimated in the past and are still being underestimated today.

1.4 A PRECIS OF SUBSEQUENT CHAPTERS

The remaining chapters in this book represent a summary of our attempt to begin to redress the neglect of niche construction as an evolutionary agent. In chapter 2 we begin with definitions of niche construction, ecological inheritance, and other important terminology. Chapter 2 also presents a systematic collation and categorization of examples of niche construction, as well as of traits that appear to have evolved as a consequence of selection pressures modified by niche construction. These empirical data illustrate the ubiquity of niche construction.

Chapter 3 discusses previous attempts to handle aspects of niche construction, including frequency- and density-dependent selection, habitat selection, coevolution, indirect genetic effects, maternal inheritance, and various other approaches. We show that, while each of these separate bodies of theory has features germane to niche construction, none of them captures all of the pertinent characteristics. Thus, aside from our own analyses, there has been no attempt to explore the evolutionary consequences of niche construction in a systematic and general manner. Nonetheless, findings from these disparate approaches strongly suggest that niche construction is likely to be an important evolutionary process.

Chapter 3 goes on to investigate the likely evolutionary consequences of niche construction by presenting theoretical population-genetic models that explicitly incorporate the process of niche construction into the evolutionary dynamic. If niche construction is as important an evolutionary process as we claim, then its inclusion should make a significant difference to the behavior of theoretical models and should generate some unusual and hitherto unpredicted dynamics. In the text of chapter 3 we describe the findings of our formal analyses, with all technical and mathematical details relegated to the appendixes. The results of these analyses clearly demonstrate that there are myriad ways by which niche construction is likely to have an evolutionary impact.

There is one prerequisite of evolutionary theory that is often taken for granted. Natural selection can obviously only work when it is fed with a continuous supply of organismal diversity. Superficially, however, organisms appear to violate the second law of thermodynamics merely by staying alive and reproducing, since this law dictates that net entropy always increases and that complex, concentrated stores of energy will inevitably break down. In chapter 4 we ask what characteristics any organism must have *merely to live*. Drawing from theoretical developments in physics and thermodynamics, which offer a description of the Maxwell's-demon-type properties any agent needs to drive a system out of equilibrium, we identify the universal properties that niche construction must have if organisms are not to violate physical laws. As some characteristics of niche construction are universal, it follows that some aspects of the impact that niche-constructing organisms have on their environments will also be universal. Moreover, we suggest that, like natural selection, niche construction is a selective process and that, distinct from other evolutionary processes (e.g., drift, mutation), it introduces directedness to the evolutionary process.

If there are universal and characteristic features of niche construction then it follows that the evolutionary process must have universal and characteristic impacts on the local environments of evolving species. This raises the possibility that niche construction may have implications for ecosystem-level ecology and that a niche-construction perspective may shed light on problems traditionally considered within the domain of ecology. We spell out these implications in chapter 5, where we draw heavily on the insights of ecosystem-engineering researchers. We also illustrate how, with niche construction, evolutionary theory can help describe ecosystem dynamics in spite of the fact that ecosystems include abiotic components. An extended evolutionary theory that takes account of how evolving organisms affect both biota and abiota can provide an integrative evolutionary framework for ecology.

In chapter 6 we address the repercussions of the niche-construction perspective for the human social sciences. A focus on niche construction has important implications for the relationship between genetic evolution and cultural processes. By integrating developments in niche construction and gene-culture coevolutionary theory and explicitly recognizing the guiding role of learning and cultural processes in the niche construction of complex organisms, we develop a new evolutionary framework for the human sciences. This conceptual model is designed to act as a hypothesis-generating framework around which human scientists can structure evolutionary approaches to their disciplines.

In the final section of chapter 6 we illustrate how aspects of this new evolutionary framework can be translated into formal models that illustrate how cultural niche construction may have driven genetic evolution throughout the last two million years. Many results characteristic of gene-based niche construction are also found for cultural niche construction, although cultural niche construction may well have been, and may continue to be, even more potent. Any bias in cultural transmission, or differences in the rate at which alternative behavior patterns are acquired, can increase the impact of niche construction over and above that resulting from genes. Where cultural transmission and natural selection conflict, there is a broad range of circumstances under which cultural transmission can overwhelm natural selection. This is one reason why maladaptive behavior is possible among humans (Cavalli-Sforza and Feldman 1981).

We maintain that these proposed extensions fundamentally alter evolutionary theory. If we are correct, then there should be a set of empirical predictions that would generate data consistent with the niche-construction perspective and inconsistent with more conventional evolutionary perspectives. We acknowledge that unless and until we, or others, generate data that are irreconcilable with conventional neo-Darwinism, or at least are more consistent with the niche-construction perspective, the revisions to evolutionary thinking that we suggest are unlikely to become accepted by the biological community. Consequently, in chapters 7–9 we describe how our hypotheses concerning the evolutionary role of niche construction may be tested and suggest methods for doing so.

Empirical methods and predictions for evolutionary biology, ecology, and the social sciences, respectively, are spelled out in chapters 7, 8, and 9. These methods range from experiments that investigate the consequences of canceling or enhancing a population's capacity for niche construction, to comparative analyses that explore the phylogeny of trait evolution across related species, to directly testing the predictions of our theoretical models. In these chapters we also suggest areas in which our perspective may stimulate empirical study. There is a rich array of possibilities for testing the evolutionary credentials of niche construction, and we hope that this new perspective will stimulate empirical research in the biological and social sciences.

Finally, chapter 10 integrates these findings to make the case that niche construction should be regarded as a significant evolutionary process in its own right; part of an "extended evolutionary theory." For readers without the time or inclination to read all the preceding chapters, this final statement summarizes the contents of the book and our overall argument. It describes why we believe not only that the niche-construction perspective is a more accurate depiction of the evolutionary process than the conventional view, but that it will eventually prove to be a more useful evolutionary framework. We suggest that niche construction is not just an important addition to current evolutionary theory; it requires a reformulation of evolutionary theory. When evolutionary biologists and researchers in related disciplines start using niche construction as a means of formulating hypotheses and generating insights in their fields, then niche construction will be seen to earn its keep.

CHAPTER 2

The Evidence for Niche Construction

2.1 INTRODUCTION

A basic feature of living organisms is that they take in and assimilate materials for growth and maintenance and eliminate or excrete waste products. It follows that, merely by existing, organisms must change their local environments to some degree. Niche construction is not the exclusive prerogative of large populations, keystone species, or clever animals; it is a fact of life.

That organisms engage in niche construction is beyond dispute, but whether this niche construction makes any substantive difference to the world in which they live, or to ecological and evolutionary processes, is open to debate. If the influence that organisms exert on their environments is trivial, a mere drop in the ocean compared with the action of physical, chemical, or meteorological processes, then niche construction is arguably best ignored. While there is widespread acknowledgment that some species, such as the beaver (*Castor fiber*), should be regarded as dominant species and important ecosystem engineers, it is not clear whether such species should be regarded as special cases, or as the more visible examples of a general phenomenon. Whether or not niche construction should be referred to as a significant process in evolutionary and ecological systems is an empirical issue, dependent upon its prevalence and impact.

The goal of this chapter is to provide evidence that organisms do make a significant difference to the structure of their own world. We also detail the evidence that this niche construction has fed back to affect the evolution of these species as well as that of other species. A secondary goal is to organise these data into logical categories and provide some key definitions.

We begin by giving working definitions of the terms "niche" and "niche construction." We then present a simple categorization of different types of niche construction, which is used to organize our examples. As collations of data may often be indigestible, some readers may prefer to skip most of this chapter. However, we recommend that section 2.2 be read now, as it contains definitions that will be needed in later chapters.

2.2 NICHES

First we will consider what ecologists mean by "niche" on the basis of recent reviews by Schoener (1989) and Leibold (1995). Then we will adjust this ecological concept to make it evolutionary as well as ecological.

2.2.1 The Ecological Niche

Schoener (1989) begins his review by contrasting two historical approaches to niches in ecology, those of Grinnell (1917, 1924, 1928) and Elton (1927), which he describes as "similar except in minor ways" (p. 79), versus Hutchinson's (1944, 1957, 1978), which he describes as "revolutionary" (p. 79). Another analysis, by Leibold (1995), differs from Schoener's (1989) principally in treating Grinnell's and Hutchison's approaches as more similar to each other and Elton's as different. We will follow Schoener more closely, but we will also refer to Leibold to amplify some points.

Grinnell did not give a precise definition of his niche concept, which is one reason why it has been interpreted in so many ways (Schoener 1989). He refers to a niche in the figurative sense of a "recess" or place in the community, characterized by a set of environmental conditions including foraging opportunities. Grinnell's niche is primarily concerned with a species' habitat and with what a species requires for survival in its habitat. Contrary to some oversimplified textbook versions of it, however, Grinnell's niche is more than just a habitat and encompasses both spatial and dietary dimensions. A second important property of Grinnell's niche is that it is distinct from its occupant. While every species occupies a niche, niches can exist in the absence of occupants or be vacated by their occupants, implying that niches are provided by environments and are independent of their occupants. Yet Grinnell did not expect vacant niches to remain vacated for long: "if a new ecologic niche arises, or if a niche is vacated, nature hastens to supply an occupant" (Grinnell 1924, p. 227).

Elton's niche is also a recess but differs from Grinnell's primarily by centering on trophic properties and by emphasizing what organisms are *doing* to their communities. Hence, compared with Grinnell, Elton focused more on the role in communities of species or higher-order taxonomic groups. In doing so, Elton (1927) introduced a problem to ecology that has been central ever since, namely, how to derive community properties, such as community structure, from the interactions of diverse species in communities. Leibold (1995) highlights the same point by suggesting that Grinnell and, later, Hutchinson (see below) were concerned primarily with what Leibold calls the "requirement niche," whereas Elton was primarily concerned with the "impact niche." The latter term is used by Leibold to describe the per capita effects of species on their environments, which he says correspond to Elton's "roles." Schoener (1989) acknowledges these differences between Grinnell and Elton but still argues that the similarities between them are "more important than the differences" (p. 87). For example, Schoener points out that Grinnell and Elton "both see niches . . . as largely immutable 'places' or 'recesses' in the community" and "both include dietary and microhabitat factors" (p. 87).

In contrast, for Hutchinson (1944, p. 20): "The term niche . . . is . . . the sum of all the environmental factors acting on the organism; the niche thus defined is a region of an *n*-dimensional hyperspace." Later Hutchinson (1957) described the "*n*-dimensional hypervolume" of each different species as its "fundamental niche," and he described the "realised niche" of a species as that portion, if any, of its fundamental niche not overlapping the fundamental niches of other species, plus that overlapping portion within which the given species can survive. Schoener points out that a major difference between Hutchinson's approach and the earlier approaches of both Grinnell and Elton is that, for Hutchinson, a niche can only be defined "with respect to its occupant . . . and not at all with respect to the place or 'recess' in the community" (Schoener 1989, p. 90). Hence, a niche cannot exist without an occupant. Hutchinson also understood the niche properties of occupants to be "mutable" (Schoener 1989, p. 90), thus making it possible to think about niche evolution.

Schoener argues that Hutchinson's relativistic niche concept was both a considerable advance on its predecessors and a fundamental concept in modern niche theory. However, it introduced several empirical and conceptual difficulties. One difficulty is that, although Hutchinson's multidimensional niche concept is potentially precise, it is, nonetheless, an abstraction that cannot be made operational until the "abstract space" of a Hutchinsonian hypervolume has been mapped onto real habitats, in real space and time. Another closely associated problem concerns niche breadth and niche overlap. For Hutchinson, realized niches cannot overlap where species coexist and compete, making it difficult to understand either competition or coexistence. In the second half of his review Schoener describes how modern niche theory surmounts, or at least circumvents, these difficulties.

Like Hutchinson's conceptualization, the modern niche is associated with a species rather than a recess. Today, Hutchinson's *n*-dimensional hypervolumes are typically made operational by replacing them with "utilization distributions," comprising frequency histograms of the resources used by specific populations, with as many resource axes as are needed in any given case. Space utilizations are determined by observing where individuals of a population are found over some activity period, and temporal utilizations are typically determined on a daily or a seasonal basis by counting the relative number of active individuals at various times of day or at various seasons. Modern niche theory also relates niche overlap to competition coefficients, a conceptualization that both allows for and determines the degree of the coexistence of species. Mathematical formulations derive the conditions for the coexistence of species, niche separation along one or more dimensions, and niche changes such as character displacement (MacArthur and Levins 1967; May 1973, 1974; Schoener 1982; Tilman 1982; Grover 1997).

2.2.2 An Evolutionary Niche

To set the scene for niche construction all we need do is translate this modern ecological concept of a niche into one that is also evolutionary. We will do so by adopting a simple, pragmatic, and minimalist definition. We will treat the niche of any population as the sum of all the natural selection pressures to which the population is exposed. This includes both selection pressures that are likely to cause the occupant population to evolve further, as happens in directional selection, and selection pressures that are likely to stop it from evolving further, as in stabilizing selection. Like Hutchinson's, this definition is strongly relativistic in that selection pressures are only selection pressures relative to specific organisms. It is also consistent with Hutchinson's fundamental niche. It differs only in that the fundamental niche is now treated as a set of "*n*" natural selection pressures relative to its occupant, in addition to being a hypervolume of resources and tolerance limits relative to its occupant, the former being merely the evolutionary aspect of the latter. In principle, it would be possible to relate each selection-pressure dimension to a specific utilization distribution, such that the resource frequency corresponds to the intensity of selection that would be acting on the population. Also like Hutchinson's ecological niche, our evolutionary niche has a duality. It refers to natural selection pressures relating to the "lifestyles" of organisms, and therefore to the many different ways in which different organisms survive and reproduce by actively interacting with their environments, or to what Ehrlich and Roughgarden (1987) call the "occupations" of organisms. It also refers to the real habitats of organisms in real space and time, or to what Odum (1989) calls their "address." We will again assume that the addresses of organisms can be translated from abstractions to physical realities by observing where and when organisms are found in their environments. Thus, for most practical purposes the niche of a population will be equivalent to its realized niche: that part of its fundamental niche from which it is actually earning its living, from which it is not excluded by other organisms, and in which it is able either to exclude other organisms or to compete with other coexisting organisms.

We also want to capture the relativity of this niche concept by defining at least putative units in terms of which we can describe both how organisms relate to their environments and how environments relate to their organisms. For this descriptive purpose only we will use a scheme originally proposed by Bock (1980), who assumed that any organism can be decomposed into arrays of subsystems (traits or characteristics) which he called *features*, and that, similarly, any organism's environment can be decomposed into arrays of subsystems called *factors*. However, to correspond with conventional use of the term "adaptation" (Rose and Lauder 1996), here we regard a feature of an organism as an adaptation only if it was favored by prior natural selection arising from one or more environmental factor. For example, the feature may previously have increased the fitness of the organism by permitting more efficient acquisition of a food resource or avoidance of a predator. Thus, natural selection can be described as promoting a matching of features and factors, a correspondence that Bock described as "synerg." As we describe below, niche construction can also result in synerg. Bock's factors thus correspond to the multiple sources of selection pressures acting on the organism in its niche.

2.2.3 A Definition of Niche Construction

We can now define niche construction and ecological inheritance. *Niche construction* occurs when an organism modifies the feature-factor relationship between itself and its environment by actively changing one or more of the factors in its environment, either by physically perturbing factors at its current location in space and time, or by relocating to a different space-time address, thereby exposing itself to different factors.[1] This basic definition needs several qualifications.

(i) First, niche construction is typically expressed by individual organisms, but natural selection is a process that operates within pop-

[1] Our use of the term *construction* refers to a physical modification of the selective environment or actual movement in physical space and not to the perceptual processes responsible for constructing a mental representation of the world from sensory inputs (as used in psychology), nor to the creation of scientific theories or facts (as used by postmodernist philosophers).

ulations. Therefore, the modification of an environmental factor by individual niche-constructing organisms is likely to require some temporal persistence, and often some kind of accumulation in an environment, before the factor's modification can affect natural selection over successive generations of a population. (ii) When they construct niches, individual organisms may modify a natural selection pressure in their own selective environment, or they may modify a natural selection pressure in the environments of one or more other populations, or both. (iii) In many instances, when niche-constructing organisms modify a selection pressure, the modified selection pressure may not only affect selection in the niche-constructing generation itself, but also in the next generation and in subsequent generations, by acting on the parent generation's offspring and subsequent descendants. We define *ecological inheritance* as any case in which organisms encounter a modified feature-factor relationship between themselves and their environment where the change in the selective pressures is a consequence of the prior niche construction by parents or other ancestral organisms. Moreover, these ancestral organisms may be genetic ancestors, or they could be the ancestors of other species in their communities or in shared ecosystems.

These definitions can be formalized in terms of a "niche function" $\mathcal{N}(t)$ that builds on the pair of coupled equations 1.3 and 1.4 that Lewontin (1983) used to summarize his original criticisms of standard evolutionary theory (see chapter 1). We can envisage that $\mathcal{N}(t)$ represents the niche of the population of organisms O at time t, and that

$$\mathcal{N}(t) = h(O,E), \tag{2.1}$$

where the dynamics of $\mathcal{N}(t)$, which we can regard as "niche evolution," are driven according to equations 1.3 and 1.4, by both O's prior niche-constructing acts and natural selection, including selection stemming from sources that have previously been modified by O's niche-constructing activities, and sources that are independent of O's niche construction and that may change for independent reasons. In this book we shall be focusing primarily on the evolution of populations, and therefore on equation 1.3, although chapters 4, 5, and 8 are concerned with ecological matters that relate more to equation 1.4. Of course $\mathcal{N}(t)$ is specified by O and E according to equation 2.1. Thus equations 1.3 and 1.4, which give the dynamics of O and E, form a natural foundation for theoretical developments.

Niche construction occurs whenever a population O changes its relativistic niche by changing a factor in E relative to its own features. If, by modifying a factor, O also modifies a natural selection pressure for itself, then subsequently the change in the niche caused by O's prior niche construction may feed back to O either to select for a change in O's features or to counteract an independent change in E's factors that would otherwise have selected for a change in O. It may thereby either create, preserve, or destroy a synerg or matching relationship between O's features and E's factors.

Niche construction enriches evolutionary theory by providing the second route to an O-E match that we introduced in chapter 1. Synergistic relationships between O's features and E's factors may now be achieved by natural selection, which changes O to match E. Or synergistic relationships between O's features and E's factors may be achieved by O actively changing one or more of E's factors in its own niche in ways that cause E's factors to match O's features.

Niche construction also occurs when one population, O_1, changes the relativistic niche of a second population, O_2, by changing a factor in O_2's niche. In this case if the changed factor in O_2's niche also modifies a natural selection pressure for O_2, then the consequences of O_1's niche construction will feed forward to O_2, and perhaps also to other populations in O_1's community or ecosystem. These modified natural selection pressures may then select for different features in O_2. Subsequently any change in O_2, particularly any changes in O_2's own niche-constructing activities, may eventually feed back to O_1 in the form of a changed factor, and/or a changed natural selection pressure in O_1's niche, via n other intermediate components. Here n may vary from zero to many, for example in a food web comprising other populations in O_1's community (Cohen 1978; DeAngelis 1992) or via a biogeochemical cycle, comprising both abiota and biota in an ecosystem (O'Neill et al. 1986; Jones and Lawton 1995).

These alternative consequences of niche construction, and alternative feedback routes, are of interest because they have the potential to link Grinnell's approach to niches, which are primarily concerned with how organisms survive and persist in their worlds, to Elton's approach to niches, which are primarily concerned with how organisms contribute to their communities and ecosystems, through the niche-constructing activities of organisms. In Leibold's terms, they potentially link the requirement niche to the impact niche, a combination he calls the "total niche." Focusing on niche construction may therefore lead to a fuller understanding of both communities and ecosystems. It also helps us to spell out when niche construction has evolutionary consequences, and when the consequences of niche construction are exclusively ecological. If there is no detectable modification of any natural selection pressure for any population in a community or an ecosystem as a consequence of a focal population's niche construction, then there will only be ecological consequences of niche construction. In this case the only consequence of niche construction is ecosystem engineering. If, on the other hand, as a consequence of the focal population's niche construction, one or more natural selection pressures are modified with a resulting evolutionary response anywhere in a community, or in an ecosystem, then niche construction will result in more than ecosystem engineering. It will then lead to evolutionary as well as ecological outcomes. At present, some of this evolutionary feedback, for example, that between coevolving populations in communities, is already well modeled by standard models (Futuyma and Slatkin 1983; Thompson 1994). However, because standard evolutionary theory does not yet incorporate niche construction, it is still not usual to model other kinds of feedback in evolution, for example, feedbacks that incorporate abiota in ecosystems (O'Neill et al. 1986). We will return to these issues in chapters 3 and 5.

2.2.4 Categories of Niche Construction

Two ways that organisms can change environmental factors in their niches and the selection pressures to which they are exposed are *perturbation* and *relocation*. Perturbation occurs if organisms actively change one or more factors in their environments at specified locations and times by physically changing them. It pertains to the causal impact that organisms may have upon their world. For example, organisms secrete chemicals, exploit resources, and construct artifacts. Relocation refers to cases in which organisms actively move in space, not only choosing the direction and/or the distance in space through which they travel, but sometimes also choosing the time when they travel. In the process, relocating organisms expose themselves to alternative habitats, at different times, and thus to different environmental factors.

In practice, most cases of niche construction are likely to involve some degree of both perturbation and relocation. For example, when animals construct artifacts such as nests and burrows, they commonly select the location of their abode with some care. Similarly, every time an organism moves, it necessarily alters the habitat from which it moved by depriving it of its presence and the habitat into which it moves by introducing itself. Moreover, as relocation is likely to result in the modification of multiple selection pressures, it may subsequently result in new trade-offs being made by organisms in response to the fitness effects associated with each modified selection pressure at their new location. Alteration of life-history strategies may ensue. A possible example is provided by the cuckoo family, where enhanced migration to exploit more open and seasonal habitats and a change in diet seemingly created selection pressures for reducing the costs of reproduction, leading to brood parasitism (Kruger and Davies 2001). In spite of these interactions, we retain our perturbation-relocation distinction for two reasons. First, we believe it to be a useful and logical dimension by which to organize and categorize the myriad cases of niche construction. Second, the distinction may be empirically important (see chapter 3).

A second dichotomous category of niche construction focuses on whether organisms initiate or respond to a change in an environmental factor. The changed factor might be either a conventional resource or a conventional condition. For example, the factor could be a food or a water source, the incidence of a particular chemical, the presence of a predator, the amount of detritus, a pH level, a humidity level, or any other significant biotic or abiotic component that is a source of selection in a population's niche.

We describe as *inceptive* niche construction all cases in which organisms initiate a change in any factor, through either perturbation or relocation. More precisely, organisms express inceptive niche construction when by their activities they generate a change in the environmental factors to which they are exposed, thereby changing the relationship between their own features and their environment's factors. Inceptive niche construction can occur either relative to an environmental factor that is not changing, or relative to a factor that is currently changing, provided the niche construction is not merely counteracting that change. Inceptive niche construction may be obligate, for example, if a population is forced to perturb its environment by polluting it with detritus, or facultative, for example, an opportunistic switch to a new habitat. By allowing established traits to be coopted for a different function, inceptive niche construction is also one process that could be responsible for converting previous adaptations and exaptations (Gould and Vrba 1982) into novel adaptations.

Conversely, if an environmental factor is already changing, or has changed, organisms may oppose or cancel out that change, a process we describe as *counteractive* niche construction. More precisely, organisms express counteractive niche construction when they either wholly or partly reverse or neutralize a prior change in an environmental factor. They thereby restore a match between their previously evolved features and their environment's factors. The prior change in the factor could be due to independent processes in the organisms' environments, to the organisms' own prior niche-constructing activities, or to the prior niche-constructing activities of other organisms. Counteractive niche construction is therefore conservative or stabilizing, and it generally functions to protect organisms from shifts in factors away from states to which the organisms have been adapted.

These two binary categories of niche construction are orthogonal. It follows that we can describe all cases of niche construction as *inceptive perturbation*, *counteractive perturbation*, *inceptive relocation*, or *counteractive relocation* (table 2.1). In the following section we illustrate each of these four categories of niche construction with examples. However, it is important to bear in mind that all niche construction is relative to specific natural selection pressures on specific populations. In reality, perturbational or relocational niche construction is likely to involve many different modified natural selection pressures, and not just a single factor. In the case of niche construction through perturbation, any full account of an instance of niche construction would also have to consider the direction in which an environmental factor is altered by niche construction, the size, rate, and persistence of the change caused by the niche construction, and whether the change in the factor is unidirectional, reversible, or

TABLE 2.1. Examples of the Four Categories of Niche Construction

	Perturbation	*Relocation*
Inceptive	*Organisms initiate a change in their selective environment by physically modifying their surroundings.*	*Organisms expose themselves to a novel selective environment by moving to or growing into a new place.*
	e.g., emission of detritus	e.g., invasion of a new habitat
Counteractive	*Organisms counteract a prior change in the environment by physically modifying their surroundings.*	*Organisms respond to a change in the environment by moving to or growing into a more suitable place.*
	e.g., thermoregulation of nests	e.g., seasonal migration

Note: Niche construction may be inceptive or counteractive and may occur through perturbation of the environment or through relocation in space.

oscillatory. Similarly, in the case of niche construction through relocation, we would have to consider the direction, distance, and rate of travel, and also whether the relocation is a migration (two-way) or dispersal (one-way). These variables could potentially have a profound impact upon the extent of ecological inheritance, and hence on the significance of niche construction.

Finally, we define as *positive niche construction* phenotypic activities that change environmental factors into states that on average increase the fitness of the niche-constructing organism, while *negative niche construction* refers to niche-constructing activities that change environments in such a way as to reduce fitness. For instance, positive niche construction might refer to processes that increase the frequency of a valuable resource, such as the fungal agriculture of leaf-cutter ants, or to the construction of a nest that buffers the builder from the extremes of temperature. Negative niche construction might result from severe depletion of a prey resource, leaving the population with a scarcity of food, or to a buildup of detritus that pollutes the local environment. Positive niche construction increases the fitness of niche-constructing organisms, relative to non-niche-constructing organisms, by changing one or more factors in the environment to enhance the match between features of the organisms and factors in their environments. Negative niche construction will typically reduce fitness by modifying factors in the populations' environments in a manner that leads to a weaker match between the features of the organisms and the factors in their environments. There are two further qualifications. One is that every niche-constructing act is bound to incur the metabolic cost of the act itself. Therefore, whether a niche-constructing act is positive or negative will be determined by its net effect. The second is that in the short run we expect almost all niche construction by individual organisms to be positive, as few organisms are likely to niche-construct in ways that reduce their immediate fitness. In the long run, however, the direct or indirect consequences of past niche construction may sometimes accumulate in ways that become negative for populations. The buildup of pollution by niche-constructing organisms is an example.

The impact of niche construction on the dynamic organism-environment relationship is depicted schematically in figure 2.1. Here, for illustrative purposes only, we assume that each organism at any point of time is described by a set of features or traits, for example, traits relating to binocular vision, an arboreal lifestyle, or a frugivorous diet. These traits of each organism are represented by an array of lower-case letters (*c, n, h, k, q, j*). Similarly, the organism's environment could hypothetically be decomposed into an array of factors, for instance, the local temperature, the presence of trees, or the presence of a predator, here represented by upper-case letters (*A, B, N, H, K, Q, Z, L*).

The usual description of an organism as adapted to its environment corresponds to a complementary matching of the organism's features and its environment's factors. In the highly simplified depiction of figure 2.1 a match between a feature of the organism and a corresponding factor in the environment is represented by the use of the same letter, while different letters represent a mismatch. Thus at time t in figure 2.1, the organism is well adapted to its environment according to the high level of feature-factor matching (*n-N, h-H, k-K, q-Q*), but there are also some mismatches (*c-B, j-Z*). At time $t + 1$, as a consequence of the action of natural selection, the match has

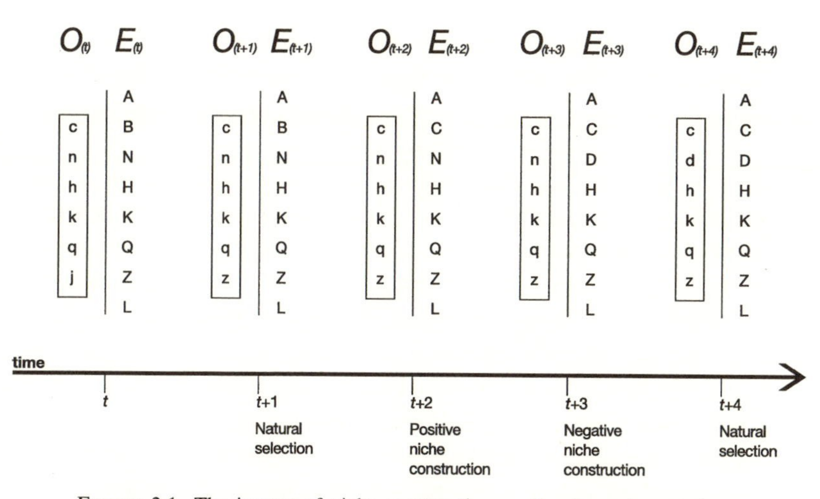

FIGURE 2.1. The impact of niche construction on the dynamic organism-environment relationship depicted schematically. It is assumed that, at any point in time, each organism **O** can be described by a set of traits (lower-case letters) and that the organism's environment **E** can be decomposed into an array of factors (upper-case letters). A trait of an organism is only an adaptation if and when it is matched to a specific selection pressure arising from an environmental factor as a direct result of prior natural selection. A match is represented by the use of the same letter in both **O** and **E**. A mismatch is represented by the use of different letters. Thus at time *t* there is a high level of matching between **O** and **E**. However, there are also some mismatches (*c-B*, *j-Z*). At time $t + 1$, as a consequence of natural selection, the (*j-Z*) mismatch is replaced by the match (*z-Z*). At time $t + 2$ the match has been further enhanced, but this time by positive counteractive niche construction. The organisms modifies environmental factor *B* to become *C*, generating a (*c-C*) match. At time $t + 3$ through negative and inceptive niche construction, the organism modifies environmental factor *N* to *D* generating a new mismatch (*n-D*). Subsequently at time $t + 4$ this (*n-D*) mismatch is replaced by a (*d-D*) match as a consequence of further selection.

been enhanced (resulting in *z-Z*) through the selection for individuals with feature *z* at the expense of those with feature *j*. At time $t + 2$ the match has been further enhanced, but now in a different manner, through positive and counteractive niche construction. Here the organism modifies environmental factor *B* to become *C*, generating the *c-C* match. For instance, the population of organisms offsets a scarcity of food in their environment (*B*) by relocating to a new environment that has more food (*C*). At time $t + 3$ through negative and inceptive niche construction, however, the organism modifies environmental factor *N* to *D*, generating a new mismatch, (*n-D*). For instance, the excrement of a population of burrowing mammal pollutes their burrows to the point where they become uninhabitable. This modification of the environment subsequently at time $t + 4$ leads to natural selection that favors individuals with feature *d* at the expense of those with feature *n*. For instance, in our mammal example, natural selection now favors individuals that deposit their feces at a latrine site away from the burrow.

Clearly, this scheme is a very simplistic representation of the constantly evolving organism-environment relationship, and many caveats and complications would have to be addressed if it were to be operationalized into a useful account of organism-environment coevolution. However, the rudimentary scheme in figure 2.1 allows us to illustrate two points. First, the complementary match between organism and environment can be brought about by two processes, and not just one, that is, by niche construction as well as through natural selection. Second, not all niche construction increases the fitness of the constructors. Negative niche construction may result in discordance between organisms and their environments, triggering a decrease in population numbers, subsequent natural selection, or further counteractive niche construction.

2.3 THE UNIVERSALITY OF NICHE CONSTRUCTION

We can now return to the question whether high-profile examples of niche construction, such as birds' nests and beavers' dams, are atypical, or whether all organisms, albeit to varying degrees, construct important components of their worlds. Examples of niche construction occur in all taxonomic groups, so ideally, to answer this question we should demonstrate the extent to which organisms in each of life's kingdoms construct niches (table 2.2). That task, however, is beyond the scope of this book. We will therefore do no more than make some general comments and offer in table 2.2 a few representative examples only for the first four kingdoms of life. Thereafter,

TABLE 2.2. The Universality of Niche Construction (illustrated with examples from all kingdoms)

Kingdom	*Niche Construction*
Bacteria (Archaebacteria and Eubacteria)	Decomposition of vegetative and animal matter, photosynthesis and production of oxygen, and ammonia, dimethylsulfide and nitrogen fixation. Bacterial allelopathy. Manipulating infected hosts.
Protista	Physical and chemical weathering of rocks, soil production, photosynthesis and production of oxygen; initiation of primary ecological succession.
Fungi	Decomposition of vegetative and animal matter. Lichens responsible for physical and chemical weathering of rocks, soil production, moisture retention, creation of environments favoring plants and microorganisms. Soil fungi extract minerals that travel to the roots of plants.
Plantae	Photosynthesis and production of oxygen; physical and chemical weathering of rocks, stabilization of bedrock; alteration of hydrology, chemical environment, pattern of nutrient cycling, temperature, humidity, stability, and fertility of soils, soil production; modification of nutrient cycles, importation and exportation of nutrients, storing of food and water; allelopathy; initiation and perpetuation of ecological successions; scattering and absorption of light, creation of shade; modification of wind speed.
Animalia	Construction of nests, burrows, protective cases, food stores, tools, defense, maintenance, and regulation; provision and protection of offspring and nursery environments; consumption of resources, selection of resources, import and export of nutrients, production of detritus; storage of food and water, farming of food resources; erosion of substrates; modification of chemical environment, pattern of nutrient cycling, temperature, humidity, stability, and fertility of

TABLE 2.2. (*continued*)

Kingdom	*Niche Construction*
	soils; generation and interception of light; generation and dissipation of heat; decomposition of vegetation and animal matter; aggregation; migration, dispersal; creation of paths; pollination of plants; transportation of other organisms; emission of signals.

we will concentrate on the animal kingdom,[2] from which we provide many more examples both in the text and in tabular form. Inevitably, our data must be regarded as massively incomplete, but they illustrate the points we need to make and may provide researchers with testable hypotheses related to niche construction.

2.3.1 Microorganisms: Bacteria and Protists

The first two kingdoms in table 2.2, the bacteria and the protists, consist entirely of unicellular microorganisms. Here the most striking point about all these organisms is that their "per capita" capacity for changing their environments by niche construction is tiny, and we might therefore be tempted to ignore microorganisms. In practice this is not possible because the ecological and evolutionary consequences of niche construction by microorganisms are enormously amplified by their superabundance, by their universality, and by the gradual accumulation of their effects in environments, as follows.

[2] Note that we concentrate on the Animalia because it is here that we have the greatest expertise, and we do not wish to imply that there is more niche construction by animals than by members of the other kingdoms. For instance, we anticipate that under close examination at least as many good examples of niche construction are likely to be identified at a local scale for plants. These might include soil modification; control of local gradients (e.g., in light availability or microclimate); provisioning of structurally complex habitats for many species (e.g., bark lice, epiphytes); providing resources for mutualists, such as Beltian bodies to feed protective ants, or pollen and nectar for pollinators; and the ability to detect appropriate time and conditions for germination.

(i) To the extent that microorganisms share the same genes, they express the same niche-constructing activities and affect their environments in the same ways. For example, all the organisms in a particular species typically take approximately the same resources from their environment and dump the same detritus in it. Populations of microorganisms can therefore act as powerful, unidirectional biological pumps for specific nutrients and gases, for example, CO_2, and they can significantly affect many environments by doing so even though this may take a long time (e.g., Falkowski 1997; Emerson et al. 1997). (ii) In favorable environments, microorganisms reproduce extremely rapidly. They have very short generation times, and they can become vastly abundant at any place in a very short time. Their large numbers also allow them to disperse, or be dispersed, widely, thereby spreading the potency of their niche construction. (iii) The short generation time of microorganisms also allows them to closely track environmental changes, by evolving rapidly in response to changing selection pressures, including selection pressures that may have been changed by the prior niche construction of the same or a different species. That potentially introduces a vigorous feedback to the evolution of microorganisms, and it may be one reason why they have been so conspicuously successful. The bacteria had the planet to themselves for approximately the first two billion years of life on earth, and during that time, as well as since, they changed many fundamental components of the biosphere through their collective niche construction. In doing so they made the subsequent evolution of more complex forms of life possible. Thus, even though the per capita capacity of microorganisms for niche construction is tiny, it appears to have had a huge influence on evolution on earth.

A good example of bacterial niche construction on the local scale is the production of allelopathic agents, or bacteriocins, by bacteria, that kill or inhibit the growth of other competing bacteria (Chao and Levin 1981; Levin 1988; James et al. 1996; Tan and Riley 1997). Colicins are an extensively studied group of bacteriocins. Under conditions of stress, a fraction of colicinogenic bacteria release colicin into the environment, killing competitors. An apparent evolutionary consequence of this bacterial niche construction is that colicinogenic cells produce an immunity protein that provides protection against their own colicin. The immunity protein recognizes its own colicin and inhibits killing. However, it is only cells carrying the same colicin plasmid that survive under conditions of colicin production (Tan and Riley 1997). Invasions by a novel colicin into an ancestral population have also been studied. Some colicin-encoding plasmids naturally possess an additional immunity gene, which expands their host's immunity and killing functions. Experiments have shown that when these "superkiller" strains are allowed to compete against their own ancestors, they rapidly displace the ancestral strain (Tan and Riley 1997). Some of the evolutionary consequences of these kinds of competition among bacteria have been modeled by Levin (1988), Durrett and Levin (1997), Iwasa et al. (1998), and Kerr et al. (2002).

Other examples of microorganismal niche construction occur on a large scale. Some are widely known, such as the probable contribution of microorganisms to the replacement of the earth's early anaerobic atmosphere by an aerobic atmosphere, which today comprises approximately 21% oxygen (O_2). This process is thought to have begun with the evolution of photosynthesizing cyanobacteria more than two billion years ago (Lovelock 1979; Stanley 1989; Odum 1989), and photosynthesizing microorganisms continue to influence the earth's atmosphere (H. Holland 1995). For a long time O_2 was slow to build in the atmosphere because "oxygen sinks" in the earth's crust, in the form of elements that combine readily with oxygen, notably sulfur and iron, were filled first. When atmospheric O_2 did start to accumulate, however, it must have fed back in the form of modified natural selection pressures to affect both the subsequent evolution of bacteria and, eventually, the evolution of innumerable other organisms too. One bacterial example of this feedback was the evolution of nonphotosynthetic, marine sulfide-oxidizing bacteria between 1.05 and 0.64 billion years ago, which Canfeld and Teske (1996) argue was probably driven by an increase in the concentration of atmospheric O_2 from less than 5% to 18% of its present level. At approximately the same time different kinds of photosynthesizing microorganisms, Protista from the second kingdom in table 2.2, the eukaryotic algae, for example, also began to contribute to both atmospheric and oceanic O_2.

Microorganisms also make numerous other large-scale contributions to the biosphere. For instance, marine phytoplankton scatter and absorb light in the upper layers of water columns, enhance the warming of surface waters, initiate the development of thermoclines, and play an important role in nitrogen fixation (Jones et al. 1994; Capone et al. 1997). Marine phytoplankton and zooplankton contribute significantly to the global carbon cycle by acting as biological organic carbon pumps in subtropical oceans, and by affecting the exchange of O_2 and CO_2 between those oceans and the atmosphere (Emerson et al. 1997; Doney 1997). Microorganisms contribute to other gas exchanges between the oceans and the atmosphere as well, for example, those involving dimethylsulfide (DMS), a gas that influences cloud nucleation (Kiene 1999). In this case it is apparently not just phytoplankton but whole food webs that release DMS as a by-product of complex food-web interactions (Simo 2001). It has also been argued that the release of DMS may contribute to climate regulation by increasing cloud albedo (Charlson et al. 1987), but that idea is more contentious (Simo 2001).

Microorganisms, including both bacteria and protists, build large microbial mats in shallow seas and have done so for billions of years (Reid et al. 2000). Stromatolites, which are fossilized microbial mats, are among the oldest fossils on earth (Golubic 1992a,b). Bacteria affect muds and sediments. Turner (2000) describes how different species of bacteria sort themselves according to the redox potential gradient in muds, with aerobic bacteria near the surface and anaerobic bacteria below them, and how by doing so they build complex mutually dependent systems. Bacteria in the waterlogged anoxic portions of wetland soils contribute to another atmospheric gas, methane (Bubier and Moore 1994). Finally, microorganisms in the form of protozoal, bacterial, and in this case viral pathogens, too, generate diseases in multitudes of other organisms, and when they do, they often alter the anatomy, physiology, or behavior of their hosts in ways that may have the effect of increasing their own transmission between hosts (Ewald 1994; Diamond 1998).

2.3.2 Multicellular Organisms: Fungi and Plants

The three remaining kingdoms in table 2.2 (two of which we deal with in this section) refer to multicellular organisms which, because of their larger size, have greater per capita capacity for niche construction than microorganisms. Individual multicelullar organisms can also express a greater diversity of niche-constructing activities. There are, however, no conceptual differences between the niche-constructing activities of multicellular organisms and those of micro-organisms, and the logic that describes the evolutionary and ecological consequences of niche construction by organisms in every kingdom is the same.

The niche-constructing activities of fungi primarily consist of their contributions to the processes of decomposition. In this respect fungi are obvious ecosystem engineers. One simple example, cited by Jones et al. (1997), is rot fungi, which make holes in trees that are subsequently used by other species (Kitching 1971; Bradshaw and Holzapfel 1992). A larger-scale example is the sequence of fungal species that colonize pine needles in the litter generated by forests of Scots pines (*Pinus sylvestris*) (Kendrick and Burges 1962; Richards 1974). Begon et al. (1996) cite this sequence as a good example of degradative succession. The sequence starts before the pine needles have been shed by the trees, with the presence of *Coniosporium* on approximately 50% of living pine needles, and it ends approximately ten years later when the pine needles are no longer recognizable, after they have been slowly degraded for several years by basidio-mycetes to humus in the soil. An ecological feedback occurs here because humus changes the structure and properties of the soil, which affects subsequent plant growth and the production of litter.

In other cases evolutionary feedback is more visible (Jennings and Lysek 1999). Consider the production of antifungal compounds by fungi that have already invaded a host and that subsequently defend their host from rival fungi. For instance, the trunks of beech trees that have been invaded by *Oudemansiella mucida* are apparently never hosts to any other wood-decaying fungus because the trees are protected by the antifungal compound mucidin. Similarly, the alkaloid production of the ergot fungus *Claviceps purpurea* deters animals from feeding on its plant hosts. A different kind of example, also from Jennings and Lysek, concerns the ability of the dry-rot fungus *Serpula lacrymans* to relocate by spreading around buildings. *S. lacrymans* is very susceptible to dry conditions, but it can partly offset them by transporting water through its mycelium and losing water at its growing mycelial front, causing its local environment to favor its growth and propagation.

Another large-scale example of niche construction by fungi that almost certainly has had evolutionarily significant consequences is provided by mycorrhizae. Mycorrhizae form mutually beneficial associations with the roots of woody plants. They come in two broad classes, endomycorrhizae, which penetrate the root tissue of plants and form full symbiotic relationships with them, and ectomycorrhizae, which grow sheaths over the surfaces of plant roots. We will focus on the commonest type of endomycorrhizae, vesicular-arbuscular (VA) mycorrhizae, which benefit plants by providing them with enhanced supplies of mineral nutrients, notably phosphorus, while benefiting themselves by taking carbon from the plants. It is likely that the evolution of these symbiotic relationships originally involved some coevolution, and therefore some prior niche construction (see chapter 3).

Law (1985) suggests that for symbioses, in general, genetic change should be slower for the "inhabitant" species, by which he means the internal partners in symbioses, here the fungi, because they should be selected for relative constancy, compared with their "exhabitants," here the plants, which must continue to adapt to an external world. Law (1985) provides some evidence that in VA mycorrhizae this is true. Similarly, he argues, and provides some supporting evidence, for less sexual reproduction, and more asexual reproduction, in inhabitants than in exhabitants. Some of Law's assumptions can be challenged; for instance, he supposes that there is a general tendency for antagonistic interactions between species to be replaced by mutualistic ones, an idea that is opposed by Hibbett et al. (2000), who found that ectomycorrhizal symbioses are evolutionarily unstable. In addition, Law himself discusses alternative possible interpretations of the data. However, all of these interpretations involve feedback from prior niche construction by fungi.

One more example may clinch this point; the colonization of polluted soil by a plant plus its mycorrhizal symbiont (Sharples et al. 2000). The pollution consists of arsenate contamination in the vicinity of an arsenic and copper mine. Mycorrhizal fungi usually enhance both phosphate and arsenate uptake in their plant hosts. However, under these contaminated conditions, Sharples et al. found a plant host, *Calluna vulgaris*, and its mycorrhizal symbiont, *Hymenoscyphus ericae*, that appeared to have evolved in parallel to obtain phosphate and exclude arsenate. The crucial adaptation seems to have been the evolution of enhanced arsenic efflux by *H. ericae*. Sharples et al. suggest that this fungus dominates arsenate-phosphate accumulation and acts as a filter to maintain low plant arsenic levels through arsenic efflux, while enhancing the plants' phosphorus status. In this way the fungus modifies its host's environment, and hence its own.

The fourth kingdom in table 2.2 comprises the plants themselves (Wilson and Agnew 1992; Mora et al. 1996). Plants photosynthesize too, so along with photosynthesizing bacteria and protists, they also contribute oxygen to the earth's modern atmosphere, and at such a rate that the entire inventory of atmospheric O_2 could be produced in as little as 4,000 years (H. Holland 1995). Most of the O_2 produced by plants, however, is consumed almost immediately in respiration. Hence, the production of O_2 by terrestrial plants is probably less important ecologically than another closely related function, their capacity to act as sinks for carbon and to contribute to the carbon cycle. The role of tropical rain forests as sinks for CO_2 is topical because of human-induced deforestation and associated fears about the accumulation of CO_2 in the atmosphere and its anticipated consequence, global warming. These fears are an implicit acknowledgment of the role of niche construction by forests. Remove the trees, and their niche-constructing activities become noticeable by their absence (Houghton et al. 2000; Carvalho et al. 2001).

Shukla et al. (1990) point out that forests also contribute to another equally vital ecological cycle, the hydrological cycle, through the retention and evapotranspiration of water, and that by doing so they may affect their own weather (Shukla et al. 1990; Holling 1992). Other kinds of plants, for instance, species of bog-forming *Sphagnum* mosses, can also profoundly affect local hydrology (Tansley 1949; Jones et al. 1994). In the sea, niche construction by marine plants may take forms not observed on land, but it too can have significant consequences. One example here is the formation of kelp forests, which benefit from, and modify, the actions of waves near shorelines, and by doing so create relatively protected three-dimensional submerged habitats for many other species, for instance, nursery grounds for lobsters (Barnes and Hughes 1999).

These are examples of ecological consequences of plant niche construction, but their evolutionary consequences are again harder to pin down. It may therefore help to consider other phenomena on smaller scales. One candidate is allelopathy in plants. Many plants change both their own and other species' local environments by emitting toxins that kill competitors (Rice 1984). Rasmussen and Rice (1971), for example, found that a small grass species (*Sporobolus pyramidatus*) can spread rapidly into heavy sods of Bermuda grass and buffalo grass by exuding toxins from its living roots and decaying shoots, thereby inhibiting seed germination and growth in other species. As in bacteria, the most obvious evolutionary feedback here occurs in the plants' protection of themselves from their own toxins. Williamson (1990) discusses several ways in which plants can do this. He also points out that allelopathy in plants is difficult to study because the conditions in which the toxins act are often the same as those in which many other competitive interactions occur as well. Nevertheless, Williamson concludes that there is now compelling evidence that allelopathy occurs.

A second example of an evolutionary response by plants to prior plant niche construction is fire resistance and flammability. Mutch (1970) suggested that if species have evolved a capacity to survive periodic fires, then fire-dependent plants might also evolve traits that enhance the flammability of their own communities. Mutch's hypothesis was criticized by Snyder (1984) and Troumbis and Trabaud (1989) on the grounds that it is evolutionarily implausible. More recently, analyses by Bond and Midgley (1995) and Zedler (1995) have shown that it could occur, while Kerr et al. (1999) demonstrate how Mutch's hypothesis becomes more plausible if niche construction is explicitly included in the scenario. Regardless of these theoretical disputes, there is now considerable empirical evidence to indicate that fire resistance and adaptations to fire covary across plant species, with niche construction implicated in the evolution of these traits. For example, pine and chaparral trees accumulate oils and litter and in the process encourage fires (Mount 1964; Allen and Starr 1982). Many plants, for instance, indigo species, smaller, shorter plants, and nitrogen-fixing legumes, are known to have evolved a resistance to fires, and some pines, for example, require fire before their own seeds will germinate (Tilman 1997; Kerr et al. 1999). Williamson and Black (1981) found that specific litter deposition patterns by different species of trees (oaks and pines) are a major cause of different intensities within fires, and they argued that, since these litter traits are themselves subject to natural selection, different species can be regarded as relatively fire-facilitating or fire-retarding. They further suggested that early successional plants that are poor competitors may exhibit fire-facilitating traits, while late successional plants that are superior competitors may show fire-retarding traits. Finally, Williamson (1990) suggested that allelopathy may have evolved to reduce fire risk and that the same mechanism, protection from fire, may also have resulted in selection for allelopathy by shrubs in the Californian chaparral. If so, that would constitute a further evolutionary feedback loop.

2.4 ANIMAL NICHE CONSTRUCTION

Animals are the fifth kingdom in table 2.2, and for them we can give many more examples. For most of the hundreds of examples of animal niche construction that we present in this section, more than a single species is represented, and in many cases, tens, hundreds, or even thousands of species are involved. We will also indicate how these examples may be categorized according to the taxonomy in table 2.1, which shows four main kinds of niche construction. The reader may find reviews by Wcislo (1989) and Rossiter (1996) relevant here, too.

2.4.1 Inceptive Perturbation

Examples of inceptive perturbatory niche construction in animals are so well known that we tend to take them for granted. Animals construct nests, burrows, and pupal cases, provision nursery environments for their offspring, generate detritus, and much more. While the fact that these processes occur may be no surprise, the number and range of species involved is astonishing. For instance, there are 9,500 known species of ants and 2,000 known species of termites (Isoptera), almost all building nests, which are frequently elaborate structures with complex features or associated behavior to regulate temperature, humidity, and gas exchange. Ants and termites dramatically change the mineral and organic composition of the surrounding soil, alter local hydrology and drainage, and consume vast amounts of terrestrial litter (Hölldobler and Wilson 1994; Pearce 1997). There are 20,000 species of solitary bees, with immensely varied nests, and many social bees construct nests too (Gullan and Cranston 1994). There are at least 13,000 species of gall-forming insects. These chemically manipulate the morphology of plant parts to construct the gall, which serves as food, shelter, or protection from natural enemies. There are more than 7,000 species of caddis fly (Trichoptera), most of whose larvae use their silk, vegetation, and stones to build fixed or portable shelters, or even to construct foraging tools, thereby defending themselves, and exploiting their habitats in other ways (Hansell 1984; Gullan and Cranston 1994). Some 1,800 species of earwigs (Dermaptera) build nests (Gullan and Cranston 1994). Moreover, there are 140,000 described species of butterflies and moths, most of which construct a pupal cocoon (Gullan and Cranston 1994). Almost all of the 34,000 or more species of spider construct a silk shelter or sac to enclose and protect their eggs, while those that do not spin webs commonly dig burrows or make nests (Preston-Mafham and Preston-Mafham 1996).

There are more than 9,000 species of birds, the vast majority of which construct nests (Forshaw 1998), and probably as many species of fish that do the same or construct elaborate spawning sites and bowers (Paxton and Eschmeyer 1998). While the burrow complexes of land mammals such as moles and rabbits are well known, the burrowing behavior of the 156 species of caecilians (order Gymnophiona), the 234 species of wormsnakes (Anomalepididae, Typhlopidae, and Leptotyphlopidae), the 900 species of gecko (Eublepharidae and Gekkonidae), and the 1,300 species of skinks (Scincidae) is less familiar. There are 135 species of chameleons that bury their eggs in simple nests, while all 250 species of turtles, terrapins, and tortoises (Testudinata) construct a nest chamber. Fossil evidence suggests that vertebrates were digging burrows as early as the Devonian, more than 400 million years ago (Benton 1988).

Many organisms construct niches in ways that significantly alter the environment of their offspring, or subsequent generations. The nests of insects, birds, and fish, the nest chambers built in reptile, amphibian, and mammal burrows, and the egg sacs of spiders all serve as constructed nursery environments for offspring. Frequently the nest, burrow, or some other constructed entity is inherited from one generation to the next. In many birds, such as swallows, weaverbirds, and virtually all cooperative breeding birds, nests are reused, or nest sites inherited, by offspring (Hansell 1984; Skutch 1987). The same may be said of many mammals. For instance, the dens of many foxes, arctic foxes, and wood rats, the nests and burrows of ground squirrels and prairie dogs, the runways of American harvest mice, and the dams of beavers are all inherited from one generation to the next, for up to tens or hundreds of years (Nowak 1991). Insects also inherit resources. As we noted in chapter 1, prior to her nuptial flight, the leaf-cutter ant queen takes into her mouth some of the fungus that these ants cultivate, and when she reaches her new chamber, she spits it out (Hölldobler and Wilson 1994). Here the fungus constitutes an ecological inheritance.

2.4.2 Counteractive Perturbation

Much counteractive perturbatory niche construction is itself an evolved response to ancestral niche construction. For example, the nests of social bees, wasps, ants, and termites are the source of selection for many nest regulatory behaviors, which counteract fluctuations in temperature, humidity, or other factors. Vespoid wasp workers maintain the comb at an appropriate temperature, heating it through muscular activity or moistening the cells so they are cooled by evaporation (Spradbery 1973; Mathews and Mathews 1978). Similarly, many insect, reptile, and mammal species construct elaborate nests or burrows with complex structures to regulate hygiene, temperature, humidity, and gas exchange (Frisch 1975; Hansell 1984). Examples of evolutionary responses to niche construction are treated in greater detail in the following section. Here we merely describe some broad classes of counteractive perturbation.

Many niche-constructing behaviors have the effect of moderating humidity. Several species of burrowing frogs (Hemisotidae, Rhinophrynidae, Pelobatidae, *Cyclorana* spp.) and swamp eels (order Synbranchiformes) burrow into river beds or lake bottoms in order to estivate (remain dormant) until the next rainy and breeding season (Cogger 1998). Many species of turtle also exhibit such burrowing and estivation behavior, for example, side-necked turtles (suborder Pleurodira) and some softshell turtles (family Trionychidae). Some beetles (family Tenebrionidae) that live in the arid coastal region of the Namib desert construct shallow trenches that trap dew (Hansell 1984) as an alternative strategy to estivation.

Animals also go to great lengths to avoid extremes in temperature. Many mammals construct nests for hibernation (see Lyam et al. 1982 for a review). The winter hibernation of tortoises (family Testudinidae) in temperate regions is well known, but in warmer regions they utilize the burrow daily to avoid the midday sun (Cogger 1998). The beaded lizards (family Helodermatidae) also spend much of their time in burrows avoiding extremely high temperatures. Conversely, macroteiid lizards (family Teiidae) retreat to a burrow during periods of inclement weather and during the colder months (Cogger 1998). Some animals are able to take advantage of the niche construction of others in order to counteract environmental temperature changes. Simply through excavating a nest cavity in an unoccupied termite mound, the Nile monitor (*Varanus niloticus*) provides her eggs with a suitable and carefully regulated microclimate of humidity and ventilation (Cogger 1998). Grass snakes (*Natrix natrix*), in colder regions, lay their eggs in manure heaps of livestock as they require high temperatures to incubate (Cogger 1998).

Birds and mammals also damp out natural fluctuations in food availability by hoarding or storing food at appropriate times. Food storing is common in many birds, such as corvids (Corvidae) and tits (*Parus* spp.) (Forshaw 1998), as well as among many mammals (Nowak 1991). For instance, as autumn approaches, the beaver (*Castor fiber*) cuts down saplings and sinks them in a lake for retrieval throughout winter (Frisch 1975). The red fox (*Vulpes vulpes*) also has food-storage holes that can be reached through tunnels from the den (Nowak 1991).

2.4.3 Inceptive Relocation

Examples of niche construction through relocation are as plentiful as those by perturbation and are equally familiar. In all such cases, when organisms move, they potentially expose themselves to different selection pressures. Many species of animals relocate in the wake of ecological successions initiated and perpetuated by plants and algae. Animals select habitats, nesting sites, and developmental environments for their offspring. They also disperse from one environment to another, frequently when their own perturbation has significantly exploited the resident resources. Most parasites move through space until they find their hosts, which not only serve as food or a dwelling, but more frequently as a nest site or food source for their offspring. Through their movements, organisms also import and export nutrients to and from local environments, playing vital roles in the dynamics of resource flow through ecosystems (Jones et al. 1994, 1997).

We have described nest building and web spinning as examples of bird and spider niche construction that involve perturbation, but virtually all birds carefully choose the location of their nests (Attenborough 1998; Forshaw 1998), and there is evidence that spiders actively select the location of their webs (Turnbull 1964). Eusocial bees from temperate zones tend to select enclosed nest sites in cavities of around 40 liters in volume, while tropical bees choose smaller cavities or construct nests outside, for example, hanging them from a tree (Hansell 1984). Many colonies of ants are concentrated beneath rocks or in the bark of decaying stumps and logs, locations that warm in the sun faster than does soil (Hölldobler and Wilson 1994). Thousands of insect parasites of ants (mites, silverfish, millipedes, flies, beetles, wasps, and other small creatures) relocate to actively chosen nests of particular ant hosts (Hölldobler and Wilson 1994).

Numerous species of organisms also select a location for their nest, eggs, or offspring and in the process change the environment for their offspring and create an ecological inheritance. The vast majority of the 5 to 80 million species of insects are oviparous, and usually the eggs are deposited on or near the food required by the offspring upon hatching (Gullan and Cranston 1994). This is probably one of the most frequently documented cases of ecological inheritance. The offspring of virtually all insects inherit from their mother the legacy of an appropriate source of larval food. There are hundreds, if not thousands, of species of mouthbrooder fish that use their own mouths as a safe environment in which to relocate their offspring (Paxton and Eschmeyer 1998). In some cases a dispersal is triggered by ecological factors, rather than being a fixed feature of the organism's lifecycle. For example, lemmings (*Lemmus* spp.) that exhaust their food supply may aggregate in huge populations and exhibit large-scale dispersals, heading down mountainsides into valleys and crossing rivers.

2.4.4 Counteractive Relocation

Migration refers to cases where organisms relocate and then later return, in the process avoiding, and thereby counteracting, temperature or other climatic extremes, or food or water shortages, or, in general, transitory environmental states to which they are not well adapted. By moving, organisms negate the selection pressures that, because of the occurrence of fluctuations, would have acted on them had they not moved. Organisms may thereby generate selection pressures in favor of enhanced relocatory ability, rather than adapting to fluctuations in ecological variables, for example, in food availability or temperature, in a single location. Many pelagic fish (fish that live in the midwaters between 200 and 1000 m deep) undertake a daily migration from the daytime depths of 500–1000 m to the upper 200 m at night, to feed in the rich surface water (Paxton and Eschmeyer 1998). Both dragonfishes (order Stomiiformes), a group of approximately 250 species of mostly mid- or deep-water fish, and lanternfishes (family Myctophidae), the most widely distributed, the most species diverse, and the most abundant of all fishes in the deep ocean seas, migrate nightly from resting areas in mid- or deep waters, up to the more productive areas near the surface, where food is plentiful (Paxton and Eschmeyer 1998).

Much animal migration is on an annual or seasonal time scale. Numerous species of birds migrate from their breeding grounds to avoid the harsh weather conditions and food shortages of the winter and return to take advantage of the plentiful supply of insects, seeds, fruits, and other foods available in the spring and summer. For example, European swallows (Hirundinidae) and swifts (Apodidae) migrate to Africa and return each spring. Similarly, hirundines breeding in North America winter in Central and South America. In the northern hemisphere, migratory birds typically move south for the winter, with birds that nest in the southern hemisphere typically migrating north, for the same reasons. However, some mountain birds merely move down to the valleys in winter. The relevant consequence of these migrations is that migratory birds can avoid environmental changes that they would encounter if they did not migrate. In this way they can counteract unfavorable seasonal conditions to which they are poorly adapted. They, and their offspring, may also continue to encounter environments where the seasonal conditions are favorable and to which they are better adapted.

Probably for similar reasons, some fish are known to frequent particular spawning sites, and their migrations are so reliable that predators and human fishers alike have learned some of these localities and spawning times and are able to take a heavy toll. The advantage of this seasonal trek is that both adults and developing larvae can each live where they have the richest feeding grounds. Some species of Clupeiformes seasonally migrate up river to spawn in fresh water (Paxton and Eschmeyer 1998). A similar seasonal migration is found in some freshwater fish, such as the Characiformes (Paxton and Eschmeyer 1998). Tunas and their relatives (suborder Scombroidei) engage in unparalleled long-distance migrations (Paxton and Eschmeyer 1998). In one case a northern bluefin tuna traveled nearly 5,000 miles across the Atlantic Ocean in just 119 days!

Seasonal migration is also found in many crocodiles, iguanas, and turtles, the latter migrating up to 5,000 km to reach nesting beaches where they bury their eggs (Cogger 1998). Many species of bats (Chiroptera) also migrate, apparently for reasons similar to birds

(Nowak 1991). Plains zebras (*Equus burchelli*), wildebeest (*Connochaetes taurinus*), and Thomson's gazelles (*Gazella thomsonii*) make an annual trip across the African plains, following the rains, in search of fresh grass (McFarland 1987). Whales swim thousands of kilometers to return to traditional mating and calving areas (McFarland 1987).

For some species, migration is a once-in-a-lifetime event. The most famous case is that of the salmon (family Salmonidae), many species of which are born in freshwater, migrate to sea as juveniles to live and grow to adulthood, and then return to the home stream for spawning (Paxton and Eschmeyer 1998). Smelts (family Osmeridae), galaxids (family Galaxidae), lampreys (order Petromyzontiformes), and many freshwater gobies (suborder Gobioidei) exhibit a similar life cycle. Freshwater eels (family Anguillidae) exhibit the reverse pattern.

Many insects make long journeys at particular stages in their life cycle. Butterflies are renowned for their seasonal migrations. Every autumn the monarch butterfly (*Danaus plexippus*) makes a southward migration from Canada and the United States to the southern states and Mexico, where it overwinters. Similar patterns of migration are found in Europe in the red admiral (*Vanessa atalanta*) and the painted lady (*V. cardui*) butterflies (McFarland 1987). Locusts (Acrididae) gather in swarms and migrate downwind into areas of low pressure, where they are most likely to encounter rain, which stimulates them to stop and reproduce (McFarland 1987).

2.5 EVOLUTIONARY CONSEQUENCES OF NICHE CONSTRUCTION

The previous sections illustrated how organisms construct niches and in so doing change their world. From the flatworm (*Dendrocoelum lacteum*) that secretes a mucous sticky patch to trap passing isopod prey (Hansell 1984) to the "diving bell" woven by spiders of the genus *Argyroneta* that contains a bubble of air allowing the spider to hunt underwater (Turner 2000); from the moss (*Sphagnum* spp.) that transforms the hydrology, pH, and topography of peat bogs (Jones et al. 1994) to the vast urban metropolises built by humans, organisms across all known taxonomic groups construct important components of their local environments. Yet if organisms significantly change environmental factors, through either perturbation or relocation, it follows that they may also change the pattern of natural selection to which they and subsequent generations will be exposed.

In this section, we focus our attention on the traits that may have evolved as a direct consequence of that modified natural selection. We document these traits because they represent plausible cases of *recipient traits* that might not have evolved without prior niche construction. By recipient traits we mean features of an organism that are exposed to selection pressures apparently modified by this prior niche construction. These recipient traits may have coevolved with the niche-constructing traits that were responsible for their selection, thereby representing potential instances of the kind of unusual evolutionary dynamics predicted by theoretical analyses (Feldman and Cavalli-Sforza 1976; Kirkpatrick and Lande 1989; Laland et al. 1996, 1999; Robertson 1991). Here we focus on the evolutionary consequences of niche construction for the niche-constructing species itself. In the penultimate section we describe cases with important evolutionary consequences for other species as well.

The evolutionary consequences of niche construction are inferred from correlations between niche-constructing traits and other characters. It is important to note that in most cases there is as yet no direct evidence that the traits that we regard as likely evolutionary responses to the selection pressures modified by niche construction are indeed derived traits. Thus we cannot rule out the possibility that some such traits predated the evolution of the pertinent niche construction. For a number of reasons, however, these characters are worth investigating.

First, in a significant number of cases there is either a well-resolved phylogeny or strong comparative evidence suggesting that the niche construction preceded the traits that we describe as recipient traits. Second, in many cases, the niche construction has set up selection pressures favoring an elaboration of the constructed resource or a more efficient construction, indicating that the original niche construction must predate the elaboration. Third, while we accept that it is possible, even likely, that some of the traits that we describe as evolutionary responses to niche construction may predate the niche-constructing trait, it is unlikely that they all do. Fourth, and perhaps most important, the correlations between niche-constructing traits and other characters that may have evolved as a consequence of this niche construction constitute data that can be employed in empirical tests of niche-construction theory. We discuss how this can be done in the final section and in chapters 7–9.

2.5.1 Evolved Responses to Perturbational Niche Construction

We begin by considering the likely evolutionary consequences of niche construction that involved the prior perturbation of an environment. These include selection for anatomical and behavioral adaptations that allow organisms to carry out their niche construction with greater efficiency (table 2.3), selection favoring the further elaboration of a previously constructed resource or artifact (table 2.4), selection favoring regulation of a constructed resource or artifact (table 2.5), and perturbations that have generated selection for modified courtship, mating, and parental behavior (table 2.6).

2.5.1.1 Anatomical and Behavioral Adaptations to Perturbation

Many animals that dig burrows or build nests exhibit characters that appear to be anatomical or behavioral adaptations to their ancestors' perturbatory niche construction (table 2.3). The larvae of many soil insects have well-developed legs to permit active movement through the soil, and pupae frequently have spinose transverse bands that appear to assist their movement to the soil surface for eclosion (Gullan and Cranston 1994). Many soil-dwelling insects have reduced eyes and their wings are either protected by hardened forewings, reduced, lost altogether, or shed after the dispersal flight, as in the reproductives of ants and termites.

Web spiders appear to have evolved responses to the threat of predation on the web and an ability to communicate on the web (Preston-Mafham and Preston-Mafham 1996). Spiders in the genus *Cyclosa* (also *Gasteracantha* and *Ulborus*) construct dummy spiders out of silk and prey remains, presumably to divert the attention of

TABLE 2.3. Anatomical and Behavioral Adaptations That May Be Evolutionary Responses to Prior Perturbational Niche Construction

Organism	*Perturbational Behavior*	*Resource*	*Anatomical and Behavioral Adaptations to Perturbation*	*Reference*
Insects				
Soil insects	Dig burrow in soil	Burrow/nest	Well-developed larval legs, spinose transverse bands on pupae, wings protected by hardened forewings, reduced, lost, or shed after dispersal	Gullan and Cranston (1994)
Ant lions (Myrmeleontidae)	Dig conical pit to trap ants	Pit	Jerk to create landslides, flick sand at prey	Gullan and Cranston (1994)
Potter wasps (Eumenidae)	Construct mud cell/clay pot	Mud nest	Moisten mud with regurgitated water, nest camouflage, and provisioning	Frisch (1975)
Parasitoid insects	Construct burrow to contain young/prey	Burrow	Plug burrow, provision burrow	Evans and West-Eberhard (1970), Gullan and Cranston (1994)
Mole crickets (Gryllotalpidae) and some cicadas and beetles	Construct nest for young	Burrow/nest	Forelimbs modified for digging, lick eggs preventing mold infection	Gullan and Cranston (1994), Preston-Mafham and Preston-Mafham (1996)
Bark beetles (Passalidae and Scolytidae)	Construct nest/pupal case	Nest/fungus	Fungus cultivation as beetle predigests wood	Preston-Mafham and Preston-Mafham (1996)

Arachnids				
Trap-door spiders (Ctenizidae) and giant trap-door spiders (Liphistiidae)	Dig tubelike burrows lined with silk	Burrow	Rakelike modified basal cheliceral segment, trap-door holding and tying behavior, trip lines, sensitivity to frequency of trip-line vibration	Preston-Mafham and Preston-Mafham (1996)
Water spider (*Arggroneta aquatica*)	Construct underwater nests	Nest	Low metabolic rate	Preston-Mafham and Preston-Mafham (1996)
Web spiders	Construct funnel or sheet webs	Web	Less powerful eyes and legs, extra claw on each leg, throwing silk	Preston-Mafham and Preston-Mafham (1996)
Some *Cyclosa, Gasteracantha*, and *Ulborus*	Web building	Web	Build dummy spiders	Preston-Mafham and Preston-Mafham (1996)
Other *Cyclosa*	Web building	Web	Build stabilimentum	Preston-Mafham and Preston-Mafham (1996)
Theridion, *Tetragnatha*, and *Areneus*	Web building	Web	Drag leaves and other material onto webs for protection	Preston-Mafham and Preston-Mafham (1996)
Segestria, *Mallos*, and others	Web building	Web	Communicate by plucking messages on web	Preston-Mafham and Preston-Mafham (1996)
Ogre-faced spiders (Deinopidae)	Web building	Web, net	Improved/enlarged eyes, marks ground with white feces to ease detection of its prey	Preston-Mafham and Preston-Mafham (1996)

TABLE 2.3. (*continued*)

Organism	*Perturbational Behavior*	*Resource*	*Anatomical and Behavioral Adaptations to Perturbation*	*Reference*
Fish				
Some conger eels (Congridae)	Burrow in sea bed	Burrow	Enlarged eyes, binocular vision, upturned mouths, specialized tail tips, lateral line and mating-behavior modifications, mucus-secreting organs	Paxton and Eschmeyer (1998)
Snake eels (Ophichthidae)	Burrow in sand or mud	Burrow	Loss of all fins, movement of the posterior nostril to within the mouth	Paxton and Eschmeyer (1998)
Weeverfishes (Trachinoidei)	Burrow into sand	Self-burial	Eyes positioned on top of head and directed upward	Paxton and Eschmeyer (1998)
Amphibians				
African shovel-nosed frogs (Hemisotidae)	Dig burrows	Burrow	Pointed snouts with a hardened tip for digging, nest guarding, channel-tunnel digging enabling tadpoles to reach water, estivation (torpid state)	Cogger (1998)

Frogs with spadelike tubercle (Pelobatidae, Microhylidae, etc.)	Dig	Self-burial	Keratinized spadelike tubercle on the inner edge of each hind foot, estivation (torpid state)	Cogger (1998)
Water-holding frog (*Cyclorana platycephala*) and others	Dig	Skin/burial chamber	Loose baggy skin, sheds its skin, lining of shed skin with mucus to form a waterproof barrier that prevents desiccation	Cogger (1998)
Reptiles				
Pipesnakes and shield-tailed snakes (Aniliidae and Uropeltidae)	Burrow	Burrow	Skull bones solidly united, enlarged ridged or spined scale near tip of tail creating soil plug versus predators, ridged scales reduce friction against soil	Cogger (1998)
Some alligator lizards, galliwasps, and glass lizards (Anguidae)	Dig (or steal) burrows	Burrow	Loss of limbs, loss of external ears, burrow-stealing behavior	Cogger (1998)
Worm snakes (Anomalepididae, Typhlopidae, Leptotyphlopidae)	Burrow	Burrow	Blunt head, short tail tipped with a small spine to assist movement in tunnels, ant trail following, chemical defenses against ants	Cogger (1998)

TABLE 2.3. *(continued)*

Organism	*Perturbational Behavior*	*Resource*	*Anatomical and Behavioral Adaptations to Perturbation*	*Reference*
Some pythons (Pythoninae)	Burrow/coil around eggs	Nest/ warmth	Crush prey against side of burrow as they cannot constrict prey, extend range into colder regions	Cogger (1998)
Worm lizards (Amphisbaenia)	Burrow in sand/soil	Burrow	Heavily ossified spade- or keel-shaped skull, recessed mouth. Loose skin, muscular adaptations for tunnel movement, cone-shaped tail that attracts dirt. Auditory system of modified throat and scales	Cogger (1998)
Caecilians (Gymnopphiona)	Burrow in sand/soil	Burrow	Ramlike head, additional jaw muscles, reduced eyes, baggy outer skin, reduced scales and skull bones, recessed mouth, sensory tentacles	Cogger (1998)
Many limbless lizards (dibamids, teiids, skinks, some pygopodids)	Dig burrows	Burrow	Limb loss and shortening, loss of eardrums and middle ears	Cogger (1998)

Skinks (Scincidae)	Dig burrows	Burrow	Limb loss, loss of eardrums and middle ears, protected ears, small eyes, head shields, countersunk lower jaw, very smooth scales	Cogger (1998)
Brazilian spiny-tailed lizards (*Hoplocercus spinosus*)	Dig shallow retreats in soil	Nest retreat	Spiny tail blocks entrance to predators	Cogger (1998)
Gopher tortoises (*Gopherus polyphemus*)	Dig burrows	Burrow	Extended foremost plate of lower plastron, used for burrowing	Attenborough (1990)
Lizards (Sauria)	Dig burrows to avoid extreme temperatures	Burrow	Countersunk lower jaws, valved nostrils and modifications of ears to prevent sand entrance. Smooth scales, larger feet, webbed feet	Cogger (1998)
Birds				
Woodpeckers (Picidae)	Peck holes	Nest hole/ sap	Lengthened, strengthened, sharply pointed bill, elongated sticky tongue, stiffened tail feathers, hammering communication, food storage	Kaufman (1996)
Barbets (Capitonidae)	Excavate cavity in tree	Nest hole	Short heavy bill for biting or gouging	Hansell (1984)

TABLE 2.3. (*continued*)

Organism	*Perturbational Behavior*	*Resource*	*Anatomical and Behavioral Adaptations to Perturbation*	*Reference*
Bowerbirds (Ptilonorhynchidae)	Build a bower	Bower/ nest	Drab plumage combined with increased bower complexity	Hansell (1984)
House martin (*Delichon urbica*) and white-winged chough (*Corcorax melanorhamphus*)	Construct nests from mud	Nest	Vibration of mud until liquified for ease of construction	Hansell (1984)
Lovebirds (*Agapornis* spp.)	Build nests	Nest	Nest-material-carrying methods (bill/back feathers)	Hansell (1984)
Some hummingbirds (Trochilidae)	Construct nests of spider silk	Nest	Gathering silk, rotatory style of flying aids construction	Attenborough (1990)
Swifts (Apodidae)	Build nests using saliva	Nest	Enhanced saliva quantity and quality, oral mucus, gluing egg to nest	Attenborough (1990)
Kingfishers (Alcedinidae)	Excavate burrow nest in soil	Burrow, nest	Powerful bills	Hansell (1984)
Manx shearwater (*Puffinus* spp.)	Excavate nest in soil of seashore	Burrow	Digging claws on feet	Hansell (1984)
Megapodes	Build nest of soil/vegetation	Stable nest	Temperature-sensitive organ in mouth	Frisch (1975)

Mammals				
Beavers (*Castor fiber*)	Build dam/lodge	Dam, lodge	Aquatic mite-extractor claws, dam maintenance, food storage	Hanney (1975)
Moles (e.g., *Talpa europaea*)	Build burrow complex	Burrow	Digging legs, poor eyesight, hoarding of live, decapitated earthworms	Hansell (1984), Nowak (1991)
Mole rats (Bathyergidae)	Build burrow complex	Burrow	Poor thermoregulatory capacity, increased burrow complexity, food storage	Nowak (1991)

avian predators (Bristowe 1958). Other *Cyclosa* build a vertical line of debris, or stabilimentum, in which they are camouflaged. Spiders in the groups *Theridion*, *Tetragnatha*, and *Areneus* drag material onto their webs under which they hide in bad weather or if predators appear. *Segestria* and *Mallos* spiders communicate by tapping out messages on the web. That the web life style has affected selection acting on many aspects of the phenotype is also suggested by the observation that web spiders have an extra claw on each leg to aid movement on the web, and many have the ability to throw silk over their prey (Preston-Mafham and Preston-Mafham 1996). In addition, free-roving spiders (e.g., wolf spiders, jumping spiders) tend to have better developed eyes and more powerful legs than web spiders, while trapdoor spiders have evolved teeth that facilitate digging (Preston-Mafham and Preston-Mafham 1996).

Numerous burrowing vertebrates exhibit what appear to be adaptations for underground life. Several groups of animals have evolved anatomical characters that function as effective digging tools. African shovel-nosed frogs (Hemisotidae) are so named because of their pointed snouts with a hardened tip, which they employ to burrow by vertical motions of their heads. The frogs may bury themselves to seek daytime retreats or for long periods of estivation, and their eggs are also laid in an underground cavity (Cogger 1998). Logically equivalent cases are the frogs that have large keratinized, spadelike tubercles on the inner edge of each hind foot, which act like little trowels to scrape away earth (table 2.3). All burrowing frogs tend to have short legs (Cogger 1998), and in many limbless lizards (dibamids, teiids, skinks, and some pygopodids), too, limb shortening or loss appears to correlate with the occupation of underground habitats (Cogger 1998). Many skinks (Scincidae) also have head shields that are used in burrowing, small eyes, protected or no external ears, a countersunk lower jaw, and very smooth scales, all apparently adaptations to life underground (Cogger 1998).

The heads of caecilians (order Gymnopphiona), wormlike burrowing amphibians, are so powerfully constructed that they can be used as rams, allowing the animals to force their way through loose mud or moist soil and create tunnels. They also have an extra set of jaw muscles, their eyes are poor and reduced in size, and they can move their body inside their baggy outer skin, which helps them to burrow by allowing them to press their skin against the tunnel wall while at the same time pushing forward with their skulls. Some caecilians also have reduced numbers of scales and skull bones and a recessed mouth. In addition, caecilians have evolved their own unique sensory organ: tentacles that more than compensate for their poor eyesight by allowing them to sense the tunnel environment and detect their earthworm or insect prey (Cogger 1998). Worm lizards (Amphisbaenia) are a group of 152 species of predatory reptiles with a similar set of anatomical features that appear to be adaptations for underground existence.

Pecking in the 215 species of woodpeckers (Picidae) is correlated with morphological and behavioral adaptations, such as a lengthened, strengthened, and sharply pointed bill that is a particularly effective pecking tool, an elongated sticky tongue, stiffened tail feathers for support when pecking, and possibly also food storage and communication through their drumming (Frisch 1975; Kaufman 1996). Their cousins the barbets (Capitonidae) also excavate a nest in a tree, and some dig in the ground or in termite nests, but unlike woodpeckers they do not peck at the wood or earth, but rather bite and gouge (Hansell 1984). This difference in behavior appears to have selected for a very different-shaped bill, which is much shorter and heavier (Hansell 1984). The same point is made equally effectively by the Galápagos woodpecker finch, which uses a tool, such as a cactus spine, to grub for insects. As a result, rather than evolving a bill resembling that of a woodpecker, it has a bill that is well shaped for tool manipulation. Additional examples are given in table 2.3.

2.5.1.2 Elaboration of the Niche-Constructed Resources or Artifact

Perhaps the clearest examples of perturbational niche construction leading to selection for further perturbation involve the elaboration of the niche-constructed resources. Many animals that dig burrows, build nests, or construct artifacts appear to have evolved improvements or elaborations on an original structure (table 2.4). In many spiders (e.g., Segestriidae) it would seem that the construction of a burrow has established selection pressures favoring the evolution of a number of elaborations to the simple burrow structure, transforming it into an effective foraging tool and enhancing its security. For

TABLE 2.4. Additional Elaborations of Niche-Constructed Resources That May Be Evolutionary Responses to Prior Perturbational Niche Construction

Organism	*Resource*	*Elaboration of Pertubational Niche-Constructed Resource*	*Reference*
Insects			
Termites (Isoptera)	Mound (nest)	Conical/mushroom-shaped structures added to mound to push rain away from the nest	Hansell (1984), Attenborough (1990)
Fungus-growing termites (Macrotermitidae)	Mound (nest)	Ventilation system of vertical channels in thick outer walls utilizes metabolic heat of fungus to power air conditioning and gas exchange. Spiral cooling vanes in basement or thick outer walls permeated with a labyrinth of fine galleries	Hansell (1984), Frisch (1975), Attenborough (1990)
Caddis fly larvae (Hydropsychidae, Philopotamidae)	Case (house)	Spin a silken net to capture small prey swept down in current	Frisch (1975)
Caddis fly larvae (*Ceraclea neffi*, Helicopsychidae, Limnephilidae)	Case (house)	Adapt shape of house to act as ballast against displacement (e.g., lateral wings, spiral-shaped, tapering, curved tube)	Frisch (1975)
Caddis fly larvae (*Macronema transversum*)	Case (house)	Build funnel to direct water into chamber containing a net. Larva lives in tube alongside chamber, thus can feed at anterior and defecate at posterior opening of tube	Frisch (1975)

Sand wasps (genus *Bembix*)	Provisioned burrows	Plug entrance to protect egg. Creation of diversions, e.g., false burrows, kick sand over burrow, misleading tracks away from burrow	Hansell (1984)
Mole crickets (*Gryllotalpa vineae*)	Burrow	Add two horn-shaped tunnels (sweeping to surface) to chamber to enhance and direct male song	Hansell (1984)
Ants (Formicidae)	Nest	Regulate temperature by plugging entrances and adding stones (heat up quickly) to nest's surface. Regulate humidity through thick outer crusts reducing evaporation, absorbent materials placed at entrance, wallpaper over surface to keep dry	Hölldobler and Wilson (1994), Gullan and Cranston (1994)
Arachnids			
Tube-web spiders (Segestriidae) and trap-door spiders (Ctenizidae, Liphistiidae)	Burrow	Silk and twig trip lines radiate from the burrow. Plug entrance with stones, create stone trap doors using silk or gravel weights	Henschel (1995), Preston-Mafham and Preston-Mafham (1996)
Corolla spider (Segestriidae: *Ariadna*)	Burrow	Place selected stones (up to 100 times body weight) around burrow (resembling flower's corolla), to which are attached trip lines that amplify vibrations of passing prey. Turrets reduce flooding	Henschel (1995)
Orb-web spiders (*Cyclosa*)	Web	Drag material onto web, providing shelter from predators/weather. Build a line of debris (stabilimentum) in which camouflaged	Edmunds (1974), Bristowe (1958)

TABLE 2.4. (*continued*)

Organism	*Resource*	*Elaboration of Pertubational Niche-Constructed Resource*	*Reference*
Reptiles			
Green sea turtle (*Chelonia mydas*)	Nest (pit in sand)	Excavates small flask-shaped egg chamber at base of pit	Hansell (1984)
Tuatara (Rhynchocephalia)	Burrow	Females dig nest chamber and block entrance once eggs laid	Cogger (1998)
Birds			
African chat (*Myrmecocichla nigra*) and others	Nest	Cup of nest softened with grass and plant fibers	Hansell (1984)
Thrush (*Turdus* spp.)	Nest	Cup of nest softened with mud	Frisch (1975)
Bearded tit (*Panurus biarmicus*)	Nest	Cup of nest softened with flower petals	Frisch (1975)
Honey eater	Nest	Cup of nest softened with hair it often plucks itself	Frisch (1975)
North American house wren (*Troglodytidae* spp.)	Nest	Cup of nest softened with snake skin	Frisch (1975)
Eider duck (*Somateria mollissima*)	Nest	Line nest with downy feathers plucked from own body to maintain heat	Jones et al. (1994)
Cape penduline tit (*Anthoscopus minutus*)	Nest	Nest entrance has a closable flap. Prominent false entrance (blind pocket) to divert would-be predators	Hansell (1984)

Hornbills (Bucerotidae)	Nest	Males close females and eggs inside by narrowing entrance with mud	Attenborough (1990)
Magpies (*Pica pica*) and others	Nest	Build roof of cross beams and corbelling (using twigs)	Hansell (1984)
Weaverbirds (Ploceidae)	Nest	Add long tube entrance to protect against snakes. Improve roofing techniques (thatching) to protect chicks from weather	Frisch (1975)
Bee eaters (Meropidae)	Nest (in rock)	Dug at end of rainy season (rock softer) ready for use months later	Hansell (1984)
Motmots (Momotidae)	Burrow (in soil)	Dug at end of rainy season (soil softer) ready for use months later	Hansell (1984)
Ovenbirds (Furnariidae)	Burrow (in soil)	Dug in bouts coinciding with mild or wet weather when mud plentiful	Hansell (1984)
Mammals			
Blind mole rats (*Spalax* spp.)	Burrow	Nest chambers, store rooms, and specific latrine sites away from burrow	Jones et al. (1994)
African mole rats (*Tachyoryctes* spp.)	Burrow	Add a (grass-lined) nest chamber, a nearby bolt hole, food store, sanitary area, and a series of foraging tunnels (up to 50 m long)	Nowak (1991)
Badgers (Melinae)	Den	Complex system of tunnels, linking nursery, sleeping sites, latrines. Moss and ferns used as bedding material (regularly aired aboveground)	Frisch (1975)
Foxes (*Vulpes* spp.)	Den	Tunnel network connecting dens to hunting, food storage, and resting sites	Nowak (1991)

TABLE 2.4. (*continued*)

Organism	*Resource*	*Elaboration of Pertubational Niche-Constructed Resource*	*Reference*
Hamsters (*Cricetulus* spp.)	Burrow	Separate rooms for nest, food storage, excrement, hibernation	Nowak (1991)
Hedgehogs (*Erinaceus* spp.)	Burrow	Contains leaf-lined nest	Nowak (1991)
Lemmings (*Lemmus* spp.)	Burrow	Contains nest, latrine, and retreat chambers	Nowak (1991)
Marmots	Burrow	Different chambers (with bedding). Blocking of entrances for hibernation	Frisch (1975)
Prairie dogs (*Cynomys* spp.)	Burrow	Tunnels up to 30 m long. Two entrances to each tunnel to aid ventilation. Entrances cone shaped to protect from rainfall and aid ventilation	Jones et al. (1994), Hansell (1984)
Jerboas (Dipodidae)	Burrow	Side tunnels act as emergency exits from predators	Nowak (1991)
Muskrats (*Ondatra zibethicus*)	Mound-shaped nest (in marsh/swamp)	Underground tunnels to access water. Constructs dome of frozen vegetation over hole in ice where it can come up for air or to eat	Nowak (1991)
Porcupines (*Hystrix* spp.)	Burrow	Several entrances, grass-lined nest chambers	Jones et al. (1994)
River otters (*Lutra* spp.)	Den	Contains nest chamber lined with dry grass and rolling/grooming areas	Nowak (1991)
Wombats (Vombatidae)	Burrow	Shallow resting place excavated near mouth of burrow for sunbathing	Nowak (1991)

instance, most tube-web spiders spin a series of trip lines that radiate out from their burrows, and if an insect touches one the spider rushes out and grabs it. The corolla spider (Segestriidae: *Ariadna*) attaches its trip lines to stones, which amplify the vibrations, while some trap-door spiders (Ctenizidae) use twigs in a similar manner to extend their sensory range (Henschel 1995). The original function of spider silk is thought to have been the lining of burrows (Preston-Mafham and Preston-Mafham 1996), so the use of silk for the construction of webs, egg sacs, and drag lines may ultimately have been a consequence of spider burrow construction.

The simplest birds' nests are complex artifacts, requiring numerous distinct behavior patterns in their construction and the selection of relevant material. The simplest nests are constructed by large tree-nesting birds, such as rooks, storks, and eagles, and are commonly described as piles of twigs and sticks (Hansell 1984). However, on closer inspection, even these nests can be quite sophisticated, require careful selection of appropriate materials, and have clearly identifiable layers. Many smaller birds with delicate eggs are obliged to build more complex structures, frequently using moss, roots, grass, fur, spider silk, and hair as material to soften the cup.

Comparative studies by Crook (1963) and Schnell (1973) have used nest structures to construct phylogenies for weaverbirds. Such studies provide compelling evidence that relatively simple nest construction preceded, and set up, the selection pressures favoring nest elaboration. Elaborations include an improved nest roof that prevents the chicks from getting wet during heavy rain (Hansell 1984). Such elaborations are not restricted to birds, and many species of termite that live in humid areas build conical structures on or above the mound that drain the rain away from the nest (Hansell 1984). Many other species exhibit modifications to nest structure that confer protection from predators (table 2.4). For example, weaverbirds construct a long tube entrance to their nests apparently to protect against snakes (Hansell 1984).

The cases of caddis fly larvae, constructed of sand grains, stones, organic fragments, and silk, may have originated in response to the physical environment of flowing water but most likely also have a defensive and in many cases a foraging function (Gullan and Cranston 1994). Many primitive caddis fly species spin a silken tunnel attached to substrate, open at both ends. In other species the shelter is modified in a variety of ways: *Agapetus fuscipes* builds a temporary tube-shaped house before a larger permanent, barrel-shaped one. In Limnephilidae the basic plan is a tapering, slightly curved tube. In *Ceraclea neffi* the tube design is modified by the addition of lateral wings and a hood to protect the head. Other species build a simple house by gluing two, three, or four leaf pieces together. The Helicopsychidae build spiral houses resembling small snail shells. Hydropsychidae and Philopotamidae spin a net that captures small items of food swept down by the current. However, members of the *Macronema* genus build the most impressive houses, with a funnel facing into the current that collects water, which is directed into a chamber divided across the middle by a silken net. In *M. transversum*, the larva lies in a tube alongside the feeding chamber; the anterior opening allows it to feed at the net, while the posterior opening allows feces to be flushed away.

The species *Nothomyrmecia macrops* is regarded as the most primitive extant species of ant and reveals something of the ancestors of ants (Hölldobler and Wilson 1994). The colonies are less than 100 individuals, the nests are simple chambers, the eggs are left scattered on the floor, there is little contact between adults, they forage in a solitary manner, and they only have two chemical signals. These contrast with the large size, complex nest structure, and sophisticated communication and coordination of behavior that characterize ant colonies. Ant evolutionary history appears to reflect increasing elaboration of the nest, and an important source of the selection favoring this elaboration is probably the nest itself, with divergent evolutionary responses to such selection pressures responsible for the adaptive radiation seen across ant species (Hölldobler and Wilson 1994). The same point can be made with regard to termites. Schmidt (1964) constructs a phylogeny for *Apicotermes* termites based largely on nest structure. He proposes elaboration of an ancestral structure of simple chambers connected by corridors, through a series of intermediates, to the varied and intricate modern forms.

Numerous mammals (including badgers, blind mole rats, foxes, gophers, ground squirrels, hamsters, hedgehogs, lemmings, marmots, moles, mole rats, muskrats, opossum, prairie dogs, rabbits, and rats) construct burrow systems that are much more than simple holes in the ground, with underground passages, interconnected chambers with different functions, and multiple entrances (Nowak 1991). Here too there is evidence that burrow and nest-site elaboration has evolved in response to selection pressures that were initiated by prior niche construction (Hansell 1984; Nowak 1991). A good example is provided by the evolution of specific latrine sites positioned apart from the nest sites of numerous species, such as blind mole rats, lemmings, and rats. That the construction of a nest has been the source of selection favoring latrine sites is suggested by observations of nest-building mammals without latrines. For example, deer mice (*Peromyscus* spp.) are forced to make several nests each year as soiled nests must be abandoned when excretory products build up to noxious levels (Nowak 1991).

2.5.1.3 Regulation of the Niche-Constructed Resource

One important class of evolutionary consequences of perturbational niche construction includes what appear to be adaptations for regulating the constructed resource, for example, regulating temperature. This regulation, typically a response to inceptive perturbation, is itself one of the more common forms of counteractive perturbatory niche construction. We describe some of the more compelling instances in the text, with other examples in table 2.5.

There are numerous cases of organisms for which nest or burrow building seems to have led to selection for temperature regulation. Ants function poorly below 20°C but can survive in temperate zones by regulating the temperature of their nests (Hölldobler and Wilson 1994). Colonies are commonly concentrated beneath rocks or the bark of decaying stumps and logs, which warm in the sun faster than soil, and materials that heat up rapidly are commonly added to the nest surface. Ants frequently plug entrances to their nests at night or in the cold, and by adjusting the height or slope of the mound, appear to enhance intake of heat from the sun or, in hotter environments, to prevent overheating.

Termites, too, regulate the temperature in their nests by plugging entrances at night and in the cold and by building thick outer walls or cooling vanes to dissipate heat (Hansell 1984; Pearce 1997). Noirot (1970) showed that the internal temperature of a *Cephalotermes rectangularis* nest was constantly higher and appreciably less variable

TABLE 2.5. Physical or Behavioral Adaptations for Regulating Niche-Constructed Resources

Organism	*Resource*	*Physical/Behavioral Regulation of Niche-Constructed Resource*	*Reference*
Insects			
Ants (Formicidae)	Nest	Regulate temperature by plugging entrances, adding stones which heat up quickly, and adjusting height or slope of mound to maximize solar-heat intake. Regulate humidity through outer crusts reducing evaporation, water-absorbent materials placed at entrance, wallpaper over surface to keep dry, and dispatch workers to collect dew. Eggs and larvae kept in moist rooms and pupae in drier rooms. Nest material recycled to prevent mold formation	Hölldobler and Wilson (1994), Gullan and Cranston (1994)
Leaf-cutter ants (*Atta* spp.)	Nest	Special castes forage for leaves, chop them up, crush to pulp, weed the growing fungal garden, and take the crop for food	Hölldobler and Wilson (1994)
Compass termites (*Amitermes meridionalis*)	Mound (nest)	Tall and flattened nest presents large surface to sun in early morning and evening and small surface at midday	Hansell (1984)
Termites (Isoptera)	Mound (nest)	Plug entrances and spiral cooling vanes in basement. Thick outer walls permeated with a labyrinth of fine galleries or vertical channels (buttresses). Function for air conditioning, temperature control, and gas exchange	Hansell (1984), Frisch (1975)

Fungus-growing termites (Macrotermitidae)	Mound (nest)	Cultivate fungi on which they are nutritionally dependent in specially constructed chambers. Feces used to nurture fungus gardens	Hansell (1984), Frisch (1975)
Eusocial bees (subfamily Apinae and others)	Comb/nest	Workers collectively fan if carbon dioxide too high. If too hot, collect water and fan to evaporate. If too cold, heat by vibrating flight muscles. Maintain at 35°C	Matthews and Matthews (1978)
Honey bees (*Apis mellifera*)	Comb/nest	Storage of honey/pollen in cells of comb enables avoidance of hibernation as provides energy for activity and heat generation. Hygienic behavior, whereby workers uncap sealed cells and remove dead pupae	Matthews and Matthews (1978), Frisch (1975)
Social wasps (Vespinae)	Nest	Workers maintain nest at appropriate temperature, heating it by stretching/contracting abdominal muscles, cooling with drops of water, or directional fanning. Hygienic behavior	Spradbery (1973), Matthews and Matthews (1978), Frisch (1975)
Tiger beetle larvae (*Cicindela willistoni*)	Burrow	Construct turrets above burrow which, as they are cooler than soil surface, enable prey capture during midday heat	Hansell (1984)
Bark beetles (Passalidae, Scolytidae)	Nest (in timber)	Infect nest with specific fungus that predigests the wood into an edible form that they can eat	Preston-Mafham and Preston-Mafham (1996)
Amphibians			
African tree frog (*Chiromantis* spp.)	Nest (foam of body fluids)	Prevents nest dehydration by absorbing water through skin (in pond) and urinating on the nest	Frisch (1975), Cogger (1998)

TABLE 2.5. (*continued*)

Organism	*Resource*	*Physical/Behavioral Regulation of Niche-Constructed Resource*	*Reference*
Reptiles			
Crocodiles and alligators (Crocodilia)	Nest (vegetation/soil mound)	Soil/sand dampens temperature extremes. Decaying vegetation generates heat. Female splashes nest with water aiding vegetation fermentation	Frisch (1975)
Green sea turtles (*Chelonia mydas*)	Nest	Egg chamber in nest filled in so that sand does not fall between eggs. Vital for gas diffusion and development/survival of embryos	Hansell (1984)
Pythons (Pythoninae)	Nest (vegetation)	Coil around nest generating heat by shivering, maintaining egg temperature	Cogger (1998)
Birds			
Megapodes (e.g., turkey)	Nest	Add or remove fermenting vegetation to maintain stable temperature	Frisch (1975)
Merganser (*Mergus serrator*) and others	Nest	Damps out cold temperatures by lining nest with feathers	Hansell (1984)
Weaverbirds (Ploceidae)	Nest	Thatching technique/improved roof protects from rain and temperature extremes	Frisch (1975), Attenborough (1990)
Eider duck (*Somateria mollissima*)	Nest	Lines nest with downy feathers plucked from own body to maintain heat	Jones et al. (1994)
Rufous castle builder (*Synallaxis erythrothorax*)	Nest	Builds elaborate roof to keep out rain and maintain temperature	Hansell (1984)
Woodpeckers (Picidae)	Holes (foraging)	Regulate food supply by hoarding food in pecked holes	Frisch (1975)

Mammals			
Blind mole rats (*Spalax* spp.)	Burrow	Excavate deeper parts of burrow to avoid extreme temperatures of summer or winter. Also hoard food in separate storerooms	Nowak (1991)
Hamsters (*Cricetulus* spp.)	Burrow	Excavate deeper tunnels for hibernation and food caching during winter	Nowak (1991)
Marmots	Burrow	Deeper chambers with bedding and blocking of entrances for winter hibernation. Clearing of dirt in spring	Frisch (1975)
Hedgehogs (*Erinaceus* spp.)	Burrow	Leaf-lined nest enables maintenance of temperature during winter hibernation	Nowak (1991)
Jerboas (Dipodidae)	Burrow	In summer, plug burrow entrances to keep heat out and moisture in	Nowak (1991)
Western harvest mouse (*Reithrodontomys* spp.)	Nest	Summer nests lined with thistle down to buffer extreme heat	Nowak (1991)
Beaver (*Castor fiber*)	Lodge/dam	Constant maintainance of lodge or dam due to seasonal changes in water flow and cache food underwater for winter	Frisch (1975)
Prairie dogs (*Cynomys* spp.)	Burrow	Two entrances to each tunnel to aid ventilation from overground breeze. Entrances cone-shaped to protect from rainfall and aid ventilation	Jones et al. (1994), Hansell (1984)

than the external temperature. Compass termites are so called because they position their tall thin mounds to point north-south, exposing the large flat surface to the rising and setting sun, but leaving only a small surface exposed to the midday heat.

Honey bees keep their nests at 35°C. If the nest should become too hot they collect drops of water, place them on the nest, and fan it to cool the nest by evaporation. If it is too cold, they heat the nest by vibrating their flight muscles. The social wasps do the same (Spradbery 1973). Unlike wasps, honey bees are typically active all winter, and they are able to escape hibernation by using stored honey as energy to allow activity and generate heat in the nest.

Temperature regulation is a characteristic of many reptiles, particularly those for whom the sex of offspring is temperature-dependent, such as the Crocodilia, who use soil, sand, and decaying vegetation to manipulate temperatures (Cogger 1998). Female pythons coil around their clutch and keep the eggs at a high and stable temperature, generating heat by shivering. This adaptation has apparently allowed pythons to extend their range into relatively cold regions (Cogger 1998) and constitutes a case of counteractive perturbatory niche construction leading to relocation. Birds' nests can frequently be well described as insulating cavities that shelter the eggs from rain and cooling winds and allow the body heat of the parent bird to be used to best advantage (Forshaw 1998). It is characteristic of many burrowing mammals, and an essential survival mechanism in arid or cold regions, for them to plug the entrances to their burrows during the day, keeping the heat or cold out (Nowak 1991). For example, naked mole rats have the poorest capacity for thermoregulation of any known mammal, but the temperature and humidity within their burrows is maintained as a result of their entrance plugging and other regulatory activities (Nowak 1991).

Animals also regulate other characteristics of their nests. Many termites construct air-conditioning systems, complete with ventilation chambers, air ducts, chimneys, and other structures that aid gas exchange, without which the entire colony would suffocate within several hours (Hansell 1984; Pearce 1997). Some ants and many mammals exhibit similar structures (Nowak 1991; Hölldobler and Wilson 1994). For instance, prairie dogs (*Cynomys* spp.) build elaborate burrows, with grass-lined chambers, tunnels up to 30 m long, and volcano-shaped cone entrances that keep water out of the burrow. Each tunnel has two openings, one at each end. One is a chimney, and because the air moves faster above the ground a breeze sucks out stale air from the burrow (Hansell 1984). Many species employ a variety of strategies to regulate the humidity in their nests. African tree frogs (*Chiromantis* spp.), like many tropical frogs, whip up a foam nest from a mucous secretion containing eggs and seminal fluid, with a kicking motion of the legs (Frisch 1975; Cogger 1998). The female guards the nest and ensures that it does not dehydrate by repeatedly going down to a pond, absorbing water through her skin, returning, and urinating on the nest.

Numerous species of animals regulate their food supply. Leaf-cutter ants tend extensive fungal food gardens, and many Asian and African termite species also cultivate fungi (Gullan and Cranston 1994). Bark beetles in the families Passalidae and Scolytidae inhabit tiny galleries in timber, which they infect with a particular species of fungus. The fungus both predigests the wood into an edible state and is food for the beetles. Fungus-cultivating scolytids even have specific pouches in which they ferry a supply of the fungus to new trees. Countless mammals (including blind mole rats, field mice, gerbils, hamsters, moles, mole rats, and voles) also store food within their nests or burrows (Nowak 1991). Up to 18 kg of potatoes and sugar beet have been found in a single burrow of the blind mole rat (Nowak 1991).

2.5.1.4 Courtship, Mating, and Parental Behavior

In many cases nest building has apparently led to the evolution of elaborate courtship, mating, and parental behavior (table 2.6). Adult bark beetles (Passalidae and Scolytidae) help to construct the complex pupal cases needed for a new generation of brothers and sisters and respond to the larva's needs by placing fresh supplies of fungal food at the mouth of its burrow and removing its droppings (Preston-Mafham and Preston-Mafham 1996). In some species of scarab dung beetles (Scarabaeidae) parental care is well developed, with mothers removing fungus from the eggs and fathers defending against ants (Gullan and Cranston 1994). Parental behavior within the nest, such as tending and protecting the eggs, keeping the eggs free from fungi, maintaining appropriate conditions for egg development, and herding

TABLE 2.6. Courtship, Mating, and Parental Behavior That May Be an Evolutionary Response to Prior Perturbational Niche Construction

Organism	*Niche Construction*	*Evolved Courtship, Mating, and Parental Behavior*	*Reference*
Insects			
Ants (Formicidae), termites (Isoptera)	Nest	Structure enables workers to move immatures to best chambers for growth (temperature, humidity)	Hölldobler and Wilson (1994), Pearce (1997)
Bark beetles (Passalidae, Scolytidae)	Nest	Adults construct pupal cases for next generation and supply fungal food to the larvae while removing its droppings	Preston-Mafham and Preston-Mafham (1996)
Scarab dung beetles (Scarabaeidae)	Dung removed underground (tunnels)	Eggs laid in dung, tended by female who removes fungus, and male who defends eggs against invading ants	Gullan and Cranston (1994)
Earwigs (Dermaptera)	Nest	Eggs licked to apply fungicide, egg guarding, female assists nymphs to leave the egg, young fed via regurgitation and cared for up to third instar	Preston-Mafham and Preston-Mafham (1996)
Treehoppers (Membracidae)	Nest (in tree)	Female guards egg for weeks in order to assist nymphs in penetrating bark	Preston-Mafham and Preston-Mafham (1996)
Social species (e.g., Blattodea, Hymenoptera, Orthoptera)	Nest	Tending and protecting eggs, removal of fungi, maintaining appropriate conditions for development, herding and feeding young	Gullan and Cranston (1994)

Arachnids			
Nursery web spider (Pisauridae)	Egg sac	Egg sacs carried around in mouth, silken tent constructed and guarded when eggs ready to hatch	Preston-Mafham and Preston-Mafham (1996)
Spider mites (*Tetranychus urticae*)	Webbing attaching preadult female to leaf	Males locate and guard dormant female until she molts and emerges to mate. Web strands may act as a signal to the male	Hansell (1984)
Wolf spiders (Lycosidae)	Egg sac	Eggs sacs carried by female attached to spinnerets, leaving jaws free at all times	Preston-Mafham and Preston-Mafham (1996)
Fish			
Three-spined stickleback (*Gasterosteus aculeatus*)	Nest	Males lead female to nest, point at nest, zigzag dance, prod protuding tail of female to induce spawning. Once eggs laid, ventilate and maintain nest. Retrieve fry and defend them against predators	Paxton and Eschmeyer (1998)
Cichlids (Cichlidae)	Nest/spawning site	Males perform an elaborate courtship display before spawning. Eggs guarded and when hatched protected in the mouth by mouth brooders	Paxton and Eschmeyer (1998)
Bowfin (*Amia calva*)	Nest (circular mat)	Mating ritual of nose bites, nudges, chases until female lays eggs on nest. Nest and fry guarded by male against females	Paxton and Eschmeyer (1998)

TABLE 2.6. (*continued*)

Organism	*Niche Construction*	*Evolved Courtship, Mating, and Parental Behavior*	*Reference*
Catfish (Siluriformes)	Nest	Nest guarding, fry guarding, mouth brooding	Paxton and Eschmeyer (1998)
Garibaldi (*Hypsypops rubicundus*)	Nest site (red algae)	Males continuously defend the eggs from marauding predators and fan them with fins to keep them free of sediment and well oxygenated	Paxton and Eschmeyer (1998)
Toadfish (Batrachoididae)	Nest	Attraction of gravid females by calling, nest guarding	Paxton and Eschmeyer (1998)
Gobies (Gobioidei)	Egg attachment site	Male guards and keeps eggs clean. In freshwater species, larvae are carried out to sea and migrate back months later	Paxton and Eschmeyer (1998)
Paradise fish (*Macropodus opercularis*) and others	Floating nest of bubbles	Male catches and returns drifting eggs to nest. Guards nest and fry and replenishes foam	Frisch (1975)
Amphibians			
Shovel-nosed frogs (Hemisotidae)	Underground nest	Female remains with eggs until they hatch and then digs a channel to water so tadpoles can swim out	Cogger (1998)

Poison frogs (Dendrobatidae)	Nest	Tadpoles transported to water-holding bromeliad plant which is provisioned with unfertilized eggs as food	Cogger (1998)
African tree frogs (*Chiromantis* spp.)	Foam nest	Female guards nest and ensures it does not deyhydrate by absorbing water in pond and urinating on nest (tadpoles then fall into water below)	Frisch (1975)
Caecilians (Gymnopphiona)	Burrow	Female remains with eggs until hatch to protect from predation	Cogger (1998)
Reptiles			
Glass lizards and others (family Anguidae)	Burrow	Females remain with eggs until incubation and protect them from predation	Cogger (1998)
Skinks (Scincidae)	Burrow	Females remain and guard eggs protecting them from mold and predation	Cogger (1998)
Crocodiles and alligators (Crocodilia)	Nest	Males and sometimes females remain with and guard eggs. When hatched, young call and parents release them and carry them to the water	Frisch (1975)
Nile monitor (*Varanus niloticus*)	Nest in termite mounds	Do not guard eggs but may return to release hatchlings from mound	Cogger (1998)
Pythons (Pythoninae)	Nest of vegetation	Coil around eggs affording protection and maintaining temperature	Cogger (1998)

TABLE 2.6. (*continued*)

Organism	*Niche Construction*	*Evolved Courtship, Mating, and Parental Behavior*	*Reference*
Birds			
Bowerbirds (*Amblyornis flavifrons*)	Bower	Production of many elaborate courtship calls and displays	Hansell (1984)
Weaverbirds (Ploceidae)	Nest	Nests become more elaborate as used by females in mate choice	Frisch (1975)
White booby (*Sula dactylatra*)	Nest	Despite its elaboration, the nest has become vestigial and functions as a courtship ritual promoting pair formation	Hansell (1984)
European wren (*Troglodytes troglodytes*)	Nest	Males build several rough nests and the female chooses the one she prefers. The nest functions as a signal as mating cannot occur without a nest	Frisch (1975)
Long-billed marsh wren (*Telmatodytes palustris*)	Nest	Multiple nests (average of 23) built by male who sings to attract a female. She inspects nests (possibly used as indicator of male quality)	Hansell (1984), Verner and Englesen (1970)

and feeding the young, are found in some Blattodea, Orthoptera, Dermaptera, Embioptera, Psocoptera, Thysanoptera, Hemiptera, Coleoptera, and Hymenoptera (Gullan and Cranston 1994). Many spiders too will stay beside their egg sacs and protect their eggs (Preston-Mafham and Preston-Mafham 1996).

It is well established that elaborate courtship and mating rituals are associated with nest-building fish (Parrish 1998). Consider the well-known nest construction and mating ritual of the three-spined stickleback (*Gasterosteus aculeatus*). Males first defend a territory, then excavate a shallow pit, and then collect plant material and algae to build a tubelike nest. They then engage in a mating ritual with a female, which involves leading her to the nest, pointing at it, and prodding her tail as it protrudes from the nest to induce her to spawn. After fertilizing the eggs, males remain with them, ventilate and repair the nest, and care for the brood when they emerge, defending them against predators. Many aspects of the mating ritual, nest guarding, maintenance, and paternal behavior of the male are likely to have evolved in response to past nest construction. This is supported by McLennan et al.'s (1988) phylogenetic tree of the *Gasterosteus* genus, constructed using behavioral characters that are components of nest building, mating, and parental behavior. This tree, which is consistent with trees constructed using electrophoretic and molecular data, suggests that nest building evolved first, followed by the nest-show display and nest-ventilating behavior, and later by the zigzag dance, building a tunnel entrance and exit, prodding the female in the nest, and fry retrieval. The same pattern of elaborate courtship, nest construction, and parental behavior is characteristic of many of the thousands of cichlids and numerous other fish groups (Paxton and Eschmeyer 1998).

Many amphibians and reptiles also remain within or near the nest or burrow protecting and caring for their eggs. Some skinks (Scincidae), alligator lizards, galliwasps, and glass lizards (Anguidae) show a high degree of parental care; females stay with their eggs until incubation and protect them from predation or mold formation (Cogger 1998). In most species of crocodiles and alligators the parents remain with the eggs, defending them from predators and maintaining the nest at an appropriate temperature (Hansell 1984; Cogger 1998).

Where nests are constructed by males, yet are crucial for female reproductive success, it has been suggested that females base their mate choice to a large extent on the quality of the nest (Qvamstrom and Forsgren 1998). It is widely accepted that nest building plays a critical role in the mate choice of many fish (e.g., cichlids) and birds (e.g., bowerbirds and weaverbirds) (Qvamstrom and Forsgren 1998; Paxton and Eschmeyer 1998; Forshaw 1998).

2.5.1.5 Communication and Sociality

Pertubational niche construction often appears to have generated natural selection favoring enhanced communication or social behavior. Several predatory species of spider prey on other spiders by entering their web and quivering to imitate struggling prey, which leads the host spider to rush out to its doom (Preston-Mafham and Preston-Mafham 1996). Many spiders pluck out a signal on the web as an integral part of the courtship ritual, while still others imitate the courtship web plucking of female hosts and eat the male host (Preston-Mafham and Preston-Mafham 1996). The more than 40 species of social spider provide further examples of how perturbational niche construction can lead to the evolution of communication and other social behavior (Preston-Mafham and Preston-Mafham 1996). Some species, such as *Agelana consociata* in which hundreds of spiders live together on the same sheet web, have evolved an olfactory identification signal that prevents attack by conspecifics. *Mallos gregalis* is sensitive to the frequency of vibration of conspecifics and prey and rarely attacks the former. *Stegodyphus sarasinorum* taps a recruitment signal requesting help if the prey is too large to handle alone. In the case of *Metabus gravidus*, each individual builds an orb web, and these are supported by a few threads, a system that has led to both increased tolerance of conspecifics and more efficient prey capture. In the spider *Tapinillus* the construction of a social web apparently reduced cannibalistic behavior. *Philoponella republicana* establish central roosts where clusters of spiders hang out together, emerging at dawn to build their individual orb webs. These spiders cooperate to handle larger prey than any could take alone, some applying silk wrapping, others holding the prey, or applying digestive juices.

In many social insects there is evidence that niche construction has favored the evolution of more complex sociality. For example, some carpenter bees (subfamily Xylocopini) have changed the structure of the nest by removing the walls between cells, with the result that the eggs, larvae, and pupae are maintained in an "open-plan" burrow. This has allowed the rearing of larvae by continuous rather than mass provisioning (Hansell 1984). In some of these species substantial changes have also occurred in the use of nest space, again accompanied by advances in social development. In the genus *Braunsapis*, different parts of the nest space are occupied by broods at different stages, with eggs at the far end, larvae of progressively greater development nearer the entrance, and the pupae nearest it. Daughters often stay in the nest, helping with provisioning and laying a few eggs themselves. Thus, in these bees the loss of cell partitions is associated with a leap from solitary status to primitive eusociality.

2.5.2 Evolutionary Responses to Relocational Niche Construction

In this section we consider the likely evolutionary consequences of relocatory niche construction. We begin by focusing on apparent anatomical and behavioral adaptations to migration exhibited by a variety of organisms and then go on to consider adaptations to other forms of relocation. This section is brief because there are well-established databases for major categories, such as migration and dispersal (see Werner 1992; Dyer 1998), that often describe evolutionary consequences of this relocation.

2.5.2.1 Anatomical and Behavioral Adaptations to Inceptive Relocation

Oviparous insects typically deposit their eggs on or near the food required by the offspring upon hatching. Often the safety of the eggs, and the accessibility of the food to the emerging larvae, is enhanced by specialized ovipositors on the mothers that allow them to inject the eggs into a suitable location with considerable precision. Ovipositing on particular hosts appears to have generated selection for the shape and characteristics of the ovipositor. This is clearly seen in the 20,000 species of Orthoptera (Gullan and Cranston 1994). Grasshoppers and locusts (suborder Caelifera) have an ovipositor specialized to bury batches of eggs in soil chambers. Katydids and crickets (suborder Ensiferan) lay their eggs singly in plants or soil (Gullan and Cranston 1994). The saberlike ovipositor of many female katydids introduces the eggs into concealed places by probing into slits in bark, rotten wood, lichen-covered earth, cracks in the ground, or plant tissue. In tephritid flies (Tephritidae) the ovipositor is telescopic and can be extended to several times its original length, probing deep inside the stems or flowerheads of plants such as thistles and knapweeds (Preston-Mafham and Preston-Mafham 1996). Sawflies (suborder Symphyta) lay their eggs in leaves or stems, and the ovipositor has serrated edges that cut a slit into plant tissue (Preston-Mafham and Preston-Mafham 1996). In many sawfly species (e.g., *Urocerus gigas*) the eggs are accompanied by an easily digested symbiotic fungus, contained in a pair of sacs at the base of the ovipositor. In certain endoparasitic wasps in the families Ichneumonidae and Braconidae, the female injects the larval host not only with her egg or eggs but also with viruses or viruslike particles which overcome the host's immune reactions (Gullan and Cranston 1994).

Adaptations for finding an appropriate host are common in insects. The hairlike ovipositor of many ichneumonid wasps may be several times their body length, and the wasps sometimes have complex sensory methods for locating suitable hosts (Gullan and Cranston 1994). Lepidoptera frequently have to select, locate, and assess the quality of a specific host on which they lay their eggs. The initial cues are usually visual; however, volatile odors are probably also used as cues, with some scents attracting and some deterring landing and investigation (Gullan and Cranston 1994). Rather than injecting their eggs, some beeflies (family Bombyliidae) such as *Bombylius major* act like bombing aircraft, diving toward the nests of their host bees or wasps and dropping their eggs directly into the entrance (Preston-Mafham and Preston-Mafham 1996).

Many poisonous insects extract their poisons from host plants. Here, consistent habitat choice seems to have generated selection favoring the utilization of host plant chemicals. An example is provided by the monarch butterfly, which carefully selects a milkweed host where it oviposits. Emerging caterpillars eat the milkweed and become poisonous (Gilbert 1983). The same logic applies to those species for whom habitat choice has generated selection for pigment utilization.

By making the move onto land, swamp eels (order Synbranchiformes) have apparently generated selection for their lunglike vascularized mouth and pharynx (Paxton and Eschmeyer 1998). Similarly, mudskippers (genus *Periophthalmus*) spend a lot of time out of water at low tide foraging for flies and small insects, a behavior that appears to have favored selection for blood vessels that extract oxygen from the air, eyes that are high on the heads, and a muscular tail and side fins to help them skip over mud. In forming shoals (which is choosing a social environment), carp and their allies (superorder Ostariophysi) have created an environment in which selection seems to have favored production of an alarm substance that can communicate information about predators near the shoal. Lanternfishes (family Myctophidae) have evolved photocells for identification of each other, and coral reef fish have evolved cultural traditions for shoaling to particular mating sites (Paxton and Eschmeyer 1998; Warner 1988). In the ocean, small fish gather under any floating object, and the yellow-bellied sea snake (*Pelamis platurus*) is soon adopted by a small school that swarm around its tail. The snake exploits this by swimming backward, so the fish gather by its head, where it can easily grab them (Cogger 1998).

2.5.2.2 Anatomical and Behavioral Adaptations to Counteractive Relocation

Many animals that migrate to avoid extreme conditions or improve access to food and other resources exhibit adaptations to this feature of their life history. These migrations are likely to have been adaptations to prior counteractive relocation. Fish such as cod and sardines migrate to particular places to breed, so that adults and developing larvae can each live where they have the richest feeding grounds (Paxton and Eschmeyer 1998). Other species of Clupeiformes migrate to particular spawning sites, and some seasonally migrate up river to spawn in fresh water. Similarly, swallows and other migratory birds do not experience the alternative selective regimes, faced by sedentary and overwintering species, that have favored adaptations for coping with temperature extremes or low levels of food availability (e.g., torpor). Among bats (Chiroptera), some species migrate while others hibernate, suggesting that these are life-history alternatives and that by migrating they have avoided selection for hibernation.

Numerous anadromic fish, including salmon, smelts, galaxids, and sea-run lampreys, which migrate from rivers to the sea, undergo a metamorphosis entailing substantial physiological changes at critical points in their development that allow them to adjust from freshwater to saltwater environments, and then back again (Paxton and Eschmeyer 1998). Eels and sardines exhibit the reverse migration but require similar adaptations to adjust salt levels in their bodies. Other species, such as tuna (suborder Scombroidei), have evolved adaptations for speed, stamina, or endurance for their long-distance migrations (Paxton and Eschmeyer 1998). Some birds are equally impressive: for instance, some species of swift can go months on the wing (Attenborough 1990). Migratory fish, birds, and reptiles that return to the location of their birth to breed also exhibit adaptations to find their home stream or nest. Young salmon imprint on their home stream, and their extraordinary sensitivity to odor cues allows them to return to it (Paxton and Eschmeyer 1998). Female green sea turtles appear to have evolved adaptations that allow them to find their home nests, and those adults that survive the migration generally return to the same nest (McFarland 1987; Turner and Rose 1989).

2.6 EVOLVED RESPONSES TO NICHE CONSTRUCTION IN OTHER SPECIES

Finally, we consider the likely evolutionary consequences of the niche construction of one species for other species (see table 2.7). Since the coevolution of two or more species is not restricted to species from a single kingdom, we include examples from all the kingdoms in table 2.2 here. Again, this section is brief, as there are well-established databases for coevolution in which niche construction is implicit (e.g., Thompson 1994).

Most flowering plants are pollinated by animals, particularly insects, and fossil evidence indicates that these associations probably date from the Cretaceous, when the numerical dominance of insects coincided with angiosperm diversification (Gullan and Cranston 1994). Similarly, ants and termites construct nests and mounds, and in some species processed vegetation or feces are used to cultivate fungi (see chapter 1). In turn, the fungi have evolved a dependence on the ant and termite species. Nile monitors have evolved the behavior of laying their eggs inside termite mounds, exploiting the termites' thermoregulation of the nest, while other species have evolved behavior to seek out the mineral-rich termite mounds for salt and mineral licks (Hansell 1984).

Ants also have a well-known mutualism with acacia trees, with the ants destroying seedlings and attacking mammalian browsers and insect pests, and the acacia responding by evolving thorns and nectaries that house and feed the ants (Hölldobler and Wilson 1994). The chemical trails left by other ants are exploited by wormsnakes (Anomalepididae, Typhlopidae, Leptotyphlopidae) that have evolved the tendency to follow the trails, enter the nest, and eat the brood. Some species join army ants as they travel through the forest, feeding off eggs and larvae from other ants' nests taken by the army ants (Cogger 1998). Mammals, too, create and regularly use trails, and many snakes, for example, vipers, that are sit-and-wait predators have evolved the tendency to hide beside these trails waiting for a victim to ambush (Cogger 1998). The code of chemical signals belonging to the ant species *Tetramorium caespitum* has been cracked by another group of ants, *Teleutomyrmex* spp., which emit chemical signals to instruct *T. caespitum* workers to feed and look after them. Apparently as a result *Teleutomyrmex* species have evolved weak bodies, they are incapable of producing food for their own larvae, and they have a small sting and a tiny brain (Hölldobler and Wilson 1994). There are 13,000 species of gall-forming insects (mainly Hemiptera, Diptera, and Hymenoptera) and some fungi and bacteria that manipulate plant tissue through chemicals and genetic material into constructing galls, which are homes and food for themselves (Gullan and Cranston 1994). This is niche construction by proxy: the insects manipulate the host plants into constructing niches to the insects' advantage. The gall formers derive their food from the tissues of the gall, as well as protection, while the plants have evolved chemical defences against the gall formers. Many inquilines are also housed within the gall (Jones et al. 1994).

Several well-known examples of coevolution involve niche construction. For instance, the monarch butterfly selects a milkweed host on which to oviposit. Emerging caterpillars eat the milkweed, extracting the poison for their own use and imprinting on the host. This relocatory niche construction has apparently selected for Batesian

TABLE 2.7. Multispecies Coevolutionary Interactions Mediated by Niche Construction

Organism	*Constructive Behavior or Output*	*Resource*	*Evolutionary Responses to Niche Construction*	*Reference*
Insects, beetles, and flowering plants	Insects and beetles gather nectar and pollen	Nectar, pollen	In plants, evolution of flowers and other adaptations for attracting insects and facilitating pollination	Gullan and Cranston (1994)
Algae and flowering plants	Physical and chemical weathering of rocks into soil	Soil	Plants evolve adaptations for exploiting soil for nutrients, water, and physical support	Jones et al. (1994)
Pine and chaparral trees, indigo species, smaller shorter plants, nitrogen-fixing legumes	Accumulate oils and litter, encourage fires	Tinderbox environment	Many species evolved resistance to and dependence on fires, e.g., indigo species, smaller shorter plants, nitrogen-fixing legumes. Some pines require fire for seed germination	Tilman (1990)
Ants and acacia trees	Ants destroy seedlings, attack browsers	Thorns, nectar	Plants evolve thorns and nectaries to attract, house, and feed ants	Hölldobler and Wilson (1994)
Termites, fungi, Nile monitor	Termites construct mounds, feces used to cultivate fungi	Mound, feces, fungi	Fungi evolved dependence on termites. Nile monitor lays eggs in mound for thermo-regulation. Termite mounds are mineral rich, and used by other animals as salt licks	Hansell (1984)

Ants, insect, spiders, fungi	Ants build nests, forage for animal and plant foods	Nest, food resources, fungi	Other ants, spiders, other insects evolved various responses to predation by ants. Caterpillars, leaf-eating insects evolved evolutionary responses to competition for resources from ants. Fungi evolved a dependence on leaf-cutter ants. Thousands of species of mites, silverfish, millipedes, flies, beetles, wasps, and other small creatures invade ants' nests	Hölldobler and Wilson (1994)
Insects, fungi, bacteria	Insect, fungi, and bacteria gall-forming activity	Gall	Plants have evolved chemical defenses, while many inquilines exploit the gall	Gullan and Cranston (1994), Jones et al. (1994)
Monarch butterfly, milkweed, viceroy	Monarch imprints on milkweed as caterpillar, selects milkweed host as adult	Monarch's poisonous nature	Viceroy butterfly has evolved mimicry of the monarch, while birds avoid these butterflies	Gilbert (1983)
Tetramorium caespitum, *Teleutomyrmex* species	*Tetramorium caespitum* communicates via chemical signals	Chemical signals	*Teleutomyrmex* species emit the same signals to exploit their hosts, and in consequence evolved weak bodies, an inability to feed their own larvae, little protection against bacteria, frail exosketelons, small sting and poison glands, small mandibles, and a small brain	Hölldobler and Wilson (1994)

TABLE 2.7. (*continued*)

Organism	*Constructive Behavior or Output*	*Resource*	*Evolutionary Responses to Niche Construction*	*Reference*
Ants, worm snakes	Ants leave chemical trails	Chemical trail	Worm snakes follow the ant trails to the nest and eat the brood. Also follow army ants and eat prey	Hölldobler and Wilson (1994)
Mammals, snakes	Mammals make trails	Trail	Snakes (e.g., vipers) have evolved behavior of waiting by the trail to ambush mammalian prey	Cogger (1998)
Spiders, birds	Spider web building	Silk	At least 175 species of birds evolved use of spider's silk in nest construction Hummingbirds evolved a particular manner of flying when constructing nests with silk	Hansell (1984), Attenborough (1990)
Petrel (*Pelecanoides urinatrix*), fairy prion (*Pachyptila turtur*), tuatara (Rhynchocephalia), soil invertebrates	Birds excavate burrows and defecate nearby	Burrow, guano	Soil invertebrates evolved tendency to gather in bird burrows, while tuatara steal burrows and exploit the invertebrates feeding on guano	Cogger (1998)
Snapping shrimps, sand gobies	Shrimps excavate burrows	Excavated soil	Sand gobies remain near shrimp burrows, feed on invertebrates in excavated soil, and signal to shrimp with tail-waving behavior if it is safe to emerge	Paxton and Eschmeyer (1998)

Beaver (*Castor fiber*), mite (*Schizocarpus mingaundi*), beetle, other species that expoit lake	Beaver builds dam and lodge	Dam, lodge	Mite evolved to exploit beaver within lodge, and a beetle lives in the beaver's fur exploiting the mites. The beaver has evolved a claw used in mite removal	Hanney (1975)
Solitary bees	Construct nests	Nest	Other bee species reuse abandoned nests	Hansell (1993)
Nest-building birds	Build nests	Nest, parental care	Other birds evolved egg-dumping behavior, and chicks evolved fast development and egg ejection	Krebs and Davies (1993)
Potter wasps	Build nests	Nest	Barn swallows attach their nests to the old mud nests of wasps	Hansell (1993)
Insect carrion necrophages	Successional colonization, laying eggs on and feeding on vertebrate corpses	Modified corpse	Microorganisms, blow flies, house flies, sarcophagids, histerid beetles, and hymenopteran parasitoids, dipterans, all evolved tendency to exploit corpses at particular stage of decomposition, contribute further to the decay, and create an environment favoring their successors	Gullan and Cranston (1994)
Plains zebras, wildebeest, and gazelles	Migrate and feed on grasses	Modified grasses	Zebras, wildebeeste, and gazelles migrate in mixed-species herds, with the zebras feeding on coarse grass, exposing the stalks for the wildebeest, who crop the grasses further, exposing the lowest vegetation for the gazelles	Nowak (1991)

TABLE 2.7. (*continued*)

Organism	*Constructive Behavior or Output*	*Resource*	*Evolutionary Responses to Niche Construction*	*Reference*
Many plants, animals	Successional colonization, and modification of environment	Modified environment	Successional trees and plants create environment favoring their successors. Animals and birds follow in their wake	Berg (1997)
Guinea pigs (*Cavia* spp.), water rats, burrowing mice	Dig burrow, make runway	Burrow, runway	Water rats and burrowing mice have evolved tendency to use guinea pig burrows and runways	Nowak (1991)
Plants and soil fungi	Soil fungi extract minerals that travel to the roots of the plants, which in turn provide carbohydrate molecules for the fungus	Minerals, carbohydrate	Complex mutualistic associations (mycorrhizae) have evolved, and in many cases the plants have evolved a dependence on the fungus	Law (1985)
Allelopathic plants, other plants	Allelopaths exude toxins from roots and shoots	Soil quality	Other plants evolve resistance, while the allelopaths evolve high levels of disperal	Rice (1984)

Woodpeckers, owls, squirrels	Woodpeckers peck holes in trees	Hole, sap	Other species exploit abandoned holes; owls and squirrels chase woodpeckers away and use hole. Many species of mammals, birds, insects, and microorganisms exploit the exposed tree sap	Jones et al. (1994)
Earthworms, microorganisms, plants	Earthworms drag organic matter into soil, mix with inorganic and caste	Soil, nutrients in soil, casts	Earthworm casts become sites of microbial activity, modified soil environment for plant species	Lee (1985)
Bacteria, Protista, plants, animals	Bacteria, Protista, and plants photosynthesize to produce O_2, and modify chemical composition of atmosphere	Oxygen and other atmospheric gases	Animals have evolved respiratory organs used in gaseous exchange	
Lichens, mosses, and primary succession	Successional colonization and modification of environment. Lichens play a critical role in the weathering of rocks and soil formation	Modified environment	Successional lichens, mosses, and plants create environment favoring their successors	Berg (1997)

mimicry in another species, the viceroy, which benefits from the monarch's poisonous reputation, because many birds have evolved prey-selection mechanisms that result in these species of butterflies not being eaten (Gilbert 1983). Another example is the mutualism that exists between snapping shrimps and sand gobies on coral reefs. The shrimps construct extensive burrows, dumping excavated sand near the entrance. Gobies have evolved a tendency to aggregate near the burrow and feed on the small crabs and invertebrates in the excavated substrate. They signal to the shrimp by waving their tails if there are no predators around and it is safe for the shrimp to emerge (Paxton and Eschmeyer 1998). Coral itself also plays a vital role in local ecosystems, modulating current speeds and siltation rates (Jones et al. 1994). In this case crustose coralline algae overgrow and cement together detritus on the outer algal ridge of barrier reefs, in the process breaking the force of water and protecting the corals against major wave action (Jones et al. 1994).

There are at least 175 species of birds that use spider silk as a structural material in their nests (Hansell 1984). Hummingbirds have evolved a unique manner of flying when they employ the silk in nest building (Attenborough 1990). The petrel (*Pelecanoides urinatrix*) and fairy prion (*Pachyptila turtur*) excavate burrows on New Zealand islands, which the tuatara (order Rhynchocephalia) steals to live in, while their mineral-rich guano supports ground-dwelling invertebrates that the tuatara eats (Cogger 1998). Further examples are presented in table 2.7.

2.7 DISCUSSION

We have collated and categorized a large number of examples of prima facie evidence for niche construction and what appear to be their evolutionary consequences. It is clear that niche construction is a widespread property of living organisms, and the products, resources, and habitats that those organisms construct frequently cannot be regarded as trivial or inconsequential, but rather constitute fundamental components of their worlds.

There is also a substantial body of circumstantial evidence that the niche construction of organisms has modified selection pressures and generated selection for alternative traits. This includes selection for anatomical and behavioral adaptations that enhance the efficiency of their niche construction, adaptations to relocation, selection favoring elaboration and regulation of the constructed resource, and selection for modified courtship, mating, and parental behavior. Although it is not clear that all of these adaptations are actually evolutionary responses to prior niche construction, it is likely that many of them are. This means that it may frequently be appropriate to consider evolution as a process in which environment-altering traits coevolve with traits whose fitness depends on alterable sources of natural selection in environments. It is also clear that many organisms construct niches in ways that affect the local environments of other species and that generate coevolutionary dynamics. The examples in this chapter may serve as sources of data for researchers interested in the roles of niche construction in within- and between-species interactions and its impact on ecosystem functioning.

One possible criticism of our argument that niche construction plays a central role in evolution is that, in some of the examples we have given, genetic variation for the recipient trait may not have been present at the time the niche-constructing trait evolved. While, in such cases, the source of selection on the recipient trait is still the product of niche construction, the traits cannot be said to coevolve, and the evolution of each trait could be treated separately. The arguments against this are the following. First, the strong empirical evidence for the contemporary presence of genetic variation at many loci and in many organisms (Futuyma 1998) suggests that it would be unlikely that the genes underlying niche-constructing traits were fixed for eons before there appeared genetic variation underlying traits whose fitness depends on the products of niche construction. Second, there is no requirement that niche construction must result from genetic variation for it to alter the selection acting on recipient traits. Niche construction can be dependent upon learning (Bateson 1988; Plotkin 1988; Laland et al. 1996, 2000b). Moreover, the acquired characteristics of organisms may be expressed as artifacts and other niche-constructed resources, which may be inherited via nonbiological routes (Odling-Smee et al. 1996). We discuss the role of learned and culturally transmitted niche construction in more depth in chapter 6. Third, the consequences of niche construction are likely to be far more profound than just trait coevolution. Niche construction introduces feedback into the dynamics of evolution and ecosystems; for example, ecological inheritance can create time lags in the response of a trait to modified selection pressures (Laland et al. 1996, 1999). Moreover, niche construction is likely to generate indirect epistatic interactions between genes via resources in the external environment. For example, flamingos are pink not because they synthesize this color, but rather because they consistently choose environments containing their crustacean prey, and this habitat choice has generated selection for extraction of the carotenoid pigment (Fox 1979). Here the genes influencing prey choice interact with those underlying pigment extraction, indirectly via the food resources in their environment. The same logic applies to their crustacean prey.

In chapter 7 we suggest how the examples contained in this chapter could be used in empirical investigations of the evolutionary consequences of niche construction. For the moment it will be sufficient to sketch how this may be done. The effects of niche construction can be investigated by canceling out or enhancing the niche-constructing activity relative to control conditions, to create experimental comparisons between virgin, engineered, and degraded habitats in recipient traits or ecological variables (Jones et al. 1997; Laland et al. 1999). With knowledge of the appropriate phylogeny, it should be possible to establish whether the evolution of the niche-constructing behavior preceded, coevolved with, or followed the evolution of the recipient trait. This could be done by using behavioral data to construct or map onto the phylogeny (McLennan et al. 1988). Moreover, comparative methods (Harvey and Pagel 1991) could also be employed across large taxonomic groups to assess whether particular niche-constructing traits are associated with particular recipient traits. If niche construction were an important evolutionary agent, then for any clade of organisms, it should also be possible to establish which recipient traits might be adaptations in environments that have been niche constructed. Pertinent characters, and environmental states, could be measured in populations of closely related organisms that do and do not exhibit this niche construction. It might then be possible to determine whether a selected recipient character change correlates with a particular niche-constructing activity, if the niche-constructing activity is ancestral to the recipient character, and whether the recipient character in question is derived. If we are correct, there should be a significant relationship between the pertinent environmental state and the recipient character only when the niche-constructing activity is also present. The data presented in this chapter indicate that there are likely to be large numbers of species for which such empirical analyses would be feasible. Additional empirical methods, and a more detailed account of those described above, are presented in chapters 7–9.

CHAPTER 3

A Theoretical Investigation of the Evolutionary Consequences of Niche Construction

3.1 INTRODUCTION

We have seen how organisms partly construct their own niches, modifying their environments to an extent that is often nontrivial, and that this niche construction can modify selection pressures. Empirical data support our claim that niche construction is likely to change the nature of the evolutionary process. Even if they are not always aware of the extent, breadth, or scale of niche construction, ecologists and population geneticists are conscious that at least some organisms construct important components of their niches, at least some of the time. Yet, despite this, niche construction is not yet widely recognized as an important evolutionary process, and its effects are not generally incorporated into quantitative models. Niche construction is generally regarded as the product of evolution, but not as part of the process.

We begin this chapter by reviewing those bodies of population genetics and ecological theory that in some way impinge on our argument. It is important to consider how biologists have treated niche construction previously and to reflect on the extent to which these earlier treatments, individually or collectively, can be said to constitute a satisfactory theory of niche construction. We conclude that while there are areas of biological research that have explored aspects of niche construction or considered some restricted features of its consequences, there has to date been no satisfactory theoretical framework for a general exploration of the evolutionary consequences of niche construction.

In this chapter we develop our argument further with population-genetic models that explicitly incorporate the process of niche construction into the evolutionary dynamic. If niche construction is as important an evolutionary process as we claim, then its inclusion should make a significant difference to the behavior of theoretical models and should generate some unusual and hitherto unpredicted dynamics. Below we describe, in nonmathematical terms, the findings of our formal analyses. Readers interested in the technical and mathematical details are referred to appendixes 1–3, or to our earlier published work (Laland et al. 1996, 1999), which forms the basis for this chapter.

3.2 A REVIEW OF PAST THEORY

In the ecology and evolution literatures there is a considerable body of formal theory that models aspects of niche construction and its consequences. Ecological models of population growth through resource depletion and evolutionary models of habitat selection are two of the more obvious examples. In these and other bodies of theory (reviewed below) there is a feedback between organismic activity and the environment. Some of this theory includes niche construction as a process in its own right, although this is rarely acknowledged in conceptual or verbal accounts. These models could be, but rarely are, described as an explicit formalization of Lewontin's equations 1.3 and 1.4 in chapter 1. In the light of some recent developments, such as the growing interest in indirect genetic effects and maternal inheritance, it seems that standard evolutionary theory is converging on a niche-construction perspective. Yet while the basic insight that feedback between niche-constructing organisms and their environments occurs and that it can have profound effects on evolutionary dynamics is hardly novel, it is often obscured in the literature as a whole. Ecologists and evolutionary biologists have occasionally included niche construction within the framework of standard evolutionary theory, but only on a piecemeal basis. While these models may provide a satisfactory theoretical foundation for other subjects, even collectively, they do not constitute a formal theory of niche construction because some important features, for example, positive effects on environmental states, have largely been ignored. But, where this theory has been attempted, it suggests that niche construction is evolutionarily significant.

3.2.1 Ecological and Demographic Models

Ecologists have built large bodies of theory that explore the ecological consequences of niche construction. With optimization models, population-dynamic theory (e.g., Lotka-Volterra), or large-system statistical analysis, models in ecology commonly explore the consequences of organisms modifying their environments, for instance, by exploiting resources or capturing prey. These include models for the dynamics of population size under a given carrying capacity, models of density-dependent resource depletion and density-dependent predation, competition and niche theory, models of succession, models of multispecies communities, and models predicting how organisms affect the flow of energy and materials between compartments in ecosystems (Roughgarden 1979; May 1981; Begon et al. 1996).

3.2.1.1 Resource Depletion

One very basic kind of niche construction has been relatively well studied, namely, the local depletion of resources. Ecological models describe how autotrophs and other organisms in an ecosystem exert a degree of control over nutrient availability through exploitative consumption and release from biomass back into the available pool, typically with explicit tracking of resource dynamics and resulting effects on consumer abundance (reviewed in DeAngelis 1992). Models of both intra- and interspecific competition for resources are widely used to study how the intensity of competition will regulate population numbers (e.g., May 1973; Chesson 2000). Considerable attention has been given to the circumstances under which interspecific exploitative competition will result in competitive exclusion or coexistence (Tilman 1982, 1986, 1990; DeAngelis 1992; Grover 1997). The models make predictions about which species will exclude other species depending on the growth rates and population size of the species concerned, together with the amount of available nutrients and the competition for these nutrients (Tilman 1982; DeAngelis 1992; Riebesell 1974). These models explicitly track the frequency or concentration of the exploited resource, for instance, by assuming that autotroph growth rates and nutrient depletion are governed by Monod functions. Generalizations allow *n* autotrophs to compete for the same or for multiple resources in the face of different kinds of resource renewal, different levels of mortality, detrital feedbacks, and other elaborations (O'Brien 1974; Tilman 1982; Holt 1985; Abrams 1988; DeAngelis 1992; Vincent et al. 1996). Additional complexity is included with higher trophic levels, such as specialist or generalist herbivores or predators, to model idealized food webs and derive conditions for exclusion and coexistence (Levine 1976; DeAngelis 1992; Holt et al. 1994). Related models of ecological succession describe changes in the species composition of an ecological community over time, some of which are due to changes in the environment caused by the populations themselves (Armstrong 1979; DeAngelis 1992). In a similar vein to the prediction that the dominant competitor will be the species that can persist at the lowest resource level (Tilman's R^* rule; Tilman 1982), comparable rules have emerged from theoretical analyses of systems of competitors with shared predators and parasites (Levine 1976; DeAngelis 1992; Holt et al. 1994; Abrams 2000; Bohannan and Lenski 2000; Kerr et al. 2002). Typically these ecological models do not explore how interspecific competition drives evolution but focus instead on how the individuals of one species suffer a reduction in fecundity, survivorship, or growth as a result of resource exploitation or interference by individuals of another species. However, for taxa whose genetics are essentially clonal (e.g., bacteria), genetic and ecological models are to all intents and purposes equivalent; species coexistence implies genetic polymorphism, and competitive exclusion by a dominant type is similar to directional selection.

Two conclusions from these ecological analyses are relevant here. The first is that niche construction through the exploitation of resources is likely to have ramifications that affect ecosystem composition and stability. The second is more prosaic but is relevant to the modeling exercise described later in this chapter, namely, that in investigations of niche construction it is important to track explicitly the dynamics of resources affected by niche construction.

3.2.2 Evolutionary Models

Parts of population biology and related disciplines are concerned with the evolutionary consequences of the modifications that organisms bring about in their own, and in other populations', selective environments. These include frequency- and density-dependent selection, habitat selection, coevolution, maternal inheritance, indirect genetic effects, gene-culture coevolution, and others.

3.2.2.1 Frequency- and Density-Dependent Selection

Models of frequency-dependent selection capture something of the process of niche construction in evolution, in cases where the fitness of the niche-constructing trait depends on the frequency of niche constructors. Frequency-dependent selection occurs when the fitness of a genotype (or an allele) depends on its frequency (or those of the alternative types) within the population (Futuyma 1998).[1] Similarly, density-dependent selection occurs when the fitness of a genotype depends on the size of the population or the numbers of the variants in the population (Roughgarden 1998). For example, Slatkin (1979a,b) explored the dynamics and phenotypic frequencies at equilibria of populations exposed to frequency- and density-dependent selection. Wright (1984) developed a two-locus model of frequency-dependent selection in order to investigate how pleiotropy affects population growth. Frequency-dependent selection also plays an important role in models of sex-ratio evolution (Eshel and Feldman 1982; Karlin and Lessard 1986). Early models of density-dependent selection developed an interface between ecology and population genetics by making viabilities density-dependent (Roughgarden 1971; Clarke 1972), while later studies considered density-dependent fecundity and more

[1] It should be noted that there is some disagreement over when a regime should be called frequency-dependent. For example, with constant fertility differences between genotypes, the fitnesses ascribed to alleles will appear to be frequency-dependent. Such disagreements can be obviated by correct choice of the evolutionary units to be studied.

general intra- and interspecific competitive situations (Clark and Feldman 1981; Matessi and Jayakar 1976; Roughgarden 1976; Fenchel and Christiansen 1977; Christiansen and Loeschcke 1980a,b; Loeschcke and Christiansen 1984). Among population geneticists, most theoretical analyses have focused on whether frequency- and density-dependent selection can maintain genetic variability at the stable population equilibria (Levene 1953; Cockerham et al. 1972; Smouse 1976; Hoekstra et al. 1985; Eshel and Feldman 1984), or how the dynamics under frequency dependence differ from those with constant fitnesses, for example through the introduction of cycles. Theoretical ecologists have also been interested in the effects of these more complicated mechanisms of selection on competition and resource partitioning (DeAngelis 1992). The importance of frequency- and density-dependent selection has been acknowledged by evolutionary biologists and ecologists. For instance, Futuyma (1986) writes, "[I]t is likely that there is a frequency-dependent component in virtually all selection that operates in natural populations, for interactions among members of a population affect the selective advantage of almost all traits" (p. 166). There is now considerable evidence for frequency-dependent effects in natural populations, particularly in social interactions between animals (see Dugatkin and Reeve 1998, and articles therein). Other important examples of frequency dependence include self-incompatibility alleles in plant populations, mimicry in butterfly populations, and sexual selection, for example in the study of mating preferences where rare males may have an advantage (Ehrman 1968; Maynard Smith 1989). Similarly, in recent years, the combination of more powerful statistical techniques and more comprehensive data sets has produced the consensus that density dependence is just as prevalent in natural populations as most ecologists had assumed (Hanski 1999).

There are, perhaps, two main reasons why population-genetic studies of frequency- and density-dependent selection are relatively uncommon. The first is mathematical convenience. Maynard Smith (1989, p. 70) acknowledged this when he stated "there is nothing artificial about assuming that fitnesses are frequency-dependent. The artificial assumption is that relative fitnesses are constant: its justification is mathematical convenience, not truth." Frequency-dependent selection models are usually mathematically intractable, although not necessarily more so than many other population-genetic models with constant parameters. Moreover, there are countless ways to specify a frequency-dependent relationship mathematically, which makes it unlikely that truly general findings will emerge from any particular analysis. While the measurement of natural selection in terms of classical constant fitnesses has been very difficult, frequency dependence has often been invoked as a kind of default explanation, although in most cases an empirically justifiable mechanism for the frequency dependence has been hard to extract from data.

The second reason is that biologists interested in frequency-dependent relationships have often used a game-theoretical approach (Hamilton 1967; Maynard Smith 1982; Dugatkin and Reeve 1998; Hofbauer and Sigmund 1998). "Evolutionary game theory is a way of thinking about evolution at the phenotypic level when the fitnesses of particular phenotypes depend on their frequencies in the population" (Maynard Smith 1982, p. 1). Game-theoretical approaches endeavor to identify the "un-invadable" behavioral phenotype, called the evolutionarily stable strategy, or ESS, and it is this phenotype that is predicted to dominate in a particular social context. An ESS is "a strategy such that, if all the members of a population adopt it, then no mutant strategy could invade the population under the influence of natural selection" (Maynard Smith 1982, p. 10). The relationship between the ESS and the genotype frequency dynamics in classical population genetics is not always straightforward (Eshel and Feldman 1982, 1984, 2001; Eshel et al. 1998, Hammerstein 1996; Gomulkiewicz 1998). ESS and adaptive-dynamics models have been used mostly to analyze animal behavior, particularly aggression, cooperation, social foraging, and communication (Maynard Smith and Price 1973; Axelrod and Hamilton 1981; Johnstone 1995, 1998; Dugatkin and Reeve 1998; Hofbauer and Sigmund 1998; Giraldeau and Caraco 2001).

From our perspective, a striking characteristic of most models of frequency- and density-dependent selection, including those employing evolutionary game theory, is that in virtually all cases the fitnesses of alleles, genotypes, or phenotypes are functions of their own frequencies (or of the frequencies of those variants implicitly or explicitly at the same loci). In other words, models of frequency-dependent selection have focused on cases where the fitness of a genetic (or phenotypic) variant depends on the frequency only of that variant. This is, of course, only a tiny subset of the frequency-dependent fitness relationships that may play a role in evolution. What is comparatively rare is an exploration of the frequency-dependent feedback of niche construction on other loci, or other traits (although see section 3.2.2.5, below). The models below describe how the distribution of genetic variation responsible for niche construction can influence the selection acting on genetic variation at loci not responsible for the niche construction. Virtually all of the examples described in chapter 2 fit into this latter category.

3.2.2.2 Habitat Selection

Habitat selection refers to cases where individuals with a particular genotype are able to choose the habitat in which their fitness is greatest (Rosenzweig 1991). It is, therefore, a form of relocational niche construction where individuals choose pertinent resources in their environment and is another area of analysis in which a phenotypic effect feeds back to affect an individual's fitness. Habitat choice has been demonstrated in many insects, lizards, rodents, and birds (Jones 1980; Rosenzweig 1981, 1991; Jaenike 1982; Rausher 1984; Cody 1985; Morris 1994; Brown 1998; Hanski and Singer 2001). Mobile animals frequently move into especially favorable sites; for example, ectotherms move between sunny and shady spots to maintain an appropriate body temperature. Plants have the same capacity; for example, roots spread preferentially into nutritionally favorable microsites, and many plants have seed dormancy that enables their seeds to survive until conditions are favorable for growth (Bond and van Wilgen 1996; Futuyma 1998). Moreover, habitat choices can profoundly affect metapopulation structure and patterns of colonization and extinction (Hanski and Singer 2001).

Once again, for population geneticists, theoretical interest has been principally concerned with the question of whether habitat selection can maintain genetic variability in populations (Levene 1953; Maynard Smith 1966; Hoekstra et al. 1985; Garcia-Dorado 1986). However, evolutionists have begun discussing how, by circumscribing the range of habitats they experience, habitat selection by individuals can channel the direction of adaptive evolution (Rosenzweig

1987; Holt 1987), and there is a recent trend toward exploring adaptation to the chosen habitat (Holt and Gaines 1992; Brown and Pavlovic 1992; Ronce and Kirkpatrick 2001). Habitat selection has been modeled as an evolutionary game, primarily in terms of its impact on community structure (Brown 1990, 1998). Foraging theory has also considered prey and patch choice among animals, with the goal of understanding behavioral strategies (Krebs et al. 1978; Stephens and Krebs 1986; Cohen 1991; Giraldeau and Caraco 2001). Again, these analyses have focused primarily on those alleles, genotypes, or phenotypes that influence habitat or resource choice, that is, the production of the niche-constructing phenotype itself (e.g., Jaenike 1982; Maynard Smith 1962; Rosenzweig 1981, 1987). While comparatively little of this theory would apply to the kind of relocational niche construction described in chapter 2, an important exception is the growing body of theory that starts with movement patterns in heterogeneous environments and then studies the process of adaptation to chosen habitats (Holt and Gaines 1992; Brown and Pavlovic 1992; Kaweki 1995; Ronce and Kirkpatrick 2001).

3.2.2.3 Coevolution

Models of the coevolution of two or more species implicitly or explicitly take account of the fact that the niche construction of one population can affect the selection on another. This is another well-explored area of evolutionary biology and includes models of competitive, predator-prey, host-parasite, plant-herbivore, and mutualistic interactions (Futuyma and Slatkin 1983; Thompson 1994; Heesterbeek and Roberts 1995; Abrams 1996; Marrow et al. 1996; Gavrilets 1997) (see also section 3.2.1.1, above). Theoretical analyses of host-parasite interactions specify the circumstances under which parasites can evolve to become beneficial, benign, moderately virulent, or highly virulent, depending on their biological features and ecological circumstances (May and Anderson 1983, 1990; Ewald 1983, 1994; Boots and Sasaki 1999). Several "gene-for-gene" models of enemy-victim interactions have been developed in which there is a correspondence between alleles that affect the vulnerability of the victim and those that influence the effectiveness of the enemy (Futuyma 1998). Here, selection is frequency-dependent, with the fitness of each .genotype in one species depending on the allele frequencies in the other (Seger 1992). There are similar quantitative genetic models of predator-prey and host-parasite coevolution (see Futuyma and Slatkin 1983 and Thompson 1994 for reviews). A key assumption in most of these models is that there is a direct correspondence between the number of individuals, or frequency of alleles or genotypes, in one population, and the intensity of selection acting on the other. Where this interaction is indirect, for example, exploitative competition operating via a shared resource, the models have primarily been ecological or demographic (see above). However, some population-genetic coevolutionary models, including models of character displacement (Slatkin 1980) and of prey species that share predators (Holt 1977; Abrams 2000), have explored indirect interactions between coevolving species, albeit typically with comparatively simple and restricted utilization distributions of intermediary resources. For populations that coevolve via interactions in their impact on pertinent resources in their environment, the amounts of such resources are unlikely to be directly proportional to either the number of organisms or the frequency of alleles in either population, and the discordance between resource and population dynamics will be exacerbated if there is ecological inheritance. We return to this point in chapters 5 and 8.

3.2.2.4 Maternal Inheritance and Maternal Effects

In chapter 1 we introduced the concept of ecological inheritance, the legacy of modified selection pressures that ancestral generations of niche-constructing organisms create for their descendants. We described ecological inheritance as akin to a second inheritance system in evolution. Ecological inheritance shares with maternal and cytoplasmic inheritance the potential to generate interesting and unusual dynamics in evolutionary systems (Uyenoyama and Feldman 1979; Kirkpatrick and Lande 1989; Cowley and Atchley 1992; Ginzburg 1998). Maternal effects occur when a mother's phenotype influences her offspring's phenotype independently of the female's genetic contributions to her offspring (Mousseau and Fox 1998a,b). Recent reviews have discussed the ecological and evolutionary importance of maternal effects in insects (Mousseau and Dingle 1991), plants (Roach and Wulf 1987), and mammals (Bernado 1996), and many areas of current research in ecology and evolution explicitly or implicitly involve maternal effects (Wade 1998). In recent years, the term "maternal effects" has been broadly interpreted and includes the observations that mothers determine propagule size, where, when, and how propagules are dispersed, protection of young from inclement conditions or predators, and parental care and provisioning of developing young (Grun 1976; Schluter and Gustafsson 1993; Mousseau and Fox 1998a,b). Moreover, maternal affects are increasingly being regarded as a source of adaptive variation (see Mousseau and Fox 1998a, and references therein). Organisms inherit more than genes from their parents (Oyama et al. 2001); for example, properties of the cytoplasm are maternally transmitted. Although cytoplasmic inheritance is commonly thought of in terms of organelles such as mitochondria, chloroplasts, or intracellular virus particles (Grun 1976), it is now well established that a mother's experience of the environment can lead to variation in her growth, condition, and physiological state that can be transmitted to offspring via cytoplasmic factors (e.g., yolk amount, hormones, mRNA) that affect offspring development (Mousseau and Fox 1998a,b). Mathematical analyses have revealed that maternal effects may cause a response to selection that is initially in the opposite direction to that selection, or may generate an evolutionary "momentum" where the response to selection continues after the selection ceases, oscillations in the approach to equilibria, or interfamily or group selection (Feldman and Cavalli-Sforza 1976; Cavalli-Sforza and Feldman 1981; Kirkpatrick and Lande 1989; Cowley and Atchley 1992; Wade 1998; Sinervo 1998). Moreover, models of population growth based on maternal effects have been found to exhibit oscillatory dynamics with cycles of slow growth in abundance followed by dramatic population crashes (Ginzburg 1998), and this has led to the suggestion that maternal effects are a major cause of oscillations in natural population numbers (Ginzburg and Taneyhill 1995). Models of cultural evolution have also explored the behavior of maternally inherited traits, generating parallel findings (Feldman and Cavalli-Sforza 1976; Cavalli-Sforza and Feldman 1981; Boyd and Richerson 1985). So far, however, these models have explicitly limited the number of generations involved in this inheritance to two. Kirkpatrick and Lande (1989) admit that the term maternal inheritance may be misleading and accept that ancestral phenotypes of both sexes and of previous generations may influence selection on many subsequent generations. Maternal effects can thus be regarded as a subcategory of ecological inheritance, and the theoretical work demonstrating the significance of maternal effects in evolution confirms the importance of ecological inheritance as a whole.

3.2.2.5 Epistasis and Indirect Genetic Effects

Since the 1960s evolutionary biologists have devoted increasing attention to the study of epistasis, that is, interaction between genes in their effects on phenotypes or fitness (Bodmer and Felsenstein 1967; Karlin and Feldman 1970; Fenster et al. 1997; Rice 1998; Wolf et al. 2000; Christiansen 2000). Epistatic interactions between genes are now known to have a considerable influence on a number of evolutionary processes, including the evolution of sex, the evolution of recombination, mutation load, inbreeding depression, mutational meltdown, Muller's ratchet, and the evolution of reproductive isolation among species (Phillips et al. 2000). Indirect genetic effects can be viewed as a class of epistatic interactions in which "the environmental influences on the phenotype of one individual . . . are due to the expression of genes in a different conspecific individual" (Wolf et al. 1998, p. 64) and can be regarded as examples of the kind of feedback from niche construction introduced by Laland et al. (1996). The effect on the phenotype of social or biotic environments provided by conspecifics has attracted the interest of quantitative geneticists because such environmental components are heritable. If there is variation in the quality of environments provided by others that reflects underlying genetic variation, then indirect genetic effects will ensue. In our terms (Laland et al. 1996), indirect genetic effects are a consequence of niche construction, and, where parents are involved, indirect genetic effects result from niche construction and ecological inheritance. Indeed, all cases of ecological inheritance will generate indirect genetic effects. In chapter 5 we introduce the term "environmentally mediated genotypic associations" (EMGAs), which are virtually identical to indirect genetic effects, although EMGAs are broader than indirect genetic effects because they also encompass between-species genotypic interactions. Indeed, interactions between unrelated individuals may also result in indirect genetic effects in situations where there is covariance between the genes of an individual and the social environment that it experiences, because individuals both experience a particular social environment and contribute to that environment (Wolf et al. 1998). We see no reason why indirect genetic effects should be restricted to within-species interactions. Models of indirect genetic effects have generated important insights into a number of topics within behavioral ecology and evolution, including kin selection, parental investment, and mate choice (reviewed in Wolf et al. 1998). Indeed, indirect genetic effects may have wide-ranging influences on evolutionary processes, for instance, changing evolutionary rates, leading to nonintuitive (e.g., opposite) responses to selection, and generating time lags and momentum effects (Kirkpatrick and Lande 1989; Moore et al. 1997, 1998; Wolf 2000). These models of indirect genetic effects are closely related to our early genetic models for the evolution of niche construction (Laland et al. 1996, 1999). While there are clearly many parallels between indirect genetic effects and niche construction, there is one distinction that is worthy of note. Indirect genetic effects are regarded solely as "effects" or products of prior natural selection and not as part of a process of environmental modification, while niche construction is treated as a process in its own right. We return to the issue of whether niche construction is more appropriately described as "effect" or "process" in chapter 10.

3.2.2.6 Gene-Culture Coevolution

Gene-culture coevolutionary theory, also known as "dual-inheritance theory" (Boyd and Richerson 1985), is a branch of theoretical population genetics that, in addition to modeling the differential transmission of genes from one generation to the next, incorporates cultural variation into the analysis (Feldman and Cavalli-Sforza 1976; Boyd and Richerson 1985; Feldman and Laland 1996). The archaeological record documents the fact that for at least the last two million years hominid species have inherited two kinds of information, one encoded by genes, the other in cultural processes. The two transmission systems may interact because what an individual learns may depend on its genotype, and because the selection acting on the genetic system may be generated or modified by the spread of a cultural trait. Cultural processes complicate selection to the extent that the outcome may differ from that expected under purely genetic transmission

(Feldman and Cavalli-Sforza 1976; Boyd and Richerson 1985). In addition, interactions between genes and culturally influenced patterns of behavior may either speed up or slow down the response to selection.

Gene-culture coevolution is relevant here because it captures two central features of our evolutionary perspective. First, through their expression of socially learned information, humans are explicitly recognized as niche constructors, capable of modifying their own selection pressures. Second, the information underlying this niche construction is inherited from one generation to the next by an extragenetic inheritance system. Although cultural inheritance clearly differs in several important respects from ecological inheritance, the most notable being the informational content of the former, it may nevertheless generate modified natural selection pressures.

Gene-culture coevolutionary theory could be described as a body of theory designed to explore the evolutionary consequences of niche construction in those organisms capable of the stable transgenerational transmission of learned information, most obviously *Homo sapiens*. However, it is explicitly species-specific, while niche construction occurs more generally in all species of organism, including those incapable of learning or culture. The similarities between the processes of gene-culture coevolution and evolution with niche construction are such that models of the former can suggest the kind of properties that models of the latter are likely to exhibit. Whether it is based on cultural inheritance or not, niche construction transforms evolutionary dynamics, modifying the rates and trajectories of populations' evolution, the equilibria they approach, and the complexity of the dynamics they exhibit. Similarly to cultural inheritance, ecological inheritance is likely to have consequences over many generations.

3.2.2.7 Evolution in Spatially Heterogeneous Environments

There is a rich and growing theoretical literature on the role of demographic asymmetries (e.g., in directions of spatial flows of individuals, or in local abundance) in influencing niche conservatism and evolution (Holt and Gaines 1992; Holt 1996; Kawecki et al. 1997; Gomulkiewicz et al. 2000). Although this literature does not directly explore the evolutionary consequences of niche construction, some of the findings are relevant here. For instance, Holt and Gaines (1992) argue, on the basis of mathematical models, that evolution in spatially heterogenous environments is channeled toward adaptation to those regions of niche space in which abundance is greatest, rather than to other regions. Kawecki et al. (1997) put forward a similar argument. This makes evolution of the fundamental niche an inherently conservative process, leading Holt and Gaines to conclude that "in a certain sense organisms inherit their environments, as well as their genes, from their ancestors" (p. 446), clearly affirming the importance of ecological inheritance. Population size tends to be autocorrelated across generations and, with limited dispersal, most organisms will tend to be found in environments within the species' fundamental niche where local carrying capacity is high, because populations become extinct outside it. If adaptation to a local environment increases population size there, then the importance of that environment relative to other local environments over the species distribution as a whole will be increased. One consequence of this is that niche construction in a particular local environment that leads to an increase in population size there automatically biases selection toward further adaptation in that environment, relative to other environments where niche construction is absent (Holt and Gaines 1992; Holt 1996; Kawecki et al. 1997). In other words, to the extent that niche construction increases absolute fitness in a specific locality, it may inherently generate feedback leading to further adaptation to that locality, a process that would help to explain many of the data described in chapter 2.[2]

3.2.2.8 Other Approaches

Evolutionary biologists have given considerable attention to situations in which an organism's selective environment is dynamic, in the sense that forces independent of the organism may change the world to which the population adapts. Models in which the selection coefficients vary over time have been studied extensively (Kimura 1954; Haldane and Jayakar 1963; Lewontin and Cohen 1969; Hartl and Cook 1973; Karlin and Liberman 1974; Gillespie 1973). Here genotypic evolution tracks the fluctuating physical and biological re-

[2] We are indebted to Robert Holt for drawing this point to our attention.

sources in the environment (Hartl and Clark 1989). This has been described as "treadmill evolution" or the "Red Queen" hypothesis (Van Valen 1973). In these cases, a component, or components, of the external environment are thought to be changing, although not in response to the activities of the organisms under study.

Robertson (1991) conducted a theoretical analysis of the likely consequences of feedback in evolutionary systems and concluded that, because adapted organisms are both consequences and sources of natural selection, both positive and negative feedback loops should be common in evolution. These feedback loops introduce major instabilities, associated primarily with positive feedback cycles, and hyperstabilities, associated with negative feedback cycles. Feedback can produce "lock-in" effects, in which very small initial differences between alternative adaptations can be powerfully amplified by positive feedback loops resulting from a frequency-dependent fitness advantage to the most common variant. This variant may then rapidly become dominant, driving all competitors extinct. Although this work is very general and does not specifically model niche construction, it is reasonable to conclude that, if organisms were to modify the action of natural selection, they would almost certainly generate the kind of feedback in evolution that Robertson modeled.

One theoretical construct that captures some, but not all, of the consequences of niche construction is Dawkins' (1982) "extended phenotype." Dawkins argues that genes can express themselves outside the bodies of the organisms that carry them. For instance, the beaver's dam is an extended phenotypic effect of beaver genes. Dawkins emphasizes just one aspect of the evolutionary feedback from niche construction, that is, the selection acting on the genes responsible for the extended phenotype. He does not stress the feedback from those novel selection pressures created by the extended phenotype to other genes that affect other aspects of the phenotype. Moreover, Dawkins' extended phenotype requires no additional theoretical framework beyond that of any other phenotype and has not generated a body of formal theory for the evolutionary consequences of niche construction. To Dawkins, the organism itself is the constructed niche of the genes it harbors, so in terms of evolutionary processes there is no logical distinction to be made between phenotype and extended phenotype. While we also see parallels between the interactions of genes with effects inside and outside organisms (see chapter 10), niche construction is capable of changing evolutionary dynamics in ways that other aspects of the phenotype are not.

3.2.3 Conclusions

Three points emerge from our review of theories that overlap with niche construction. The first is that to date there is no body of theory that explores the evolutionary consequences of niche construction in a systematic and general manner. While many separate theoretical domains investigate phenomena with features in common with niche construction, none of these captures all of the pertinent characteristics. Not surprisingly, each has a very different and quite specific goal, focusing on issues such as the processes that may maintain genetic variation or on the consequences of maternal inheritance. While these models may constitute a satisfactory and comprehensive theoretical foundation for their own topics of interest, they provide only a limited foundation for research into the evolutionary consequences of niche construction.

Second, in spite of their different focus, these diverse theories can actually be regarded as confirming that niche construction is evolutionarily consequential. The ramifications of resource depletion on community stability, of frequency- and density-dependent selection, of active choice in habitat and sexual selection, of evidence for unusual evolutionary dynamics with additional inheritance systems and through indirect genetic effects, of the effects of feedback in evolution, and of gene-culture coevolutionary theory that explores human niche construction all attest to the importance of niche construction in evolution.

Third, much of the above theory explicitly models niche construction as a process in its own right because the environmental states in these models are fully or in part determined by the prior niche-constructing activities of organisms, while those niche-constructing activities are fully or in part determined by prior natural selection contingent on environmental states. Yet in conceptual and verbal accounts of evolutionary events, the reciprocity explicit in the models is lost, and niche construction is relegated from being an evolutionary process to merely an effect or product of prior selection.

3.3 MODELING NICHE CONSTRUCTION

Niche construction is itself an evolutionary process that can potentially codirect and regulate natural selection and other evolutionary processes. As mentioned above, niche construction involves feedback, or frequency-dependent effects, and these impose constraints on the type of theoretical analysis that is possible. The approach that we adopt involves simple idealized models that use two-locus population-genetic theory. Two-locus theory is an obvious way to begin such an investigation, as we can designate one locus as partly responsible for the population's capacity for niche construction, and a second locus for which selection pressures are modified by niche construction.

Our models include three components. First, we assume that a population's capacity for niche construction is influenced by the frequency of alleles at one genetic locus, which we label **E**, to emphasize that alleles at this locus influence the changes that the organisms bring about in their environment. Second, our models always include an environmental factor that, for convenience, we treat as a resource labeled **R**. However, in principle, **R** could refer to any biotic or abiotic source of natural selection for the population. The amount or frequency R of **R** is influenced by the niche-constructing activities of past and present generations of organisms. More specifically, R is a function of the frequency of alleles at the **E** locus over a number of generations. Hence, **R** could represent food or water, or the presence of a predator, or the amount of detritus, or any other significant component of the population's niche. R is assumed to vary between 0 (where **R** is absent) to 1 (where **R** is accessible to all members of the population). Third, the amount of this resource in the environment determines the contributions to fitnesses made by a second locus **A**. Hence, the genotypic fitnesses in part depend upon the frequency of alleles at the **E** locus over a number of generations, because these frequencies affect R. Note, for a population of organisms at time t we should write $R(t)$ to indicate the dependence of the amount of the resource on time. This will be understood in what follows and we suppress the time parameter.

This type of model minimally captures the essential dynamics of the evolutionary consequences of niche construction. Note, we are not claiming that most cases of niche construction can be well described as reflecting the expression of a single gene, but rather that the **E** locus is one of many loci that might influence niche construction and, similarly, that the **A** locus is one of many loci that might be influenced by the selection from **R**. By focusing on just two loci, we gain some insight into how genes that underlie niche construction (i.e., **E**) may coevolve with genes selected as a consequence of niche construction (i.e., **A**).

The main problem with developing a niche-construction theory that is based on a two-locus frequency-dependent selection model is that a vast number of frequency-dependent relationships are conceivable, and it is not certain that any useful general properties of two-locus frequency-dependent systems will emerge. Consequently, there is a danger that the results of the analysis will be difficult to interpret. In order to approach this problem, we have used as a baseline for our model an established fixed-fitness two-locus model well studied in the theoretical population-genetics literature, namely, the multiplicative model (Bodmer and Felsenstein 1967; Karlin and Feldman 1970; Feldman et al. 1975; Karlin and Liberman 1979; Hastings 1981; Christiansen 1990, 2000). We perturb the standard multiplicative model by introducing frequency-dependent fitnesses, and this allows us to analyze and describe the effects of niche construction relative to a comparatively well analyzed body of theory. Two-locus theory has been concerned with the effects of recombination, which occurs at rate r, and how its interaction with selection determines the dynamics of the evolving system. We will also address these topics in our analyses. In addition, we consider the consequences of manipulating the functional form of the relationship between the fitness of the genotypes at the **A** locus and the amount R of the resource **R**.

Below we describe two models. In the first and simpler model the amount of the resource present depends wholly on the niche-constructing activities of previous generations. We present two genetic versions of this model, which exhibit qualitatively similar behavior:

(a) diploid and (b) haplodiploid. The latter also includes the case where the two loci are sex-linked. Our second model extends the first to allow the amount of the resource to be influenced by independent (ecological) processes of renewal and depletion.

3.3.1 Model 1a: A Simple Model of Niche Construction for Diploids

Consider an isolated population of diploid individuals and the two genes **E** and **A**. Each gene has two alleles: **E** has alleles E and e and **A** has alleles A and a. In generation t, the frequencies of alleles E and A are $x(t)$ and $p(t)$, respectively. Genotypes at the **A** locus make contributions to the two-locus fitnesses that are assumed to be functions of the frequency R with which the resource **R** is encountered by organisms in their environment (where $0 < R < 1$). This frequency is, in turn, a function of the amount of niche construction over n generations (including the present one), relative to an implicit rate of spontaneous resource recovery or resource dissipation due to other independent factors in the environment. This treatment of the interaction between the population and the resource, while a logical framework with which to begin our investigation, is nonetheless extremely simplified. More realistic treatments would involve general distributions of the resource and simultaneous ecological dynamics of resource and population (e.g., MacArthur and Levins 1967; Schoener 1974). Further extensions would incorporate population dynamics and the implications of heterogeneity in abundance and demographic processes in space and time into analyses of niche construction.

In generation t, the amount of the resource created or destroyed as a consequence of niche construction is assumed to be a function of the frequency of allele E (x_t) at that time. The simplest such function is one in which the contribution of each generation's niche construction to the amount of the resource is weighted by a coefficient (see equation A1.1 in appendix 1). We consider three cases: (i) *equal weighting*, where the amount of the resource is equally dependent on each of the previous n generation's niche construction; (ii) *recency effect*, where the amount of the resource has greater dependence on the activities of more recent generations (for example, if the effects of earlier generations' activities have partially dissipated); and (iii) *primacy effect*, where the amount of the resource has greater dependence on the activities of earlier generations (as may be the case when a founder population moves into a new environment). Mathematical expressions for these functions are given by equations A1.2a–c in appendix 1. The key point is that in this model the effect of niche construction on the resource, and hence on selection, is reduced to a function of the frequencies of allele E in the previous n generations.

Genotypic fitnesses are given in table 3.1. It can be seen that the fitnesses have a fixed viability component and a frequency-dependent viability component. The fixed component is given by the α_i and η_i terms, which are the fitnesses of the single-locus genotypes in the standard two-locus multiplicative viability model. The frequency-dependent component is a function of the frequency R of the resource **R**, which is given by equations A1.1 and A1.2 in appendix 1. Thus the terms involving ε express the deviations from constant viabilities due to niche construction, where the size of ε scales the amount of niche construction. The frequency-dependent components of the contribution to fitness of genotypes *AA*, *Aa*, and *aa* are functions of R, $\sqrt{R(1 - R)}$, and $1 - R$, respectively, chosen so that this component of selection favors allele A when the resource is common, and allele a when it is rare. The particular functions chosen here are specified by a parameter f, which determines the power of the relationship between the contribution of locus **A** to genotypic fitnesses and the frequency of the resource, while ε determines the strength of the frequency-dependent component of selection relative to the fixed-fitness component. Positive values of ε represent resource accumulation, while negative values represent resource depletion. We shall assume $-1 < \varepsilon < 1$ and $f > 0$. The entries in table 3.1 result in the four standard gametic recursions given by equations A1.3a–d in appendix 1.

To provide a baseline for the subsequent analysis, we first describe the behavior of the model when there is no external source of selection, in which case the fixed multiplicative viabilities (α's and η's) may all be taken to be 1. Here the effects of the selection generated by niche construction are clearest. We then investigate the effects of niche construction when an external source of selection (independent

TABLE 3.1. Multiplicative Viabilities with Niche Construction for Diploid Genotypes

	EE (α_1)	Ee (1)	ee (α_2)
AA (η_1)	$w_{11} = \alpha_1\eta_1 + \varepsilon R^f$	$w_{12} = \eta_1 + \varepsilon R^f$	$w_{13} = \alpha_2\eta_1 + \varepsilon R^f$
Aa (1)	$w_{21} = \alpha_1 + \varepsilon[R(1 - R)]^{f/2}$	$w_{22} = 1 + \varepsilon[R(1 - R)]^{f/2}$	$w_{23} = \alpha_2 + \varepsilon[R(1 - R)]^{f/2}$
aa (η_2)	$w_{31} = \alpha_1\eta_2 + \varepsilon(1 - R)^f$	$w_{32} = \eta_2 + \varepsilon(1 - R)^f$	$w_{33} = \alpha_2\eta_2 + \varepsilon(1 - R)^f$

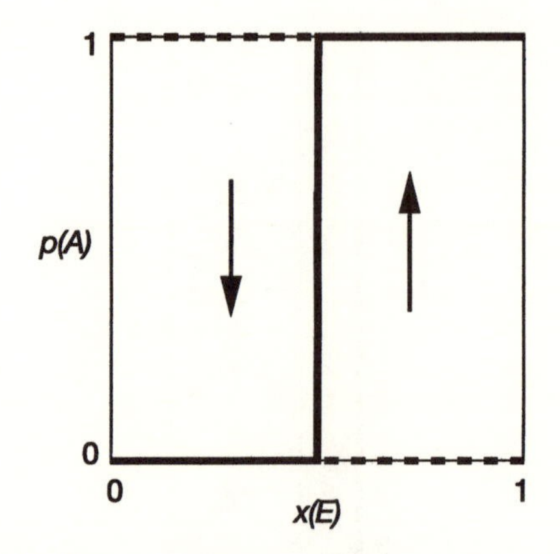

FIGURE 3.1. Model 1a. Niche construction can generate selection even when no external source of selection is acting. The arrows represent the trajectories of a population with the frequencies of allele *E* and *A* given by *x* and *p*, respectively. The heavy lines represent neutrally stable equilibria, and the dashed lines unstable equilibria. Formal details are given in appendix 1. (From Laland et al. [1996], fig. 1. Reprinted with the permission of Blackwell Science.)

of the niche construction), represented by the fixed fitness component of selection, favors one of the alleles at one of the loci. Finally, we focus on the effects of niche construction when the external selection generates overdominance at both loci.

3.3.1.1 No External Selection

With no external source of selection ($\alpha_1 = \alpha_2 = \eta_1 = \eta_2 = 1$) each genotype at the **E** locus makes the same contribution to fitness. Niche construction generates frequency-dependent selection that changes the frequency of alleles at locus **A**. If niche construction increases the amount *R* of the resource (that is, ε is positive), then in the selective environment characterized by high frequencies of *E* (i.e., *R* is large), allele *A* will be favored, while in the selective environment provided by high frequencies of *e* (i.e., *R* is small), allele *a* will be favored. The reverse is true where niche construction decreases *R* (ε is negative). Figure 3.1 illustrates that under these conditions the system exhibits a curve of equilibria, similar to that found in sexual-selection models (Kirkpatrick 1982; Gomulkiewicz and Hastings 1990). The theoretical details are included in equations A1.4–

A1.10 in appendix 1. Numerical iteration suggests that with no external selection, when ε is positive, mean fitness is highest either with A and E fixed, or with a and e fixed.

3.3.1.2 External Selection at the E Locus

We now consider the case in which external selection acts only at the **E** locus ($\alpha_1 \neq 1$, $\alpha_2 \neq 1$, $\eta_1 = \eta_2 = 1$), representing selection on a niche-constructing activity, such as nest building or resource depletion. When allele E is favored ($\alpha_1 > 1 > \alpha_2$), positive values of ε result in geometric convergence to a single equilibrium point with E and A fixed, while negative values result in fixation of E and a but at an algebraic rate. These results are illustrated for $n = 1$ in figures 3.2a and 3.2b. If allele e is favored by external selection ($\alpha_1 < 1 < \alpha_2$), then positive values of ε result in geometric fixation of e and a, and negative values cause fixation of e and A at an algebraic rate. These results are illustrated in figures 3.2c and 3.2d. Provided all alleles are initially present, numerical analysis suggests that under these conditions populations will always converge to these equilibria.

When allele E is favored, the niche-constructing behavior spreads through the population, increasing the amount of the resource and changing the selection pressure at the second locus so that allele A is favored. If the amount of the resource present depends only on the present generation's niche construction ($n = 1$), for example, if the resource is a spider's web, then the response at the **A** locus is fairly immediate. Less predictable patterns emerge when the amount of the resource R depends on multiple past generations' niche construction ($n > 1$), for instance, as occurs when earthworms generate topsoil. Under such circumstances there is a time lag between the change in frequency of alleles at the **E** locus (x), and the response to the frequency-dependent selection at the **A** locus (p). See equations A1.11, A1.14, and A1.15 and the section entitled "External Selection at the **E** Locus" in appendix 1 for theoretical details.

The effect of increasing the number of generations of niche construction on the response to selection at the second locus is illustrated in figure 3.3. The time lag between the response to the selection at the two loci creates an evolutionary inertia. The inertia occurs because, when selection favoring allele E begins, the amount of the

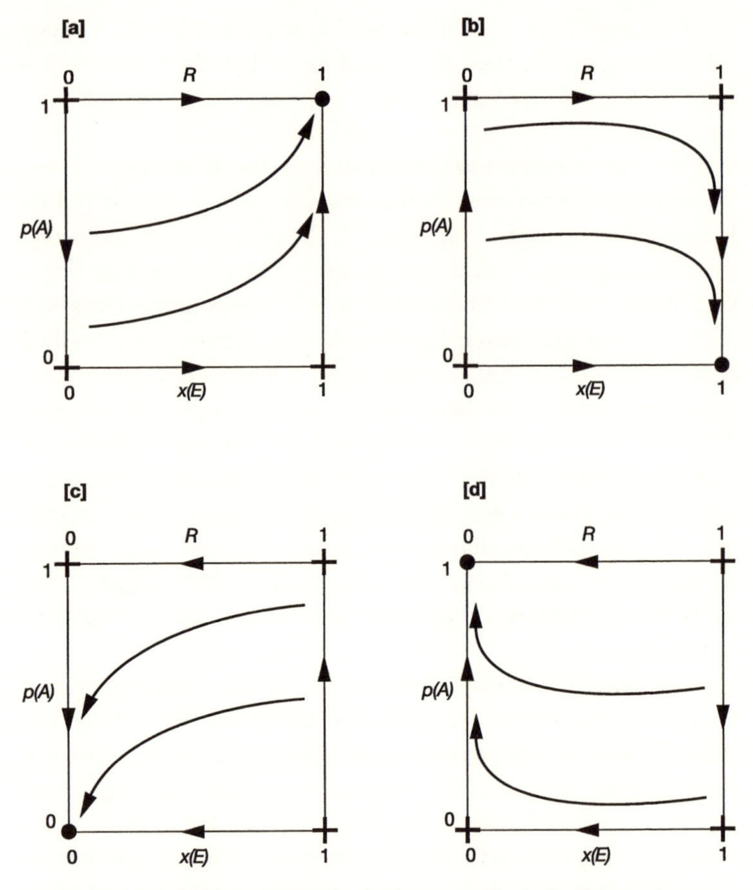

FIGURE 3.2. Model 1a. External selection acts only at the **E** locus ($\eta_1 = \eta_2 = 1$). When allele E is favored ($\alpha_1 = 1.1$, $\alpha_2 = 0.9$), (a) positive values of ε ($\varepsilon = 0.3$) result in geometric convergence to a single equilibrium point with E and A fixed ($u_1 = 1$), (b) while negative values ($\varepsilon = -0.3$) result in slow (algebraic) fixation of E and a ($u_2 = 1$). If allele e is favored by external selection ($\alpha_1 = 0.9$, $\alpha_2 = 1.1$), then (c) positive values of ε ($\varepsilon = 0.3$) result in geometric fixation of e and a ($u_4 = 1$), and (d) negative values ($\varepsilon = -0.3$) cause slow (algebraic) fixation of e and A ($u_3 = 1$). The upper horizontal boundary gives the value of R that corresponds to that value of x, and in this case $R = x$. Here, in all cases $r = 0.5$, $f = 1$, and $n = 1$.

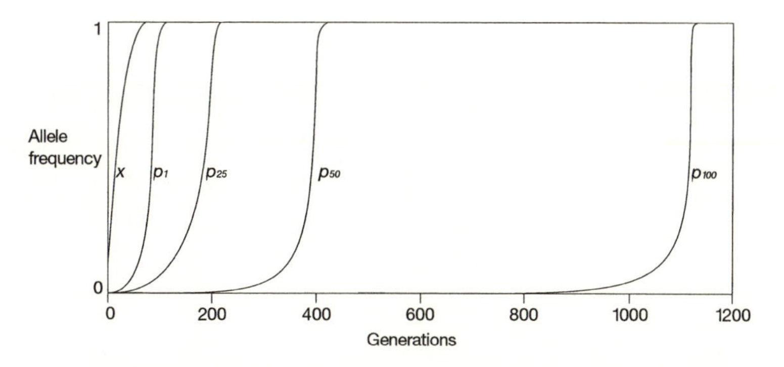

FIGURE 3.3. Model 1a. As a consequence of the spread of allele E, niche construction increases the amount R of a resource in the environment **R**, which generates selection favoring allele A. The response to selection at the second locus is dependent upon the number of generations of niche construction (n) that influence **R**. The figure plots the frequencies of E (x) and A (p) against time for $n = 1, 25, 50,$ and 100. Considerable time lags may occur between the initial niche construction and the response to the selection it generates. In all cases $\alpha_1 = 1.1$, $\alpha_2 = 0.9$, $\eta_1 = \eta_2 = 1$, $\varepsilon = 0.1$, $r = 0.5$, and $f = 1$. Here, x is only weakly dependent on n, except for small r. (From Laland et al. [1996], fig. 2. Reprinted with the permission of Blackwell Science.)

resource cannot accumulate as rapidly as allele E can spread. This means that allele frequencies for the **A** locus will take a number of generations to change from their initial values. In addition, while the resource accumulates to a level necessary to reverse selection at the second locus, the allele favored by the frequency-dependent selection will have dropped in frequency because selection will still be operating in the original direction, and this "lost ground" will have to be made up. The inertia is greatest if there is a primacy effect, since here the amount of the resource depends heavily on earlier generations, when allele E was rare and niche construction had had little effect. In contrast, the inertia is at its weakest if there is a recency effect, since in this case the amount of the resource depends on more recent generations during which allele E has increased in frequency, and hence niche construction is beginning to have some impact. Although not illustrated in figure 3.3, the time lag can also generate an evolutionary momentum. This occurs because when selection at the **E** locus stops, or reverses, the amount of the resource continues to

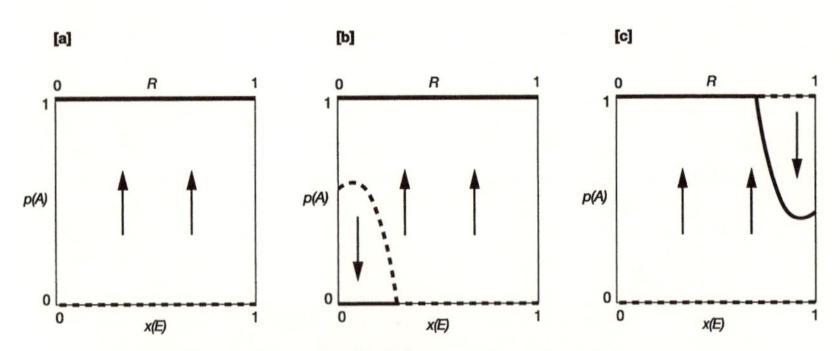

FIGURE 3.4. Model 1a. External selection favoring allele A. (a) With no niche construction ($\varepsilon = 0$), the external selection takes allele A to fixation. (b) Positive niche construction ($\varepsilon = 0.3$) generates selection that opposes the external selection, and can result in the fixation of allele a. (c) Negative niche construction ($\varepsilon = -0.3$) creates selection that opposes the external selection, and polymorphic equilibria are possible. The heavy line represents neutrally stable equilibria, and the dashed line unstable equilibria. Formal details are given in appendix 1. In all cases $\alpha_1 = \alpha_2 = 1$, $\eta_1 = 1.1$, $\eta_2 = 0.9$, $r = 0.5$, and $f = 1$. (From Laland et al. [1996], fig. 3. Reprinted with the permission of Blackwell Science.)

accumulate for a number of generations such that alleles at the **A** locus continue to change frequencies in the original direction, despite selection favoring the opposite allele. The momentum is also greater for a primacy effect, and weaker for a recency effect, for similar reasons.

3.3.1.3 External Selection at the A Locus

Niche construction creates interesting results when the selection it generates opposes external selection acting at the **A** locus (when $\alpha_1 = 1$, $\alpha_2 = 1$, $\eta_1 \neq 1$, $\eta_2 \neq 1$). This would occur if some external agent depletes the resource accrued through niche construction or accumulates the resource depleted through niche construction. The results of numerical and analytical investigations are shown in figure 3.4, which illustrates the behavior under external selection favoring A ($\eta_1 > 1 > \eta_2$). (See equations A1.11–A1.13 and A1.24–A1.29 in appendix 1 for theoretical details.) When there is no niche construction ($\varepsilon = 0$), and when the selection generated by the niche construction is very weak ($1 - \eta_1 < \varepsilon < 1 - \eta_2$), fixation on A is stable (assuming that the A allele is present) (fig. 3.4a).

The mathematical analysis in appendix 1 reveals that near the equilibrium at which chromosome ea is fixed, if $\varepsilon > 1 - \eta_2$ fixation of the chromosome is neutrally stable. That is, the niche construction can reverse the instability of this fixation state, so that fixation in both A and a may be locally stable (fig. 3.4b). In the same way, near the equilibrium at which chromosome EA is fixed, if $\varepsilon < 1 - \eta_1$, alleles a and e can increase when rare. The analysis in appendix 1 reveals that in this case a stable polymorphic equilibrium is possible (fig. 3.4c). Again, the niche construction has affected the evolutionary outcome. In this case, it appears from numerical work that selection at the **A** locus is unaffected by the number of generations of niche construction (n) influencing R. The qualitative behavior of this system is also little affected by the amount of recombination (r) or the power of the frequency dependence (f). Formal details of the separatrix and curve of polymorphic equilibria are given in appendix 1, equations and inequalities A1.24–A1.29. Note that these polymorphic equilibria are neutral in the sense that perturbations away from such an equilibrium followed by subsequent evolution will result in convergence to another point on the curve.

3.3.1.4 External Selection with Overdominance

Finally, we consider the case where external selection generates overdominance ($\alpha_1,\alpha_2,\eta_1,\eta_2 < 1$). In the multiplicative model with heterozygote advantage, an internal equilibrium with linkage equilibrium ($D = 0$) exists for all levels of recombination (r), while if the loci are tightly linked (small r), two equilibria with disequilibrium ($D \neq 0$) exist (Moran 1964; Bodmer and Felsenstein 1967; Karlin and Feldman 1970; Karlin 1975). For small values of r, with overdominance at both loci, the $D \neq 0$ equilibria are locally stable (Karlin 1975), and the internal equilibrium with $D = 0$ is unstable, although, for a small intermediate range of recombination rates, $D = 0$ and $D \neq 0$ may both be stable (Karlin and Feldman 1978). One of these $D \neq 0$ equilibria is characterized by positive linkage disequilibrium, and the other by negative linkage disequilibrium. As a convenient shorthand, we refer to these equilibria as the $+D$ and $-D$ equilibria, respectively. As the amount of recombination increases from zero, the amount of linkage disequilibrium at equilibrium decreases, and these two stable equilibria approach the internal equilibrium with $D = 0$, although they may reach it at different r values (Karlin and Feldman, 1978).

First we explore polymorphic equilibria of the form $D = 0$, which are expected to be stable for sufficiently large values of r. At an equilibrium of this form, the frequency-dependent selection does not change the equilibrium frequency of E, but $\hat{p}$ becomes a function of ε. Expressions for this equilibrium, and for the two edge equilibria that are polymorphic at the **A** locus (with allele E or e fixed) are given in appendix 1, equations A1.12, A1.13 and A1.30, A1.31. The effect of the frequency-dependent selection is to shift the internal equilibrium and these two edge equilibria in the direction of the arrows given in figure 3.1. Positive values of ε will increase the equilibrium frequency $\hat{p}$ of allele A when the resource is common ($\hat{x} > 1/2$) and decrease $\hat{p}$ when it is rare ($\hat{x} < 1/2$). Negative values of ε will have the reverse effect. The size of the change in frequency of A at equilibrium, as a consequence of the frequency-dependent selection, increases the further the frequency of E is from one-half. This is illustrated in figure 3.5, which plots the magnitude of the difference between the equilibrium frequency of p with ($\hat{p}_\varepsilon$) and without ($\hat{p}_0$) the frequency-dependent selection, against the equilibrium frequency of E ($\hat{x}$). The frequency-dependent selection appears to have its most dramatic effect on the two edge equilibria at which either E or e is fixed. As might be expected, the larger the value of ε, the stronger the shift in equilibrium frequency. Figure 3.5 also shows the effect of manipulating the power of the frequency dependence (f). For positive values of ε, as f increases the effect of ε is typically amplified because the differences between the fitnesses of the **A**-locus genotypes increase; however, for negative values of ε there is no simple relationship between f and the effect of ε on allele frequencies.

We now explore the behavior of the system in the special case $\alpha_1 = \alpha_2 < 1$ and $\eta_1 = \eta_2 < 1$, in which case the fixed fitness component of the model is the well-known symmetric multiplicative viability model (Lewontin and Kojima 1960; Bodmer and Felsenstein 1967; Karlin and Feldman 1970). For values of r below r_0, given by equation A1.33 in appendix 1, in addition to the $D = 0$ equilibrium,

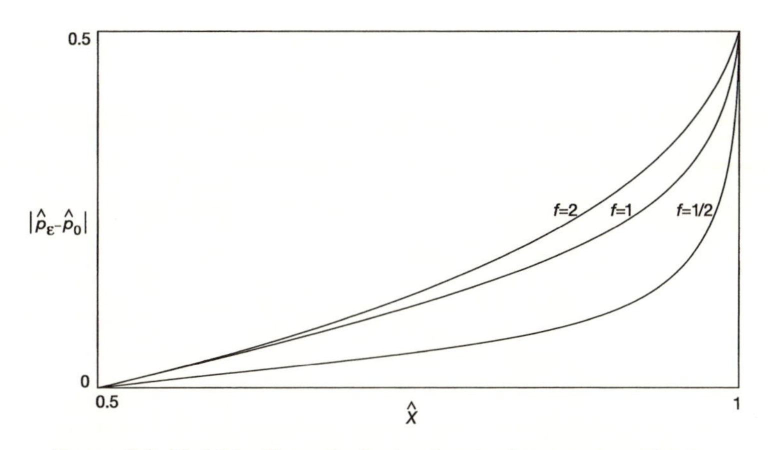

FIGURE 3.5. Model 1a. External selection favoring heterozygotes. The figure shows the difference between the equilibrium frequency of p, with ($\hat{p}_\varepsilon$, $\varepsilon = 0.1$) and without ($\hat{p}_0$, $\varepsilon = 0$) the frequency-dependent selection generated by niche construction, plotted against the E allele equilibrium frequency $\hat{x}$, for $f = 0.5$, 1, and 2. Values of α_1 range from 0.9 to very near 1.0, $\alpha_2 = \eta_1 = \eta_2 = 0.9$, $r = 0.5$, and $n = 1$. (From Laland et al. [1996], fig. 4. Reprinted with the permission of Blackwell Science.)

there are two symmetric equilibria, at which the frequencies of the gametes may be written as $\hat{u}_1 = \hat{u}_2 = 1/4 + \hat{D}$ and $\hat{u}_2 = \hat{u}_3 = 1/4 - \hat{D}$, where $\hat{D}$ is given by equation A1.32 in appendix 1. The allele frequencies at both loci are one-half at these equilibria. Nevertheless, there is an effect on the linkage disequilibrium caused by the niche construction. Here, positive values of ε decrease the amount of linkage disequilibrium at equilibrium, while negative values of ε increase it. The effect of ε on the amount of disequilibrium at equilibrium and the stability of the $D = 0$ equilibrium are illustrated in figure 3.6. An analysis of the effects of niche construction on an asymmetric viability model is also presented in appendix 1.

More generally, both positive and negative values of ε can either increase or decrease the amount of disequilibrium at equilibrium, depending on the position of the equilibrium. Numerical iteration reveals that positive values of ε increase the equilibrium frequency of gametes AE and Ae for equilibria at which the frequency of E is greater than one-half ($\hat{x} > 1/2$), and decrease the equilibrium fre-

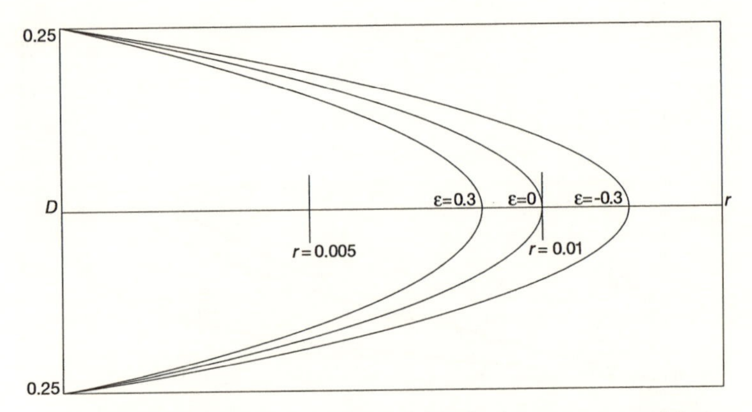

FIGURE 3.6. Model 1a. External selection favoring heterozygotes, with low levels of recombination. The figure gives the amount of linkage disequilibrium at equilibrium for the symmetric multiplicative model, plotted against r, for $\varepsilon = 0.3$, 0, and -0.3. For each value of ε, the $D \neq 0$ points are stable for r smaller than the value at which the curve cuts the $D = 0$ line, and $D = 0$ is stable for r to the right of this value. $\alpha_1 = \alpha_2 = \eta_1 = \eta_2 = 0.8$, $n = 1$, and $f = 1$. (From Laland et al. [1996], fig. 5. Reprinted with the permission of Blackwell Science.)

quency of gametes *aE* and *ae* for equilibria at which the frequency of *E* is less than one-half ($\hat{x} < 1/2$). Negative values of ε have the reverse effect. Further discussion of the effect of the frequency-dependent selection on linkage disequilibrium is given in appendix 1.

3.3.2 Model 1b: Niche Construction in Haplodiploids and for Sex-Linked Loci

We extend model 1 to the case of a population with haplodiploid genetics, because many of the best-known examples of niche construction are provided by species with haploid males and diploid females. This model also describes cases in a diploid population where both the **E** and **A** loci are sex-linked. As before, we employ a two-locus model, but here the theoretical models assume that a population's capacity for niche construction is influenced by the frequency of alleles at the **E** locus in females only. R is again a function of allele frequency at the **E** locus over a number of generations and represents any significant component of the population's niche modified by the female population. Female and male genotypic fitnesses are given in tables 3.1 and 3.2, respectively.

As before, the fitnesses are assumed to be functions of a fixed viability component and a frequency-dependent viability component. Details of the model are given in appendix 2, although the algebraic analysis of allele fixation states has proved intractable. Because the numerical analysis gave results for this haplodiploid (sex-linked) system qualitatively similar to those found for model 1a, we do not describe them in detail. The reader can assume findings similar to those given for model 1a.

Although the equilibria are not qualitatively different, typically the approach to equilibrium is slightly faster for the haplodiploid system than for the diploid system. This is because the effect of the frequency dependence on the haploid males is stronger than on diploids, which, in turn, increases the rate of response to selection of the diploid females, relative to females in a fully diploid system. Figure 3.7 considers the case in which external selection acts only at the **E** locus, representing selection on a niche-constructing activity, which generates selection on the **A** locus. In comparison with the diploid case, this haplodiploid version of the model generates shorter time lags. This is because, while selection on the females in the haplodiploid system is equivalent to that on all individuals in the diploid system, selection on the males is considerably stronger than in the diploid case. Figure 3.8 illustrates the behavior of populations under external selection favoring allele A for both (a) positive and (b) negative niche construction. The system is again similar to the diploid model, although for haplodiploids there is a slightly larger parameter space over which the selection generated by niche construction counteracts external selection to take populations to alternative equilibria. Again, this results from the stronger selection on males.

3.3.3 Model 2: An Extended Model of Niche Construction

The results of model 1 suggest that the changes that organisms bring about in their own environments can indeed be an important source

TABLE 3.2. Viabilities with Niche Construction for Male Haplodiploid Genotypes

EA	*Ea*	*eA*	*ea*
$w_{11} = \alpha_1\eta_1 + \varepsilon R^f$	$w_{31} = \alpha_1\eta_2 + \varepsilon(1 - R)^f$	$w_{13} = \alpha_2\eta_1 + \varepsilon R^f$	$w_{33} = \alpha_2\eta_2 + \varepsilon(1 - R)^f$

Note: Female viabilities are given by table 3.1.

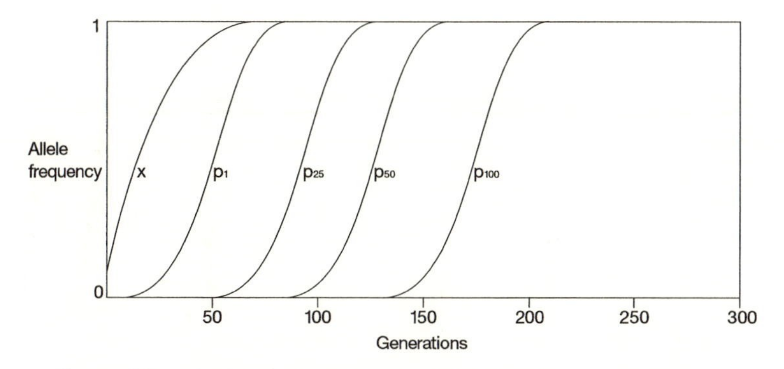

FIGURE 3.7. A haplodiploid model with diploid females as niche constructors (model 1b), which also represents a sex-linked case. As a consequence of the spread of allele E, niche construction increases the amount of **R** in the environment, which generates selection favoring allele A. It can be seen that time lags occur between the initial niche construction and the response to the selection, but these are considerably shorter than in the diploid case (see fig. 3.3). In all cases $\alpha_1 = 1.1$, $\alpha_2 = 0.9$, $\eta_1 = \eta_2 = 1$, $\varepsilon = 0.3$, $r = 0.5$, and $f = 1$. Here, x is only weakly dependent on n.

of modified natural selection pressures and can generate some novel evolutionary outcomes. However, the assumption that the frequency of the key resource in the environment depends solely on niche construction is clearly unrealistic in many cases, because the distributions of ecological resources may also depend on processes that are independent of the population living on that resource (DeAngelis 1992; MacNally 1995; Polis and Strong 1996). Here we present a model that allows the frequency R of the resource **R** to be influenced by a more inclusive set of processes. This extended model allows for R to be influenced by independent environmental processes of renewal or depletion of the same resource as well as for the population's niche construction.

The following are the extensions to model 1. In each generation, the amount of the resource at time t is given by the amount of the resource in the previous generation, diminished by two processes that deplete the resource and incremented by two processes that increase the amount of the resource. Two of these processes are positive or negative niche construction, which here refer to processes that increase or deplete the frequency of the valuable resource R. In this

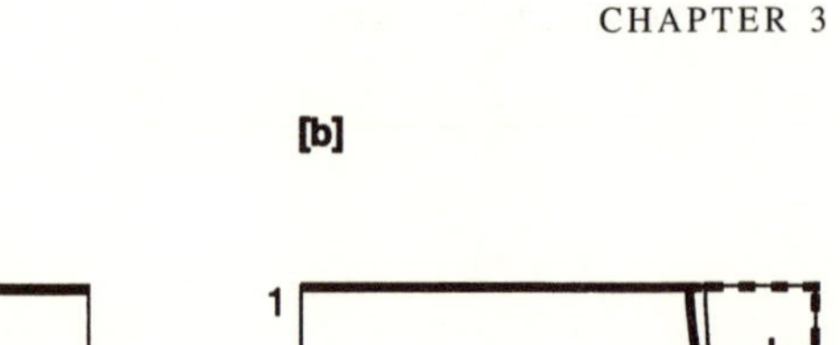
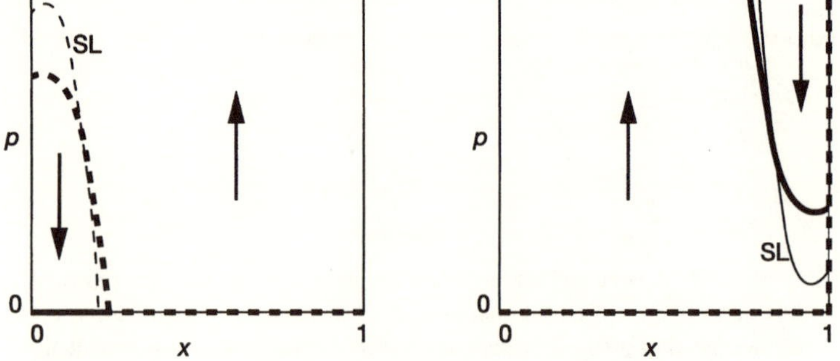

FIGURE 3.8. A comparison of the diploid model (model 1a) with a haplodiploid model with diploid females as niche constructors (model 1b), the latter also representing a sex-linked case (labeled SL). The figures show the effect of the niche construction on a population with external selection favoring allele *A*. (a) Positive niche construction $\varepsilon = 0.3$) generates selection which, for $x < 1/2$, opposes the external selection, and can result in the fixation of allele *a*. (b) Negative niche construction ($\varepsilon = -0.3$) creates selection which, for $x > 1/2$, opposes the external selection, and polymorphic equilibria are possible for *x* large enough. The two models give different results because, while selection on the females in the haplodiploid system is equivalent to that on all individuals in the diploid system, selection on the males is considerably stronger than in the diploid case. The heavy line represents neutrally stable equilibria, and the dashed line unstable equilibria. In both cases $\alpha_1 = \alpha_2 = 1$, $\eta_1 = 1.1$, $\eta_2 = 0.9$, $r = 0.4$, and $f = 1$. The upper horizontal boundary gives the value of *R* that corresponds to that value of *x*, and here $R = x$.

case λ_2 ($0 \leq \lambda_2 \leq 1$) is a coefficient that determines the effect of positive niche construction, such that the increment in *R* due to the current generation's niche construction (i.e., in generation *t*) is given by $\lambda_2 x_t$. If $\lambda_2 = 0$ there is no positive niche construction influencing the amount of the resource, while if $\lambda_2 = 1$ niche construction has its maximum impact on *R*, although the absolute impact also depends on the frequency of allele *E* (*x*). Similarly, γ ($0 \leq \gamma \leq 1$) is a coefficient that determines the effect of negative niche construction, such that the amount of the resource in the current generation is given by the amount in the preceding generation multiplied by the fraction

$(1 - \gamma x_t)$. If $\gamma = 0$ there is no negative niche construction influencing the amount of the resource, while if $\gamma = 1$ niche construction depletes the resource to a maximal degree, although the absolute impact again depends on the frequency of allele E (x).

In addition the resource may also accrue or diminish as a result of processes independent of the niche-constructing organism. In this model, λ_1 $(0 \leq \lambda_1 \leq 1)$ is a coefficient that determines the degree of independent depletion. If $\lambda_1 = 1$ there is no independent depletion, while if $\lambda_1 = 0$ the resource is completely depleted from one generation to the next. The term $\lambda_1 R_{t-1}$ represents the proportion of the resource that remains from the previous generation after independent depletion. Finally, λ_3 is a coefficient that determines the degree of independent renewal. If $\lambda_3 = 0$ there is no independent renewal of the resource, while if $\lambda_3 = 1$ then R will increase to the point where it is available to the entire population. The precise expression giving R_t as a function of R_{t-1}, x_t, λ_1, λ_2, λ_3, and γ is given by equation A3.1 in Appendix 3, or $R_t = \lambda_1 R_{t-1}(1 - \gamma x_t) + \lambda_2 x_t + \lambda_3$. In order to ensure $0 \leq R \leq 1$ we assume $0 < \lambda_1, \lambda_2, \lambda_3, \gamma < 1$, and $\lambda_1 + \lambda_2 + \lambda_3 \leq 1$. This treatment of the interaction between the population and the resource remains extremely simplified. A more realistic treatment would involve general distributions of the resource, more complex ecological dynamics of resource and population, and ecological models that take the density of niche constructors into account. The particular model we employ has properties that are likely to exemplify a broad class of models, and our assumptions concerning resource dynamics are made largely on the basis of analytical convenience.

Genotypic fitnesses are given in table 3.1, and they give rise to the standard gametic recursions A1.3 given in appendix 1. As in model 1, the fitnesses are assumed to be functions of a fixed viability component and a frequency-dependent viability component. The difference between models 1 and 2 is the function defining R. We consider four special cases of equation A3.1, representing positive or negative niche construction with independent renewal or depletion. The four cases are shown in table 3.3. For convenience we assume that the population engages either in positive or in negative niche construction, but not both simultaneously (i.e., $\lambda_2 > 0$ and $\gamma = 0$ or $\lambda_2 = 0$ and $\gamma > 0$). For the numerical analyses when $n > 1$, R_t was com-

TABLE 3.3. Expressions for the Frequency R of the Resource **R** for Positive and Negative Niche Construction, with Independent Renewal and Depletion

	Independent Renewal	*Independent Depletion*
Positive Niche Construction	*Case 1* $R_t = \lambda_1 R_{t-1} + \lambda_2 x_t + \lambda_3$ $\gamma = 0$	*Case 2* $R_t = \lambda_1 R_{t-1} + \lambda_2 x_t$ $\lambda_3 = \gamma = 0$
Negative Niche Construction	*Case 3* $R_t = \lambda_1 R_{t-1}(1 - \gamma x_t) + \lambda_3$ $\lambda_2 = 0$	*Case 4* $R_t = \lambda_1 R_{t-1}(1 - \gamma x_t)$ $\lambda_2 = \lambda_3 = 0$

Note: These represent four special cases of equation A3.1 in appendix 3.

puted by iterated substitution of the value of R in the previous generation into equation A3.1 of appendix 3 for n steps. For further details, see equation A3.4 of appendix 3.

Once again, we focus on the effects of niche construction under four regimes of external selection: no external selection, external selection acting only at the **A** locus, external selection acting only on the **E** locus, and external selection generating heterozygote advantage.

3.3.3.1 No External Selection

First we consider the dynamics of the system when there is no selection aside from that generated by the resource ($\alpha_i = \eta_i = 1$). Here the amount of niche construction remains constant and is given by x. With only positive niche construction ($R_t = \lambda_1 R_{t-1} + \lambda_2 x_t + \lambda_3$, $\gamma = 0$), for frequencies of the allele E from $x = 0$ to $x = 1$, the corresponding values of R range between 0 and 1, depending on λ_1, λ_2, λ_3. An expression for the equilibrium frequency of R is given by equation A3.2 in appendix 3. With only negative niche construction ($R_t = \lambda_1 R_{t-1}(1 - \gamma x_t) + \lambda_3$, $\lambda_2 = 0$), for values of x increasing from 0 to 1, R decreases in an interval of values contained in 0–1, depending on the values of λ_1, λ_3, and γ. An expression for the equilibrium value of R in terms of an equilibrium value $\hat{x}$ of E is given by equation 3.3 in appendix 3. It is a feature of the frequency-dependent fitnesses in table 3.1 that, for positive values of ε, selection favors a when $R < 1/2$, A when $R > 1/2$, with no selection on A when $R = 1/2$, while for negative ε, selection favors a when $R > 1/2$ and A when $R < 1/2$. For instance, in figure 3.9a, with no positive niche construction (when $x = 0$) the balance of independent renewal and depletion of resources leaves R at an equilibrium value of 1/3. Any populations containing allele E ($x > 0$) will engage in niche construction. Positive niche construction ($\lambda_2 > 0$) increases the resource by an amount proportional to x. For values of x less than 0.25, then $R < 0.5$ and the selection on the **A** locus favors allele a. However, values of x greater than 0.25 result in sufficient construction of the resource to produce $R > 0.5$, and the selection on **A** switches to favor allele A. The values of $\hat{x}$ and corresponding equilibrium values of $\hat{R}$ are shown on the bottom and top horizontal axes, respectively, in figure 3.9, and the bold lines correspond to the stable equilibria, which in figures 3.9a–3.9d and 3.9f entail fixation of A or a.

3.3.3.2 External Selection at the A Locus

We now consider the case where there is external selection at the **A** locus only ($\alpha_1 = \alpha_2 = 1$, $\eta_1 \neq 1$, and/or $\eta_2 \neq 1$). Figures 3.9b–3.9f illustrate the behavior of the model when there is directional selection favoring allele A ($\eta_1 > 1 > \eta_2$). For positive values of ε, if the equilibrium value of R is sufficiently small, the selection generated by the resource will be strong enough to counteract the external selection and take the population to the $p = 0$ boundary. With positive niche construction and weak external selection (η_2 slightly larger than 1) as in figure 3.9b, equation A3.5 in appendix 3 gives $\hat{p}$ as an approximately linear function of R, so that regions of the $p = 0$ and $p = 1$ boundaries constitute equilibria that may be simultaneously stable for a range of values of x. For stronger external selection (e.g., fig. 3.9c) $\hat{p}$ becomes more curvilinear as a function of R, and a becomes fixed for a smaller region of the parameter space. Figure 3.9d illustrates the effects of negative niche construction, with the a fixation boundary being stable from some starting conditions. In this case, since fixation on a only occurs when $\hat{R}$ is sufficiently small, independent renewal and positive niche construction make it less likely that the frequency-dependent selection will generate the kind of counter selection illustrated in figures 3.9b–3.9d, while independent depletion and negative niche construction both increase this likelihood.

Examples with positive and negative niche construction are shown in figure 3.9e and figure 3.9f, respectively. When the external selec-

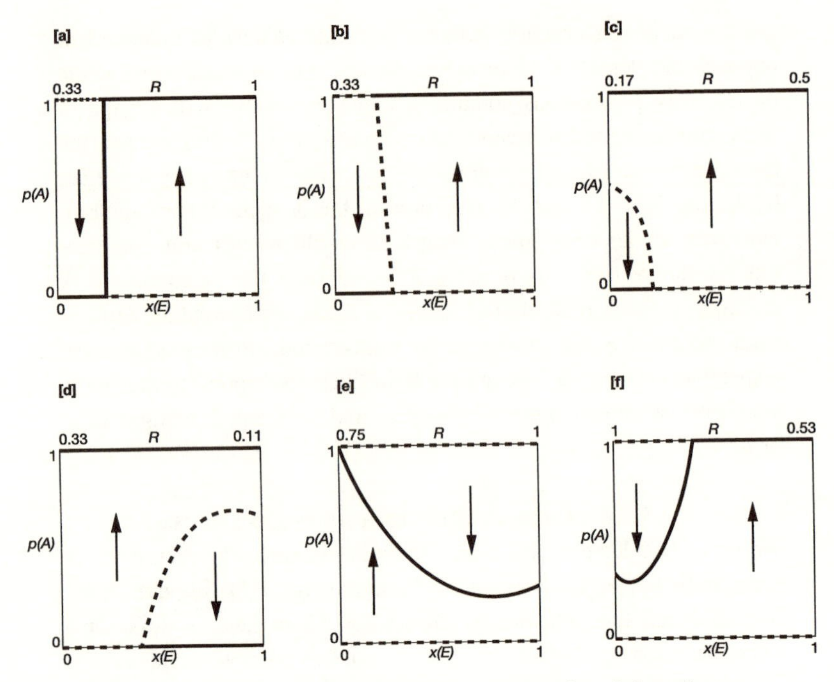

FIGURE 3.9. Model 2. (a) Positive niche construction, independent renewal, and no external selection ($R_t = \lambda_1 R_{t-1} + \lambda_2 x_t + \lambda_3$, $\gamma = 0$, $\lambda_1 = 0.7$, $\lambda_2 = 0.2$, $\lambda_3 = 0.1$, $\alpha_1 = \alpha_2 = \eta_1 = \eta_2 = 1$, $\varepsilon = 0.3$). Here the only selection is that generated by the resource. The arrows represent the trajectories of a population, the heavy line represents curves of stable equilibria, and the dashed line unstable equilibria. The direction of selection generated by the resource switches at $R = 1/2$. In (b)–(d), with independent renewal ($\lambda_3 > 0$) and external selection favoring allele A ($\alpha_1 = \alpha_2 = 1$, $\eta_1 > \eta_2$), niche construction generates selection that may oppose the external selection, taking the population to alternative fixations. The values of R at the equilibrium $\hat{x}$ are given by equation (A3.2) in appendix 3, and shown on the upper horizontal boundary. (b) shows positive niche construction with weak selection ($R_t = \lambda_1 R_{t-1} + \lambda_2 x_t + \lambda_3$, $\gamma = 0$, $\lambda_1 = 0.7$, $\lambda_2 = 0.2$, $\lambda_3 = 0.1$, $\alpha_1 = \alpha_2 = 1$, $\eta_1 = 1.01$, $\eta_2 = 0.99$, $\varepsilon = 0.3$), and (c) shows positive niche construction with strong selection ($R_t = \lambda_1 R_{t-1} + \lambda_2 x_t + \lambda_3$, $\gamma = 0$, $\lambda_1 = 0.7$, $\lambda_2 = 0.1$, $\lambda_3 = 0.05$, $\alpha_1 = \alpha_2 = 1$, $\eta_1 = 1.1$, $\eta_2 = 0.9$, $\varepsilon = 0.3$). (d) shows negative niche construction, independent renewal, and external selection favoring allele A [$R_t = \lambda_1 R_{t-1}(1 - \gamma x_t) + \lambda_3$, $\gamma = 0.9$, $\lambda_1 = 0.9$, $\lambda_2 = 0$, $\lambda_3 = 0.1$, $\alpha_1 = \alpha_2 = 1$, $\eta_1 = 1.1$, $\eta_2 = 0.9$, $\varepsilon = 0.3$]. (e) Positive niche construction when selection favors allele A and ε is negative can generate polymorphisms ($R_t = \lambda_1 R_{t-1} + \lambda_2 x_t + \lambda_3$, $\gamma = 0$, $\lambda_1 = 0.67$, $\lambda_2 = 0.083$, λ_3

tion is strong, polymorphic equilibria for both loci are possible for large values of R under conditions where inequalities A3.6 and A3.7 in appendix 3 are not satisfied. Since independent renewal and positive niche construction increase R, they increase the likelihood that there will be polymorphisms, while independent depletion and negative niche construction reduce this likelihood. Figure 3.9e and figure 3.9f illustrate how there may be a relatively large range of equilibrium values of x for which the frequency-dependent selection generated by the resource can counteract the external source of selection to generate polymorphic equilibria at the **A** locus. When the external selection is weak (η_1,η_2 close to 1) the line of polymorphic equilibria (also given by equation A3.5 in appendix 3) is approximately straight, unless the impact of the niche construction only weakly influences the amount of the resource (λ_2 or γ is small).

More generally, in the presence of independent renewal and depletion, even low frequencies of niche construction (small x), or a weak effect of niche construction on the amount of the resource (small λ_2 or γ), can switch the direction of the selection generated by the resource with the result that allele a is selected to fixation (figs. 3.9b–3.9d) or stable polymorphisms are possible (figs. 3.9e and 3.9f). The qualitative behavior of this system is little affected by the amount of recombination (r), when r is moderate to large.

3.3.3.3 External Selection at the E Locus

External selection at the **E** locus represents selection on a niche-constructing activity. When E is favored ($\alpha_1 > 1 > \alpha_2$, $\eta_1 = \eta_2 = 1$), positive values of ε result in convergence to a single equilibrium point with E and A fixed ($\hat{u}_1 = 1$) if $\hat{R} > 1/2$ at $\hat{x} = 1$, with E

$= 0.25$, $\gamma = 0$, $\alpha_1 = \alpha_2 = 1$, $\eta_1 = 1.1$, $\eta_2 = 0.9$, $\varepsilon = -0.3$). (f) Negative niche construction when selection favors allele A and ε is negative also generates polymorphisms [$R_t = \lambda_1 R_{t-1}(1 - \gamma x_t) + \lambda_3$, $\gamma = 0.9$, $\lambda_1 = 0.9$, $\lambda_2 = 0$, $\lambda_3 = 0.1$, $\alpha_1 = \alpha_2 = 1$, $\eta_1 = 1.1$, $\eta_2 = 0.9$, $\varepsilon = -0.3$]. In all cases $r = 1/2$ and $n = 1$. Formal details of these equilibria can be found in appendix 3. The upper horizontal boundary gives the value of R that corresponds to that value of x. (From Laland et al. [1999], fig. 1. Copyright 1999 National Academy of Sciences, U.S.A. With permission.)

and a fixed ($\hat{u}_2 = 1$) if $\hat{R} < 1/2$ at $\hat{x} = 1$, and with a range of polymorphic equilibria when $\hat{R} = 1/2$. Similarly, negative values of ε result in the fixation of E and a ($\hat{u}_2 = 1$) if $\hat{R} > 1/2$ at $\hat{x} = 1$, with E and A fixed ($\hat{u}_1 = 1$) if $\hat{R} < 1/2$ at $\hat{x} = 1$. The reverse pattern is found if e is favored by selection, resulting in fixation on either eA or ea. Provided all alleles are initially present, numerical analysis indicates that under these conditions a population will converge to these equilibria. Independent renewal increases the likelihood that selection at the **E** locus favoring positive or negative niche construction generates a corresponding selection pressure favoring A, and decreases the likelihood that positive niche construction favors a.

Interesting evolutionary dynamics may occur as a consequence of the difference in genotype fitnesses being a function of the amount of the resource present. For instance, when selection favors E and a positive niche-constructing behavior spreads, R will increase in value. We might predict that as the impact of niche construction on the resource (λ_2) increases, so will the rate of selection of the favored allele A. For instance, if $\lambda_1 = 0.5$ and we increase λ_2 from 0.4 to 0.5, there is a dramatic deceleration in the rate at which allele A approaches fixation. This occurs because when E is fixed, $\hat{R}$ has increased from 0.8, where the fitness difference between genotypes $AaEE$ and $aaEE$ is significant, to 1.0, where the fitness difference is tiny.

For model 1 we observed a time lag between the spread of a niche-constructing activity, represented by a change in the frequency of alleles at the **E** locus, and the response to the selection generated at the **A** locus (fig. 3.3). In model 1, R was computed as a weighted average of the previous n generations of niche construction. Model 2 extends this analysis to consider cases in which R is no longer a weighted average but, for cases of positive niche construction, is an accumulatory function that will typically increase as n becomes larger (e.g., equation A3.4 in appendix 3), while for negative niche construction R is a function that will typically decrease with increasing n. One consequence is that model 2 usually does not generate time lags in the response to selection; that is, it does not produce the momentum and inertia effects that characterized model 1. However, numerical analysis reveals that model 2 will generate time lags in two special cases: (1) if there is a primacy effect, that is, earlier generations contribute disproportionately to the amount of the resource present in the environment (λ_2 and γ decreasing with time), or (2) if we impose the constraint $\lambda_2 = 1/n$, which transforms R into a function qualitatively similar to a weighted average of the previous n generations, analogous to the assumptions of model 1. Again, the qualitative behavior is little affected by r when recombination is moderate to loose.

3.3.3.4 Heterozygote Advantage

Expressions for the equilibrium frequencies of E and A, $\hat{x}$ and $\hat{p}$ are given by equation A3.8 in appendix 3 for the case $D = 0$. As with model 1, in model 2 the selection generated by the resource can shift the position of polymorphic equilibria. The direction of the shift is in favor of a when ε is positive and $R < 1/2$ and also when ε is negative and $R > 1/2$, and in favor of A when ε is positive and $R > 1/2$ and also when ε is negative and $R < 1/2$. As before, the selection generated by the resource is strongest when R is close to 0 or 1, and absent when $R = 1/2$. By influencing the amount of the resource, positive or negative niche construction can dramatically change the position of the equilibrium and may change the direction of selection resulting from R. In figure 3.10a, the effect of positive niche construction ($\lambda_2 = 0.1$) is to increase the amount of the resource, and as a result the selection favoring a generated by the resource becomes much weaker, and A reaches a considerably higher frequency than without niche construction ($\lambda_2 = 0$). This is best understood by focusing on the effect of niche construction on the frequency-dependent components of the contributions to fitness of genotypes AA, Aa, and aa, that is, εR, $\varepsilon\sqrt{R(1 - R)}$, and $\varepsilon(1 - R)$, respectively, in the case $n = 1$. Here, with no positive niche construction ($\lambda_2 = 0$ and $R = 0.167$), $\varepsilon R < \varepsilon\sqrt{R(1 - R)} < \varepsilon(1 - R)$, and the frequency-dependent selection overcomes the fixed fitness component of selection and drives a to fixation. However, with positive niche construction ($\lambda_2 = 0.1$, which increases R to 0.47), $\varepsilon R < \varepsilon(1 - R) < \varepsilon\sqrt{R(1 - R)}$, allowing alleles A and a to coexist. In figure 3.10b, positive niche construction again increases the amount of the resource, but here the selection generated by the resource switches from favoring a to favoring A, and takes A to fixation. Now, with no niche construction ($\lambda_2 = 0$ and $R = 0.2$) $\varepsilon R < \varepsilon\sqrt{R(1 - R)} < \varepsilon(1 - R)$, but with

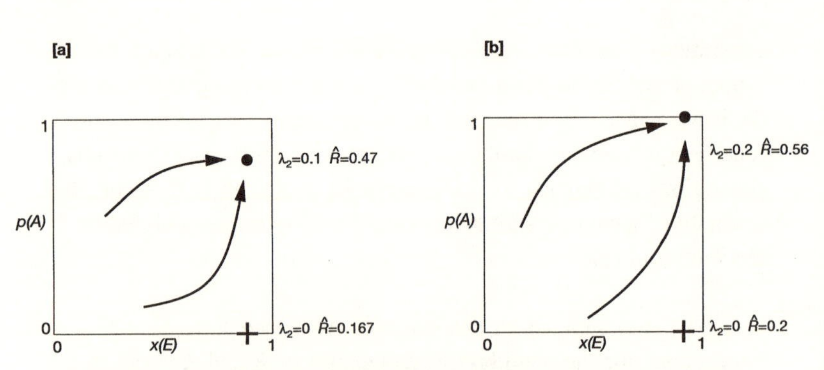

FIGURE 3.10. Model 2. Positive niche construction, independent renewal, and heterozygote advantage. Provided r is not very small, the population converges to a single equilibrium in linkage equilibrium, the position of which may be strongly affected by niche construction. The + represents the equilibrium in the absence of niche construction. By increasing R, niche construction changes the frequency-dependent component of genotype fitnesses, affecting the equilibrium frequency of allele A. In both figures, without niche construction ($\lambda_2 = 0$) allele a would be fixed, while in (a) niche construction ($\lambda_2 = 0.1$) allows A and a to coexist ($\lambda_1 = 0.7$, $\lambda_3 = 0.05$), and in (b) niche construction ($\lambda_2 = 0.2$) drives A to fixation $\lambda_1 = 0.5$, $\lambda_3 = 0.1$). The amount of the resource is given by $R_t = \lambda_1 R_{t-1} + \lambda_2 x_t + \lambda_3$. In both cases $\gamma = 0$, $\alpha_1 = \eta_1 = 0.99$, $\alpha_2 = \eta_2 = 0.9$, $r = 1/2$, $n = 1$, $\varepsilon = 0.3$. See text for further details. (From Laland et al. [1999], fig. 2. Copyright 1999 National Academy of Sciences, U.S.A. With permission.)

niche construction ($\lambda_2 = 0.2$, which increases R to 0.56) $\varepsilon R > \varepsilon\sqrt{R(1-R)} > \varepsilon(1-R)$. While the effects of niche construction are not always so dramatic, it is important that niche construction can strongly influence the pattern of selection acting on a population, even in the face of significant external renewal and depletion of the resource.

3.4 GENERAL DISCUSSION

In this chapter we have explored the evolutionary consequences of niche construction using extensions of classical two-locus models. Our analyses assume that a population's capacity for niche construction is partly influenced by the frequencies of alleles at one locus that affect the distribution of a resource in the population's environment. Selection at a second locus is a function of the amount of that resource. In model 1 the amount of this resource in the environment wholly determines the pattern and strength of selection acting on the second locus. The results suggest that the effects of niche construction can override external sources of selection to create new evolutionary trajectories and equilibria, produce polymorphism where external selection could not, and generate unusual dynamics. Moreover, these findings apply equally to diploids and haplodiploids.

Model 2 extends this analysis by allowing the key resource to be influenced also by independent processes of renewal and depletion. This is intended to enhance the ecological generality of our model. In model 1 populations were exposed to a potential conflict between the selection generated by niche construction and that from an external source. In model 2, populations may now be exposed to three-way conflicts, with selection from the resource generated by both niche construction and independent processes of renewal and depletion, and with selection from the external source acting on the same (**A** locus) alleles. Analysis of the extended model confirms that, even with independent renewal and depletion of the key resource, the effects of niche construction can be profound.

Both models explore the effects of selection favoring the spread of a niche-constructing behavior through a population, due to an increase in the frequency of allele E, as would occur, for example, in any instance of inceptive niche construction (see chapter 2). When this occurs in model 1, it changes the frequency of the resource R in the environment, and therefore the selection pressures acting at the **A** locus. Assuming that there is no external selection acting on the **A** locus, if the fitness of allele A increases with the amount of the resource (that is, when ε is positive), the resource becomes more frequent as allele E spreads, and the population eventually fixes on E and A. In contrast, if the niche construction depletes the resource (that is, ε is negative), the resource becomes less frequent as allele E spreads, so the population eventually converges to the equilibrium at which E and a are both fixed.

If the amount of the resource in the environment depends only on the current generation's niche construction ($n = 1$), then the response to selection at the **A** locus is immediate. A good example of this kind of resource is a spider's web, which is not inherited from past generations, but built by each spider. Here, locus **E** might affect the genetic variation underlying web-building behavior, while **A** might influence a web-defense behavior. On the other hand, if the amount of the resource also depends on the niche-constructing activities of past generations ($n > 1$), a time lag is generated between the change in frequency of alleles at the first locus and the response to the frequency-dependent selection at the second locus. This would be the case for the earthworm example, where **R** could represent topsoil or soil nutrients, locus **E** could influence a soil-processing or burrow-lining behavior, and locus **A** could specify some aspect of the phenotype affected by soil conditions, such as the structure of the epidermis or the amount of mucus secreted. This time lag, which can be considerably larger than n, is due first to the time it takes for the resource to accumulate (or deteriorate) in the environment, and second to the continuing fall in the frequency of the allele that is eventually favored, before the resource has had time to accumulate (or deteriorate) sufficiently.

The time lag creates both an inertia and a momentum, with at least two consequences. First, if it takes an evolving population many generations of niche construction to change the direction of its selection, the change may be too slow to prevent loss of the genetic variation on which its eventual response relies. Second, once a population reaches a stable equilibrium, it takes more time, or stronger selection, for the population to move away from it. These effects are more pronounced when there is a primacy effect (the niche construction of earlier ancestral generations has more impact on the resource than recent generations), and weaker when there is a recency effect (the niche construction of recent generations has more impact on the resource than that of their ancestors).

In model 1, R was computed as a weighted average of the previous n generations of niche construction. Model 2 extends this analysis to cases in which R is no longer a weighted average but, for positive niche construction, is an accumulatory function that typically increases as n becomes larger, while for negative niche construction R is a function that typically decreases with increasing n. One consequence of this is that model 2 does not usually generate time lags in the response to selection. In fact, the number of generations for an initially rare allele *A* to reach fixation usually becomes smaller as *n* increases. However, there are two conditions under which model 2 generates time lags in response to selection at the **A** locus: (i) if there is a primacy effect, that is, earlier generations contribute disproportionately to the amount of the resource present in the environment, and (ii) if $\lambda_2 = 1/n$, which transforms *R* into a function qualitatively similar to a weighted average of the previous *n* generations, analogous to the simpler model. The goal of model 2 is to explore the effects of niche construction in the face of independent renewal and depletion of resources, so we have chosen not to focus on the capacity of niche construction to generate time lags, which was explored in model 1. However, model 2 confirms that there are biologically meaningful circumstances under which we might predict the evolutionary momentum and inertia identified in our earlier analysis.

In their theoretical analysis of the evolution of maternal characters, Kirkpatrick and Lande (1989) found that maternal inheritance can generate time lags in the response to selection, and that as a consequence populations may continue to evolve even after selection has ceased. Gene-culture coevolutionary models with uniparental transmission of a dichotomous trait (Feldman and Cavalli-Sforza 1976) have generated similar results. Our results are clearly in accordance with these earlier findings. In fact, if we were to set our parameters so that $n = 2$, with a primacy effect, then niche construction could generate a form of maternal inheritance. That our models generate similar qualitative results to maternal-inheritance models over a broader range of parameter space serves to support Kirkpatrick and Lande's intuition that these effects could be more general than the term "maternal inheritance" implies.

Model 2 exhibits some interesting evolutionary dynamics because the difference in genotypic fitnesses is a function of the amount of the resource present. Small changes in coefficient values can result in an order-of-magnitude change in the time taken for evolving populations to approach equilibrium, with counterintuitive results. For instance, we saw examples where increasing the amount of niche construction, which in turn increases the amount of the resource, actually slows rather than accelerates the selection of the favored **A** locus allele (for instance, if $\lambda_1 = 0.5$ and we increase λ_2 from 0.4 to 0.5).

The consequences of niche construction are interesting when the selection it generates opposes the action of an external source of selection acting at the **A** locus. The examples of counteractive niche construction given in the previous chapter fit into this category, and such cases are likely to be common in nature. Many of the activities of organisms, such as migration, hoarding of food, habitat selection, or thermoregulatory behaviors, are adaptive precisely because they damp out statistical variation in the availability of environmental resources (Lewontin 1983, 2000). The results of our analysis of model 1 suggest that, in these circumstances and despite this environmental variation, niche construction decreases the probability that such populations will maintain genetic variation. If ε is sufficiently large (see details in appendix 1), then the frequency-dependent selection may overcome the external selection. If the amount of the resource in the environment accumulates as E increases (positive ε), then, while the resource is rare, the frequency-dependent selection makes it more difficult for a novel advantageous mutation to invade. Here, despite an external source of selection favoring one allele, the niche construction can generate counterselection that takes an alternative allele along a new evolutionary trajectory to fixation. If, on the other hand, the resource is common, then niche construction makes it easier for a novel advantageous mutation to invade (fig. 3.4b). In contrast, where a population (say, an herbivore) exploits a naturally replenishing resource (a novel plant species), intermediate levels of the resource may be maintained by a balance between these opposing processes. This may, in turn, maintain genetic variation at loci for which genotypic fitnesses depend upon the amount of this resource (for instance, loci affecting the herbivore's digestive enzymes). If the amount of the resource decreases as E increases (negative ε), and the resource is common, niche construction makes it more difficult for a novel selectively favored mutation to both invade and become fixed. In this case the niche construction creates new polymorphic equilibria, and consequently increases the amount of genetic variation maintained (fig. 3.4c).

The results of model 2 confirm that the frequency-dependent selection generated from the resource, and modified by niche construction, can (1) overcome the external selection and lead to the fixation of otherwise deleterious alleles and (2) support stable polymorphisms where none are expected. Our analysis illustrates how the probabilities of (1) and (2) are affected by independent renewal and depletion of resources, and the sign of ε. If an increase in the amount of resource results in an increment in the fitness of genotypes containing allele *A* (i.e., when ε is positive), then independent renewal decreases and independent depletion increases the chance that niche construction will lead to the fixation of otherwise deleterious alleles or support stable polymorphisms. If an increase in the amount of resource results in an increment in the fitness of genotypes containing allele *a* (when ε is negative), then independent renewal increases and independent depletion decreases the chances that niche construction will lead to the fixation of otherwise deleterious alleles or support stable polymorphisms. These findings hold for both positive and negative niche construction.

With overdominance at both loci, unless the loci are very tightly linked, there is usually only one stable equilibrium that is completely polymorphic (although see Karlin and Feldman 1978). The effect of the niche construction is to shift the frequency of this internal equilibrium in the direction of the arrows given in figure 3.1. That is, positive values of ε increase the equilibrium frequency of allele *A* for equilibria at which the frequency of the resource *R* is greater than one-half, and decrease the equilibrium frequency of allele *A* for equilibria at which the value of *R* is less than one-half. Negative values of ε have the reverse effect. The size of the change in frequency of *A*, as a consequence of the frequency-dependent selection, increases the further *R* is from one-half. By making the resource very common or very rare, niche construction can generate selection that results in genetic variation being lost. Also in model 2, when there is heterozygote advantage, the main effect of niche construction is to shift the position of polymorphic equilibria, and this may change the direction of selection resulting from the resource. Once again, we find that small changes in the amount of niche construction can have major consequences for the evolution of the population.

If, on the other hand, the two loci are tightly linked, then equilibria with linkage disequilibrium are possible. In model 1 both positive and negative values of ε can either increase or decrease the amount of disequilibrium at equilibrium, as well as the range of values of r over which these equilibria are stable, depending on the position of the equilibrium. As the frequency-dependent selection always changes the frequency of two gametes in ways that result in an increase in linkage disequilibrium, and two in ways which result in a decrease, its effect at any point in time depends delicately upon the gamete frequencies, as well as the sign of the disequilibrium.

It is well established that even simple models of frequency-dependent selection can generate a broad variety of outcomes depending on the choice of parameters, and hence it is important to stress which outcomes are not predicted as well as which are possible. It is in this respect that models 1 and 2 give particularly encouraging findings. Although the analyses reveal a rich array of possible outcomes, the patterns that emerge exhibit a predictable symmetry, they are easy to interpret relative to simpler models, and the dynamics are not chaotic, cyclical, or irregular. This suggests that, given sufficient understanding of the relationships between the model's parameters, it may be possible to make qualitative predictions, for example, as to the likelihood that, and circumstances under which, niche construction will generate polymorphisms. The fact that different frequency-dependent models might make alternative predictions we regard as a virtue, rather than a problem, since a comparison between the models and the data will facilitate a deeper understanding of the consequences of niche construction. For example, we noted earlier that assumptions as to how R is computed strongly affect the probability that niche construction will generate time lags. We have only begun to explore this issue, and a much more thorough analysis of the circumstances under which niche construction either leads to or does not lead to time lags is required. Ultimately, it is an empirical question to assess which of the many possible functions describing how niche construction affects R is closest to biological reality.

Chapter 2 presented empirical data suggesting that niche construction is widespread in nature. The theoretical analyses presented in this chapter suggest that the evolutionary effects of niche construction deserve serious consideration. Our models reinforce Lewontin's (1983) original intuition that the complementary match between organism and environment can be brought about by niche construction as well as through natural selection. However, it is also important that we acknowledge the limitations of our analyses. Model 1 is obviously an extremely simple representation of the world in which niche construction plays a role and can do little more than suggest the kind of qualitative behavior we might expect if niche construction were important in evolution. Nonetheless, it is striking that, when the simplest models incorporating niche construction provide us with compelling theoretical evidence, it is likely to have major evolutionary consequences.

Model 2 is an advance on model 1, in that it does allow us to begin to explore some relevant ecological issues, such as how niche construction interacts with independent processes that affect the frequency of resources. We are, nevertheless, aware of the constraints imposed by our models' two-locus, population-genetic structure. For instance, while model 2 allowed us to investigate how a population's evolutionary dynamics were affected by the proportion of niche constructors in the population (x) and the relative impact that this niche construction had on the resource (λ_2 and γ), we have not explored other demographic and ecological parameters. It would be valuable to synthesize population-genetic and ecological models in order to explore factors such as the size and density of the niche-constructing population, the scale of its impact on the environment, and how niche construction influences and is influenced by growth rates, life histories, and other relevant demographic parameters. In this respect, it would be useful to integrate the evolutionary models of niche construction with the theoretical ecological models developed to study the impact of ecosystem engineering (Gurney and Lawton 1996). In chapter 8 we make suggestions as to how such a body of theory might be developed.

Two other factors could add to the evolutionary significance of niche construction. First, in our models, locus **E** and locus **A** occur in the same species and, indeed, population, but in principle there is no reason why this should always be the case. Niche construction may also generate interactions between genes in different populations, thus providing a mechanism for coevolution in ecosystems. This may occur not only by direct effects that are already modeled by population-genetic coevolutionary theory, as, for instance, in host-parasite coevolution (Futuyma and Slatkin 1983), but also by populations affecting each other indirectly via their impact on some intermediate abiotic component in a shared ecosystem, as in competition for a chemical or water resource. This indirect coevolution is currently explored primarily by using ecological rather than evolutionary models.

For example, if niche construction resulting from genes in a plant population causes the soil chemistry to change in such a way that the selection of genes in a second population, of plants or microorganisms, is changed, then the first population's niche construction will drive the evolution of the second population simply by changing the physical state of this component of the abiotic ecosystem. The dynamics of the intermediate abiotic component may be qualitatively quite different from either the frequency changes in the genes that underlie the niche construction or changes in the number of niche-constructing organisms in the first population. Moreover, the discordance between resource and population dynamics will be exacerbated where there is an ecological inheritance. As a consequence, the coevolutionary dynamics are likely to be greatly affected by the introduction of an intermediate environmental component **R**, and this indirect feedback between species may generate some interesting and as yet unexplored behavior in coevolutionary systems. This issue is discussed in greater detail in chapters 5 and 8.

Second, there is no requirement for niche construction to be directly determined by a gene at one locus before it can alter the selection of a gene at a second locus. Niche construction can depend on learning (Bateson 1988; Plotkin 1988); the consequences for a gene at the **A** locus would still be the same, provided that the effect of the niche construction on the environmental resource was similar for learned and unlearned behavior. Moreover, at least among hominids, niche construction could also depend on cultural transmission. In chapter 6 we discuss these ideas further and present an investigation of how learning and culture may further enhance a population's capacity for niche construction.

CHAPTER 4

General Qualitative Characteristics of Niche Construction

4.1 INTRODUCTION

Before exploring the implications for ecosystem-level ecology of an evolutionary theory that incorporates niche construction, it is important to address some general qualitative characteristics of niche construction. This will bring into focus the inevitable environmental consequences of this evolutionary process and set the scene for our discussion of its ecological repercussions in chapter 5.

4.2 HOW CAN LIVING ORGANISMS EXIST?

Evolutionary biology is concerned with the complexity and diversity of living organisms, the history of this diversity, and the reasons for its change over time. Standard evolutionary theory explains the observed structure and behavior of organisms as consequences of historical contingency, chance events, and natural selection sorting between organisms that differ in their capacity to survive and reproduce in particular environments. Similarly, it explains organismal diversity as the result of natural selection in diverse environments. There is one prerequisite of the standard theory, however, that is often taken for granted. Natural selection can obviously work only when there are organisms that vary in their heritable characteristics. At first sight this prerequisite appears easily satisfied, as by definition fit organisms must reproduce and therefore resupply natural selection with fresh cohorts of organisms each generation, and variation among individuals in the parental population (and in the environments that they experience) will generate variation among offspring. But this presupposes that living organisms can persist for at least long enough to reproduce. We begin by asking what adaptations an organism must have merely to live.

The principal problem here is the sheer improbability of living organisms. Not only are organisms complicated biological entities, but they are also "far-from-equilibrium" systems relative to their physical or abiotic surroundings. Superficially, organisms appear to violate the second law of thermodynamics merely by staying alive and reproducing, as this law dictates that net entropy always increases and that complex, concentrated stores of energy will inevitably break down. In fact, in the language of thermodynamics, organisms are open, dissipative systems that can only maintain their far-from-equilibrium states relative to their environments by constantly exchanging energy and matter with their local environments (Prigogine and Stengers 1984; Eigen 1992). Organisms are like engines that must consume energy-rich sources (or free energy) in order to live and reproduce, ultimately converting this energy into less useful, more dissipated energy (O'Neill et al. 1986; Johnson 1995). Organisms feed on molecules rich in free energy and use the energy to do work, in the process generating outputs largely in the form of molecules that are poor in free energy (Turner 2000). As Schrödinger (1944) put it, an organism must "feed upon negative entropy . . . continually sucking orderliness from its environment" (p. 73). In this sense, organisms can stay alive only at their environment's expense.

Obviously organisms cannot break the second law of thermodynamics. Instead they participate with their local environments in two-way interactions that create coupled organism-environment systems that do permit organisms to stay alive without violating the second law. These two-way interactions account for the origins of obligate niche construction. To gain the resources they need and to dispose of their detritus, organisms cannot just respond to their environments. They must also act on their local environments and by doing so change them (Lewontin 1983; Odling-Smee 1988), in the process converting free energy to dissipated energy. Hence, evolution is contingent on the capacity of organisms to use their environments in ways that allow them to gain sufficient energy and material resources from their environments, and to emit sufficient detritus into their environments, to stay alive and reproduce. Variability in these processes offers the potential for the process of natural selection to operate.

It follows that biological evolution must have consequences for environments as well as for organisms. Whatever benefits are gained by organisms from their environments as a result of their niche construction must always be balanced by the costs that organisms must inflict on their environments and that ultimately take the form of a net increase in entropy in their environments.

Sometimes no practical consequences of any kind arise from these interactions with the environment and they can safely be ignored. For example, when primary producers convert radiant energy into glucose through photosynthesis, the amount of that radiant energy supplied by the sun is, for all practical purposes, limitless. Hence, there is no perceptible thermodynamic cost to any environment as a consequence of photosynthetic reactions, just perceptible gains by primary producers. Similarly, the heat loss generated by photosynthesis is ultimately returned to the universe, which in effect is infinite. Here too there is no perceptible environmental cost, only perceptible gains by organisms (Turner 2000).

More often, environments do demonstrate perceptible and measurable costs of the gain of energy and matter by organisms. For example, insofar as organisms obtain their energy from the limited material resources of the earth and its biosphere through chemical reactions, rather than from the unlimited radiant energy of the sun, they must change some of the components of their local environments, and in ways that are likely to have biologically significant consequences for themselves or for other organisms (Turner 2000). For instance, when a cow eats grass and then defecates in its pasture, there is a cost to the environment, measurable in terms of the difference between the free energy obtained by the cow from the grass and the relatively dissipated energy in its cowpat. This is so even though the cowpat may subsequently become a resource for other organisms such as dungflies or beetles (Milinski and Parker 1991). In all such cases the thermodynamic costs that organisms impose on their environments should be equal to or greater than the thermodynamic benefits the organisms receive from their environments, but, again, only after allowing for the fact that some of these costs may be imperceptible because of the photosynthesis by primary producers, here, the grass. If, in addition, organisms invest in facultative as well as obligate niche construction, for example, if a cow not only perturbs its local environments by eating and defecating but also relocates to a different field where the grass is greener, then the second law of thermodynamics still applies. The cow must still pay for whatever extra fitness benefits it may gain from its facultative niche construction by inflicting extra thermodynamic costs on its environments that are equal to or greater than the extra benefits it receives. The one exception to this rule occurs when organisms die and their dead bodies are returned to their environments in the form of dead organic matter (DOM). In this case the thermodynamic relationships in the coupled organism-environment systems may be reversed. When it dies, a cow's body gradually loses its far-from-equilibrium status, paying back part of its debt to the environment in the process, by returning some of the free energy absorbed when it was alive. However, this is still a form of niche construction. A relocation of free energy still occurs. It might be called "ghost niche construction."

As each species has a unique evolutionary history, the interactions that allow individual organisms to live at their environment's expense are usually explained in terms of the species-typical adaptations of organisms relative to their selective environments (Rose and Lauder 1996), plus the adaptive adjustments made by individual organisms to their individual environments on the basis of developmental reaction norms (Schlichting and Pigliucci 1998). For many purposes these explanations clearly suffice, but for some they are incomplete.

Two omissions are relevant here. First, by explaining niche construction in terms of species-typical adaptations, standard evolutionary theory indirectly discounts the characteristics of niche construction that might be common to all organisms. Yet some properties of niche construction should be general, if only because, regardless of species, all organisms have to live at their environments' expense, and they can do that only by acting on and perturbing their environments. The reduction of niche construction to idiosyncratic sets of adaptations obscures any universal properties of niche construction.

Second, by directing so much attention to the adaptations of organisms, and so little attention to the changes caused in environments by niche-constructing organisms, standard evolutionary theory also plays down the consequences of evolution for environments. Environmental change is seldom regarded as another aspect of the expression of biological evolution itself, and is therefore seldom included as part of evolutionary theory. Exceptions occur when environments are artificially restricted to other biota, as in population-community ecology where, for instance, coevolutionary models can be applied. However, as soon as abiotic environmental components are also included, as in process-functional ecology, it becomes difficult for the standard theory to describe environmental change in evolutionary terms.

4.3 THE UNIVERSAL CHARACTERISTICS OF NICHE CONSTRUCTION

One way to establish the universal characteristics of niche construction is in terms of physics, as the physical theory of thermodynamics already offers a simple description of the kind of properties any agent needs, whether it be a niche-constructing organism or a human-engineered artifact, to drive a system out of equilibrium and into an improbable state relative to its surroundings (Schrödinger 1944; Bennett 1987; Frautschi 1988; Layzer 1988).

In 1871 the physicist James Clerk Maxwell proposed a purely conceptual, yet apparently subversive, creature, subsequently called "Maxwell's demon." Maxwell's demon seems subversive because it appears to be able to drive a closed physical system out of equilibrium against a gradient of chance, thereby violating the second law of thermodynamics. By achieving so much, the demon appears to promise an El Dorado of free energy. To protect the second law, and to explain why free rides do not exist, generations of physicists have tried to exorcise Maxwell's demon. In doing so they have proposed a variety of reasons why the demon cannot break the second law. What makes the matter so intriguing is that most of their proposals have turned out to be wrong. Nevertheless, one fundamental objection that has always been raised turns out to be true. In any closed system the demon is ultimately forced to increase rather than decrease the entropy level in the overall system, by using, rather than storing, free energy (Bennett 1987). It follows that Maxwell's demon can only do its job if it is supplied with free energy from an external source, one that is outside the closed system, to fuel its own energy consumption. Hence the demon is not truly subversive because it can only work in an open rather than a closed physical system, and there is no violation of the second law.

The details of the many arguments among physicists are mostly of little concern to biologists (Bennett 1987; Jaynes 1996), but in one respect they are profoundly instructive. In exorcising Maxwell's demon not only have physicists established why the demon cannot work in a closed system, but they have also gone a long way toward demonstrating how the demon can work in an open system, one that is able to supply it with sufficient free energy to fuel its work. They have done this partly by describing the minimal properties the demon needs to force any system out of equilibrium, relative to its surroundings, and against the consequences of chance. Although there are continuing controversies among physicists about the extent to which these properties illuminate our understanding of thermodynamics, we propose to borrow them on the grounds that the properties assigned by successive generations of physicists to Maxwell's demon are the best available guide to the universal properties of niche construction.[1] As organisms are far-from-equilibrium entities living in open-system environments, the properties assigned to Maxwell's demon should be identical to the properties that niche construction must have if organisms are to survive.

A cartoon version of the original Maxwell's demon, based on Bennett (1987), is illustrated in figure 4.1a. Other more subtle versions of Maxwell's demon exist (e.g., Feynman 1965), but this simple version is sufficient for our purpose. Figure 4.1a shows two rooms, A and B, both of which are isolated from the rest of the

[1] A possible alternative source of guidance is von Neumann's (1956, 1966) description of self-reproducing automata, later cellular automata, which combines a universal Turing machine (Turing 1936, 1937) with a universal construction machine (Arbib 1969; Laing 1989). This approach is also relevant to any enquiry into the universal properties of niche construction.

universe by an external container and separated from each other by a wall with a small hole in it. Thus, here the demon is operating in a closed system. Both rooms are initially at equilibrium, being filled with a gas at equal temperatures and pressures. Maxwell's demon is given control of a sliding door that opens and shuts the small hole in the wall between the two rooms. The demon's job is to observe molecules of gas approaching the hole (the molecules are driven by random Brownian motion) and to open and shut the door in such a way as to allow hotter, faster-moving molecules to pass from room B to room A but not vice versa, and colder, slower-moving molecules to pass from room A to B but not vice versa. As the demon sorts between fast- and slow-moving molecules in this manner, room A gradually heats up and room B gradually cools down, thereby apparently driving the two rooms out of equilibrium in a closed system.

The minimal properties that physicists have assigned to the demon so that it can do its work are as follows. First, the demon must have a *goal*. In the example illustrated by figure 4.1a the demon's goal is the one originally assigned to it by Maxwell, namely, to drive the gas in the two rooms out of equilibrium. Second, the demon must be able to *discriminate* between fast- and slow-moving molecules and to *anticipate* their movements so that it can operate the door correctly prior to the molecules arriving at the hole. Third, the demon can only do that provided it is *instructed* by *knowledge* (Jaynes 1996). Minimally the demon must be programmed with a priori knowledge about its own goal, about how to operate the door, and about the behavior of molecules, registered in some kind of *memory* (Szilard 1929). Probably this is the demon's most controversial property and has certainly provoked many arguments (Jaynes 1996). Fourth, the demon must be capable of doing the *work* involved in operating the door. Fifth, the demon must be *designed* in such a way that it can perform this work in this environment. Finally, sixth, the demon must be supplied with *free energy*, which it cannot be as long as the demon is confined to the closed system in figure 4.1a, but it might be if it were in an open system, as in figure 4.1b.

On the assumption that far-from-equilibrium organisms can stay alive only if they are endowed with these Maxwell's-demon-type properties, we can now go back to evolutionary biology and use this

[a]

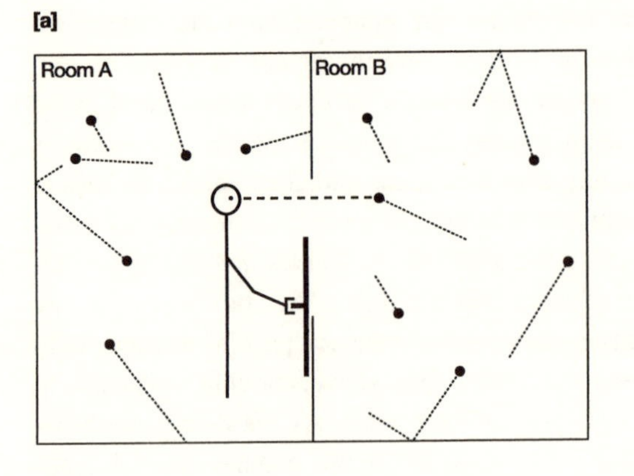
Room A
Room B

[b]

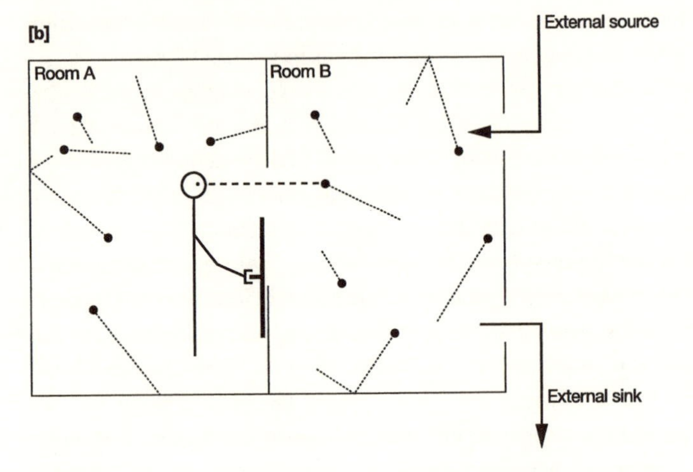
External source
Room A
Room B
External sink

[c]

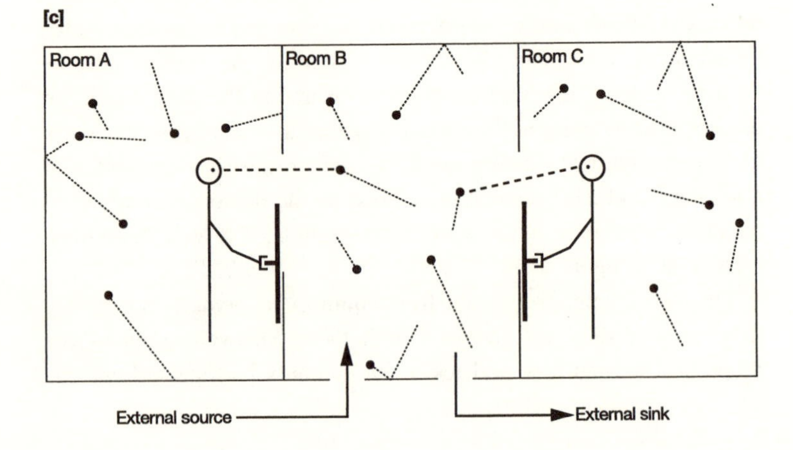
Room A
Room B
Room C
External source
External sink

list of properties to indicate the universal characteristics we should find in all niche-constructing organisms. Every instance of niche construction should demonstrate the same set of fundamental characteristics, irrespective of the species-typical properties that underlie the niche construction. A list of these universal characteristics of niche construction is given in the left-hand column of table 4.1. This list is arranged in such a way that each item in it calls for the next item.

(i) First, organisms must always be active, which means that niche construction has to do physical work. (In the case of DOM, it has to be the residue of previous physical work.) It is not sufficient for organisms to interact with their environments merely by passive response. If organisms really were just reactive, they could neither gain the resources they need from their environments, nor emit the detritus they have to get rid of, to stay alive and reproduce, and they would lose their far-from-equilibrium status by dying. In effect, organisms would be driven by environmental agents to physical states far closer to equilibrium. As niche construction must involve work, it follows that it must incur energy costs. (ii) Niche construction must be *profitable*, in the sense that it must provide sufficient energy and material resources in the short term to pay for the organism's survival and

FIGURE 4.1. A cartoon version of Maxwell's demon, indicated here by a "creature" equipped with a head that remembers, an eye that sees, and a hand that performs work. (a) Shows two rooms, A and B, that are closed from the rest of the universe by an external container, and separated from each other by a wall with a small hole in it. Both rooms are initially filled with a gas at equal temperatures and pressures. The demon's task is to observe molecules of gas approaching the hole (the molecules are driven by Brownian motion), and to open and shut the door between the two rooms selectively so as to allow hotter, faster-moving molecules to pass from room B to room A, but not vice versa, and colder, slower-moving molecules to pass from room A to room B, but not vice versa (see text). (b) The same as (a) except that the demon is now working in an open system instead of a closed system. (c) The same as (b) except that there are now two demons, in rooms A and C, interacting with each other in a shared open environmental system, represented here by room B. (Copyright 1987. Jerome Kuhl. With permission.)

TABLE 4.1. Comparison of the Two Selective Processes in Evolution, Natural Selection and Niche Construction

Niche Construction	*Natural Selection*
1. Sources of niche construction are active "fuel-consuming" agents, namely, working organisms	1. Sources of natural selection are either reactive abiotic environmental components or other niche-constructing organisms. If the latter, the source of natural selection may become identical to the prior niche-constructing activities of other organisms
2. Niche construction must be "profitable" in the short term. It must provide for an organism's survival and reproduction as well as for its niche-constructing activities	2. Natural selection need only obey the laws of physics and chemistry*
3. Niche construction must usually be restricted to fitness-enhancing behaviors or processes in the short term. Niche construction is unlikely to be random or haphazard because of prior natural selection	3. Natural selection exhibits goalless indifference*
4. Niche construction discriminates as individual organisms actively select between alternative "bits and pieces," alternative locations, and alternative interactive outcomes in their environments	4. Natural selection is selective, but it can only sort passively between fit and unfit organisms in populations*
5. Niche construction must be informed. It must be directed by semantic information whose structure and content is the result of prior natural selection	5. Natural selection is "blind." It is not directed by semantic information*
6. Niche construction must work proactively. The constraints imposed by prior natural selection inform the choices that will become available for niche construction	6. Natural selection works reactively*

*Unless the source of natural selection is the prior niche-constructing activities of other organisms.

reproductive activities, as well as for itself. Hence, the net energy benefits gained by any organism from its environment by niche construction must be greater than the net energy costs of its own niche-constructing activities. (iii) The niche-constructing behaviors or processes that are expressed by organisms must usually enhance fitness in the short term. Organisms must usually function in a manner that generates specific outcomes from their interactions with their environments that are capable of satisfying their specific survival and reproductive needs. Unconstrained, random, or haphazard niche-constructing acts could not provide any organism with a basis for staying alive because it is astronomically improbable for any far-from-equilibrium system to maintain itself by chance alone (Prigogine and Stengers 1984; Bennett 1987; Layzer 1988). (iv) Fitness differentials, however, entail that organisms respond to different environmental alternatives on the basis of nonarbitrary criteria. Hence, niche construction requires an ability on the part of organisms to discriminate; to actively sort between whatever physical bits and pieces are made available to them by their environments; to select those elements in fitness-altering ways; to change the physical state of some selected factors in their environments in ways that would usually improve their fitness; and to select between habitats of different quality in their environments. (v) Niche construction in its interactions with natural selection must therefore be regulated and controlled by information. Specifically the niche-constructing activities of organisms must be at least partly governed by and guided by *semantic information*, or knowledge, supplied by the evolutionary process as a result of natural selection (see below), and typically encoded in DNA. By semantic information we mean information that relates to the fitness of specific organisms, about their requirements, about their local environments, and about how to operate in their local environments in ways that satisfy their requirements, and that is, in this sense, "meaningful" to organisms in their local environments. (vi) Finally, because the niche-constructing acts of organisms must be selected in advance of their expression, it follows that the niche-constructing activities of organisms must either suffice, for example, because they arose for other reasons, or be oriented a priori toward targeted future outcomes of organism-environment interactions on the basis of at least rudimentary and semantically informed *search plans*. Therefore, in this limited and, in most species, entirely noncognitive sense, niche construction must be preparative or predictive in character (Odling-Smee 1988; J. Holland 1992, 1995).

The correspondence between these universal niche-constructing traits and the properties of Maxwell's demon can be seen in figure 4.1b. Here room A, including its demon, represents the far-from-equilibrium state of any living organism, while room B is its local environment, and is an open system relative to the rest of the universe. To stay alive, room A, the organism, is assigned the universal niche-constructing properties we have just listed, making its niche construction comparable to the work of the demon in figure 4.1a. These universal niche-constructing properties should then be sufficient to allow any organism to persist in its open local environment in ways that enable it to gain enough resources and eliminate enough detritus to survive and reproduce (assuming its local environment is benign enough to supply it with the resources it needs and accept its detritus).

One surprising conclusion from this exercise is that niche construction can be nothing less than a second selective process in evolution, although a very different kind of selective process from the first selective process of natural selection. Hitherto, on the rare occasions when Maxwell's demon has been applied to evolutionary biology, an analogy has usually been drawn between Maxwell's demon and natural selection, not niche construction (e.g., Kauffman 1993). This may be because both Maxwell's demon and natural selection are obvious sorting agents. However, this is clearly the wrong analogy. Natural selection shares only one property with Maxwell's demon. Like the demon, natural selection is discriminating because it can sort between organisms according to their fitness, but none of the other characteristics of natural selection correspond to any of the other properties of Maxwell's demon. This can be seen by comparing the universal characteristics of niche construction, listed in the left-hand column of table 4.1, which correspond to the properties of Maxwell's demon, with those of natural selection, listed in the right-hand column of table 4.1, which do not.

The principal contrasts between niche construction and natural selection in table 4.1 are these. (i) Niche construction is active but natural selection is reactive. (ii) and (iii) Niche construction must be profitable and in many instances may function to enhance the organism's survival and reproduction, but natural selection is goalless and is completely indifferent to all goals, even though what it selects for results in functional organisms. (iv) Both niche construction and natural selection are discriminative processes, but the targets of their discrimination are different. Niche construction selects between alternative bits and pieces in environments, alternative locations in environments, and alternative interactive outcomes between organisms and their environments. Natural selection occurs primarily between organisms that differ in their fitness. This results in selection between genotypes, and therefore between alternative items of heritable semantic information. (v) Niche construction must be an informed selective process (see below) whereas natural selection is an uninformed or *blind* selective process. By "blind" we mean that natural selection, unlike niche construction, is not guided by any kind of registration, or *memory*, of the past. Nor, unlike artificial selection, is it prescient. It cannot *prepare for* or *predict* future organism-environment interactions elsewhere or later. Natural selection can only sort between the outcomes of immediate organism-environment interactions "here and now." (vi) Niche construction is proactive. Organisms have to be informed a priori even when their information is subsequently revised. Natural selection is reactive.

Only when the sources of natural selection for one set of organisms are other organisms, as happens, for example, in frequency- and density-dependent selection and in coevolution, do these contrasts between niche construction and natural selection disappear. But this is only because in all such cases the natural selection pressures that act on one set of organisms may become identical to the prior niche-constructing activities of another set of organisms that may then express the characteristics of niche construction shown in the left-hand column rather than the characteristics of natural selection shown in the right-hand column of table 4.1. Interestingly, these considerations also suggest that natural selection pressures arising from biotic sources should have different properties from natural selection pressures arising from abiotic sources. For example, biotic sources of natural selection should be informed by naturally selected genes a priori and may therefore have some capacity for active discrimination, whereas abiotic sources of natural selection are blind and should not.

4.4 THE INTERDEPENDENCE OF NATURAL SELECTION AND NICHE CONSTRUCTION

As a direct consequence of their niche construction organisms can apparently express the full set of Maxwell's-demon-type properties. Yet niche-constructing organisms must be informed a priori if they are to support the highly improbable living systems that they constitute. So how does the evolutionary process accrue the semantic information that guides niche construction in the first place? How do organisms become informed?

Darwinian evolution, based as it is on variation, heritability, and the struggle to survive and reproduce, comprises a kind of natural algorithm by which populations of organisms gain semantic information (Campbell 1974; Odling-Smee 1983; Dennett 1995; Hull et al. 2001). The genes that are inherited by each generation of organisms include some that are naturally selected and that must necessarily reflect whatever selective biases were imposed on their ancestors by the blind action of natural selection, according to the characteristics of their ancestors' environments. Hence, in contrast to selectively neutral genes or noncoding DNA, the naturally selected genes that are inherited by offspring must indirectly register some information about whatever past adaptive successes were expressed by those same genes in earlier generations relative to the selective environments of their ancestors. The reuse of inherited genetic information that relates to past environments and its reexpression by phenotypes in the present environment amount to an inductive gamble that the selective environments of contemporary organisms are sufficiently similar to the selective environments of their ancestors to make whatever was adaptive before, adaptive again (Slobodkin and Rapoport 1974; Odling-Smee 1988).

It follows that, when organisms inherit naturally selected genes from their ancestors, their genetic inheritance is sufficient to cause them to be informed by past natural selection. It is also sufficient to allow their niche-constructing activities to be informed a priori by past natural selection. It is in this sense that genetic inheritance can be characterized as a between-generation flow of genetically registered or *remembered* semantic information. In fact, the passing on of previously selected and already meaningful information during genetic inheritance is reducible to a conventional form of information flow or communication between transmitters and receivers (Shannon 1948; Shannon and Weaver 1949). In each population a parent generation serves as a "transmitter" of DNA-encoded meaningful messages to an offspring generation that "receives" whatever semantic information the message contains, while the mechanisms of reproduction and genetic inheritance provide a "channel of communication" between them.

It is not enough, however, for each generation to inherit old semantic information from parents and ancestors as a consequence of past natural selection. The semantic information must also be repeatedly retested and updated by the acquisition of new semantic information in each generation as a consequence of current natural selection pressures. These within-generation updating processes are different from the between-generation transmission processes. For example, they are not reducible to conventional information flows between transmitters and receivers. Instead they depend on the capacity of each generation to respond to current natural selection. Hence, both the retesting of inherited semantic information and the acquisition of new semantic information by populations are determined first by the ability of each generation to supply the raw material for current natural selection to select—namely, diverse organisms, some carrying new mutations—and second by the competition among those diverse organisms to reproduce. These are the processes that ultimately convert raw organism-environment interactions into new or retested genetically encoded semantic information. Thus it is repeatedly updated, DNA-encoded semantic information that is transmitted by fit organisms between generations via genetic inheritance.

One other important aspect of what hereafter we will call the "Darwinian algorithm" also should be stressed because, even though it is so familiar, it is sometimes overlooked. Natural selection works by sorting among alternative individuals in a population. The result of this is a change in the genetic statistics of the population containing these individuals in each generation. However, no individual organism can gain new genetic information for itself during its lifetime simply by being selected. All an individual organism can do is inherit a sample of previously selected semantic information from its parents, and possibly some new mutations; it must then make do with its genetic inheritance for the rest of its life.

Of course, this does not prevent individual organisms from gaining some new information for themselves a posteriori during their lives on the basis of their individual experiences or developmental histories. To take an obvious example, many individual animals can learn from their own experience, and they can subsequently apply what they have learned to guide their present and future behavior. In this manner organisms can supplement their inherited information with some individually acquired new semantic information, which they can then use to further update and adjust their behavior. Individual organisms can only gain this extra information, however, if they possess one or more of the necessary gene-informed subsystems that allow them to do so. However, for the moment we temporarily assume that the capacity of each organism for adaptive niche construction depends exclusively on a priori information encoded in its genotype. We shall relax this assumption and examine some of these information-gaining subsystems later in chapter 6.

We must also be more precise about the nature of the semantic information that is acquired under the Darwinian algorithm. Whether or not organisms pass on information-carrying genes to the next generation depends on their differential capacity to survive and reproduce, and therefore partly on how well they are adapted to their local selective environments. Hence some, but by no means all, of the genetic information that ancestral organisms pass on to their descendants through genetic inheritance must be *about* the processes necessary to achieve adaptive matches between organisms and their environments. Information about adaptive *O-E* matches, however, is both semantic and relative, and thus niche-specific (Jaynes 1996; S. Weinstein, personal communication). It is semantic because it carries "meaning" (Maynard Smith and Szathmary 2000), namely, how to construct and maintain a viable organism. It is relative because whatever meaning it carries need be meaningful only relative to a particular selective environment. The semantic information encoded in an organism's genes and expressed in its phenotype is typically only meaningful and useful in environments similar to those of its ancestors. Semantic information about adaptive *O-E* matches is therefore a form of knowledge (J. Holland 1992, 1995) with the proviso that genetically coded knowledge is entirely noncognitive.

The trouble with knowledge is that it is such a difficult commodity to measure. One immediate problem is that the semantic information encoded in genes is not commensurate with other kinds of information that are not directly concerned with producing adaptive phenotypes. For example, semantic information is not quantifiable in terms of the various kinds of configurational information that are used by engineers and physical scientists and that are measured by binary digits, or bits. Meaning cannot be measured by bits (Stuart 1985; Eigen 1992; Hull et al. 2001). Genes themselves illustrate this point. The way in which a gene encodes the information it is carrying structurally, as a sequence of nucleotide bases in a stretch of DNA, can in principle be measured in bits, but what a gene means cannot be so measured. For example, the number of bits needed to describe the base sequence of a particular gene might change considerably through the addition or subtraction of extra meaningless sequences, such as introns or *Alu* repeats, without changing the meaning of the gene at all. The gene's function might stay the same. Conversely, a minimal structural change to a gene, such as a single-point mutation, might change its meaning completely, for instance, by rendering it lethal, without changing the number of bits needed to describe it at all.

To the extent that DNA does carry knowledge in the form of meaningful information about phenotypes in particular environments, natural selection pressures must convey information *about* the order and disorder, regularity and irregularity, or, more generally, *about* the "causal texture" of the selective environments of populations. It is only the causal texture of a population's environments that can provide the stochastic guidance that actually selects the fittest organisms. Thus, in this limited sense, natural selection leaves residues of semantic information about what characters are valuable in the environments of evolving populations, in the selected genes of organisms. Semantic information, reflecting the prior action of natural selection, must also start to flow beyond organisms and through ecosystems whenever the prior expression of semantic information by niche-constructing organisms modifies the subsequent action of natural selection on organisms. When this happens, it makes little difference whether the sources of these modifications are just other organisms or whether they are abiotic components of ecosystems. As long as subsequent natural selection is influenced by prior niche construction, and niche construction is guided by semantic information that reflects prior natural selection, there will be a flow of semantic information through both organisms and their environments via both biotic and abiotic ecosystem components.

This introduces the central issue that we will explore in the following chapter. There we will argue that it is not possible to understand how populations of organisms affect ecosystem dynamics through their niche construction (or ecosystem engineering) unless it is recognized first that niche construction is directed by semantic information, second that natural selection is a process that accrues semantic information, and third that semantic information flows between biotic components of ecosystems as a result of these two processes. We will also describe how niche-constructed environmental components, including intermediate abiotic components, can act as bridges that allow semantic information to flow through ecosystems.

In evolutionary biology, the meaningfulness of the semantic information that is encoded in naturally selected genes is currently measured by the concept of fitness. Perhaps partly because the meaningfulness of information is so tricky to quantify, there are many different working definitions of fitness (Dawkins 1982; Endler 1986; Michod 1999). Yet each definition reduces to some kind of measure of the evolutionary utility of the semantic information that is carried and expressed by organisms in their selective environments. Some additional difficulties also arise from the niche specificity of semantic information. For example, in principle a gene carried by individual organisms in one species could confer considerable fitness advantages on those individuals at a given time and place, but not on their descendants at other times and places. It could also confer fitness advantages in the selective environments of one species, but not in the selective environments of another very closely related species. In practice, this relativity of semantic information is qualified by the fact that many different species of organisms, living in many different habitats, are likely to share some selection pressures in common with other species, and consequently they may express similar adaptations. This demonstrates that even though semantic information is relative it may also be objective (S. Weinstein, personal communication).[2]

Finally, the Darwinian algorithm also incorporates the reciprocal dependence of natural selection and niche construction. This is an issue we can deal with much more briefly because we have already broached it. The starting points are items 4 and 5 in the right-hand column of table 4.1. Natural selection is blind and requires that variation exists in order to respond to selection pressures that sort between organisms in populations. But, as we have seen, if they are to supply this variation through reproduction, niche-constructing organisms must be informed by prior natural selection. They must use whatever information-based adaptations they possess to gain appropriate resources from their environments, return appropriate detritus to their environments, and select the times and places to do so.

The reason why natural selection and niche construction are so intimately dependent on each other is that neither natural selection nor niche construction by itself possesses the full set of Maxwell's-demon-type properties that are needed to build and sustain far-from-equilibrium living systems in open, energy-supplying ecosystems. In combination, however, they do possess all the properties needed to do so. Hence, evolution depends on two selective processes rather than one: a blind process based on the natural selection of diverse organisms in populations exposed to environmental selection pressures, and a second process based on the semantically informed selection of diverse actions, relative to diverse environmental factors, at diverse times and places, by individual niche-constructing organisms. The first of these processes was initially described by Darwin and is the focus of the Darwinian algorithm. The second process is not specified under the Darwinian algorithm, but it is implicit in it. In brief, the Darwinian algorithm actually depends on both the selection of

[2] Convergent evolution illustrates this point. For example, both sharks and dolphins have fins that enable them to swim. Their convergent adaptations demonstrate that the semantic information carried by these genetically distant organisms is meaningful relative to shared natural selection pressures and is "objective" rather than "subjective" in character, because its fitness value or "meaning" is decided largely independently of these organisms. Of course, we must be careful to allow for the possibility that what appears convergent may actually be the result of a single phylogenetic origin.

organisms by their environments and at least the partial selection of their environments by organisms (Lewontin 1983).

There is also a paradox implicit in table 4.1. If natural selection is *blind*, but niche construction is *semantically informed* and *goal-directed*, then evolution must comprise an entirely *purposeless* process, namely, natural selection, selecting for *purposive* organisms, namely, niche-constructing organisms. This must be true at least insofar as the niche-constructing organisms that are selected by natural selection function *so as to* survive and reproduce. This paradox is sufficiently profound that it would seem to justify Hull et al.'s (2001) claim that, when Darwin discovered the role of selection, he also discovered a new kind of causality in nature, different from any other kind of causality known to science up to that time.

4.5 THE CONSEQUENCES OF NICHE CONSTRUCTION FOR ENVIRONMENTS

The second issue that is not dealt with by standard evolutionary theory concerns the effects that niche-constructing organisms must have on their environments. Some environmental consequences of niche construction are general because they derive directly from the universal characteristics of niche construction. Others are more idiosyncratic because they depend on the particular organisms and environments involved.

We have argued that when organisms construct niches they necessarily express semantically informed,[3] discriminating activities (table

[3] Note again that there is often a difference between the kind of information that is putatively expressed by Maxwell's demon and the kind of information that must be expressed by a niche-constructing organism. For example, Bennett (1987) assumes that the demon's opening and shutting the door between the two rooms in figure 4.3a is expressing only binary digits (bits) of information, and in doing so he overlooks the semantic information in the mind of the human (Maxwell) who designed the demon in the first place. The significance of Bennett's assumption for physics is that, even if the energy cost of opening and shutting the door between rooms A and B is made negligibly small, and even if the information expressed by the demon is restricted to bits, then as Bennett shows, on the basis of work by Landauer (Lloyd 1999), the demon still has to pay an information cost that, counterintuitively, comprises the cost of erasing bits. This hitherto ignored information cost causes a rise in entropy that is equivalent to the bits erased, and it is ultimately what stops even the most reductionist demon from

4.1). These activities compel organisms to impose material and thermodynamic costs on their local environments that compensate for their own fitness gains. Hence, through their niche construction, organisms not only sustain their own far-from-equilibrium states, but they also drive some environmental components in their local environments into altered physical states. Typically, these altered states are far closer to thermodynamic equilibrium than are organisms themselves. Locally, it is even possible that some environmental components, for instance animal artifacts, may be forced into higher states of organization, and therefore to decreased rather than increased entropy states (see below). However, even when this happens, it must still be at the cost of a net increase in entropy in some wider environment.

Again, some of these points are illustrated in figure 4.1. For example, in the closed system shown in figure 4.1a, and assuming for the moment an unrealistic demon whose own work costs nothing, the demon cannot raise the temperature in room A without also lowering the temperature in room B to an extent that fully compensates for the increased temperature in room A. In reality the niche-constructing activities of organisms are not cost-free, and their local environments are open rather than closed systems. However, the same general point is true of all organisms too. Thus, given that every organism (symbolized by the demon plus room A in fig. 4.1b) is an open system relative to its local environment (symbolized by room B) and that its local environment is an open system relative to the biosphere, then any niche-constructing organisms can operate like a Maxwell's demon, without violating the second law, if and only if it imposes a balancing thermodynamic cost on its local environment, and/or on its wider environment (symbolized by the two arrows entering and leaving room B). This is true even when, in practice, some costs are ignored because they are imperceptible, or too small to worry about, for instance, the energy lost as heat in photosynthesis.

The laws of thermodynamics also determine that whatever organ-

---

breaking the second law. In biology, however, when organisms niche construct, the information that controls their niche construction is not quantifiable, in bits anyway (Stuart 1985; Jaynes 1996), but comprises the "meaning" or semantic information that underpins the fitness of organisms.

isms do to their local environments, and however much they may change them, these changes are themselves ultimately dissipated by other environmental agents. However, this does not mean that niche construction has no consequences. On the contrary, unless there is no significant difference between the times and places at which organisms generate the products of niche construction, and the times and places at which those same products are eventually dissipated by other agents in the environment, niche construction is likely to have ecological and evolutionary repercussions. Insofar as standard evolutionary theory currently recognizes that organism-environment interactions are bidirectional yet ignores the consequences of niche construction, a hidden assumption of this theory must be that this kind of negation of niche construction by dissipation must be what usually happens. Yet, as the examples in chapter 2 show, the products of niche construction often persist and accumulate in the local environments of organisms for long enough, and across large enough regions of space, for there to be important ecological and evolutionary effects. Therefore, in reality, the constraint imposed by the ultimate dissipation of all the changes that niche-constructing organisms cause in their environments is not much of a limitation.

One difference between Maxwell's demon and niche-constructing organisms is that, in figure 4.1b, room B is exclusively abiotic because it only contains nonliving molecules of a gas, whereas the local environments of organisms contain both biotic and abiotic ecosystem components. Although the material and thermodynamic costs that are imposed by niche-constructing organisms on their local environments are bound to affect both abiota and biota, the manner in which abiota and biota respond to niche-constructing organisms is different.

Abiota have no capacity to counteract the niche-constructing activities of organisms. They therefore cannot resist being driven into unusual physical states by the activities of organisms. Although these new states may not be as far from equilibrium as are the physical states of organisms, they clearly can extend well beyond the physical states those same abiota would have been in if they had only been exposed to the laws of physics and chemistry. A frequently cited example is the earth's atmosphere, which is currently in an extreme state of disequilibrium partly due to the outputs of photosynthesising and grazing organisms, as compared with the atmospheres of lifeless planets, such as Mars and Venus, that are only mildly out of equilibrium (Lenton 1998). This example illustrates how, on the largest scale, niche-constructing organisms can sometimes drive abiota into new physical states that deviate considerably from the states they would have attained in the absence of niche-constructing organisms.

In contrast, the biota in the local environment of an organism are also semantically informed and active systems. Unlike reactive abiotic systems, biota can work to protect their own far-from-equilibrium and homeostatic states relative to their own local environments in exactly the same way as can the focal niche-constructing organisms themselves. Hence, biota should resist whatever material or thermodynamic costs are imposed on them by other niche-constructing organisms, possibly by expressing counteractive niche-constructing activities of their own. Only when they die do organisms completely lose their ability to resist the costs of other organisms' niche construction.

The analogy this time is with the picture shown in figure 4.1c. Here the demon who is operating the door in room A is interacting with a competing demon, supplied with different information and operating a second door in room C. Both demons work in a shared open environmental system (room B). In this case the outcome of their interactions should depend on the relative efficacy of these two demons and on the relative fitness of the semantic information that each is using and expressing. This situation has often been modeled in terms of game theory (Axelrod 1984; Sigmund 1993; Gintis et al. 2001), and it can generate complicated consequences in ecosystems and elsewhere (e.g., O'Neill et al. 1986; Patten and Jorgensen 1995). For instance, the two demons could either assist each other or oppose each other, or they could remain neutral with respect to each other's functions. Sometimes interacting biota may therefore be expected to ratchet up the overall dissipative costs that are ultimately borne by their mutual environments as a consequence of their interactions; this may happen when diverse biota compete for the same abiotic resource. Conversely, interacting biota may become coupled to each other relative to energy fluxes and material flows in larger open environmental systems, in which case they may sometimes be able to slow or postpone the dissipation of a shared resource, for example, by recycling the resource between them. Reiners (1986) has indicated some of the theoretical roots of these possibilities in ecosystems, in a broad-based discussion of thermodynamics and stoichiometry that in many respects resembles the approach we are using here.

Because niche-constructing organisms work in open systems, they can, as we noted above, potentially drive some selected components of their environments in both thermodynamic directions, by either locally increasing or locally decreasing entropy levels. Most often, niche-constructing organisms are likely to raise entropy levels in their local environments by taking resources from them and dumping detritus in them. However, they may also build artifacts and other constructions, or store extra resources at a chosen location, or even invest in artificial agricultural practices that enhance local food production, as some ants do (Hölldobler and Wilson 1994). If niche-constructing organisms act in any of these ways, they decrease rather than increase local entropy levels. The construction of artifacts by organisms is interesting, because artifacts represent an extreme on a continuum that reflects the degree of influence of organisms on factors in their environments. They also generate a third kind of object in ecosystems that is neither biotic nor conventionally abiotic, but intermediate between the two. Artifacts are not alive, yet they can only be built by living organisms. Also, once built, they are likely to respond to niche-constructing organisms in a different way from either biota or raw abiota. Like organisms, artifacts demonstrate negative rather than positive entropy because they are usually quite highly organized; yet, unlike organisms, they have no ability to defend their own organization nor to prevent their own dissipation. Artifacts are therefore likely to demand repetitive niche construction from organisms to maintain them. However, the second law of thermodynamics still applies here. Organisms can only decrease entropy locally at the cost of raising entropy still further somewhere else in their environments. For example, to build a nest, a bird must import twigs, moss, roots, grass, fur, and spider silk from elsewhere, thereby increasing entropy elsewhere in its environment. Thus, here again, niche-constructing organisms are likely to ratchet up the overall material and thermodynamic costs they ultimately impose on their environments if they store resources or make artifacts. Organisms can impose design on some environmental components at selected locations, and they can enrich some parts of their environments at the expense of others. This suggests that when they engage in niche construction, ancestral organisms can bequeath an ecological inheritance comprising up to three different kinds of heritable components, abiota, biota, and artifacts, which are likely to respond in different ways to further niche construction and to independent dissipating agents in environments. Each component may therefore modify natural selection pressures differently in an ecological inheritance, possibly promoting different evolutionary consequences in populations.

In addition to the general consequences of niche construction for environments, there are also consequences of niche-constructing activities that are species-typical and occur because the semantic information that controls and regulates niche construction is not the same in all organisms. The semantic information that controls the niche-constructing activities of organisms is lineage-typical, species-typical, and if developmental reaction norms and ontogenetic phenotypic plasticity are added, individual-specific, too (Schlichting and Pigliucci 1998). Thus different species of organisms will modify different components of their environments in different ways when they niche construct.

This additional specificity of niche construction may complicate matters, but it does introduce one major advantage. Because the semantic information that controls niche construction is characteristic of particular taxonomic groups, and because in the limiting case it may even be reducible to information encoded in single genes, the ways in which organisms affect their ecosystems by their niche construction should be empirically discernible, regardless of whether the environmental components that are modified by niche-constructing organisms are biotic, abiotic, or artifacts. In principle, then, it should be possible for both ecologists and evolutionary biologists to investigate highly specific consequences of niche construction in ecosystems, at particular times and places, that are associated with the expression of identified genetic variation. This point is consistent with Dawkins' (1982) extended-phenotype concept. However, the extended phenotype only refers to restricted cases of perturbation of environments by living organisms, usually involving the construction of artifacts, and usually assumed to be adaptive. For us, the expression of semantic information by niche-constructing organisms includes all of this, and more. It also includes relocational niche construction, DOM, and the expression of other indirect by-products of niche construction that could be maladaptive rather than adaptive, as for instance when organisms pollute their own environments, causing negative niche construction.

4.6 SUMMARY

Standard evolutionary theory invokes only one selective process in evolution, namely, the acquisition of semantic information as a consequence of natural selection pressures sorting between diverse organisms in populations. In the absence of niche construction, these natural selection pressures are implicitly assumed to stem almost exclusively from sources of selection in environments that are independent of the organisms they select. In addition, standard evolutionary theory invokes only a single mode of expression for whatever heritable semantic information is encoded by inherited genes, namely, phenotypic expression in the bodies of descendant organisms, or under restricted circumstances in extended phenotypes.

In contrast, evolutionary theory that incorporates niche construction includes two selective processes in evolution. It assumes that the semantic information that is inherited by organisms from their parents in the form of their genes is expressed not only in their phenotypes but also in their environments. This environmental expression is in the form of further modifications by niche-constructing organisms to specific abiotic, biotic, and artifactual components at specific locations in their local environments. As some characteristics of niche construction are universal, it follows that some aspects of the impact that niche-constructing organisms have on their environments will also be universal.

Only when the two selective processes of natural selection and niche construction are combined do they provide organisms with all the Maxwell's-demon-type properties needed to support life (Schrödinger 1944). Niche construction is an ontogenetic process that allows individual organisms the opportunity to gain sufficient energy and material resources from their environments to survive and reproduce. It thereby both contributes to the building of the next generation of a population of organisms in the conventional manner and causes changes in the niche-constructing organisms' own selective environments, as well as in the environments of others. Conversely, natural selection is a phylogenetic process that can only work given diversity in a population of variant organisms. The principal consequence of natural selection is that ontogenetic processes can be informed by information encoded in naturally selected genes. Moreover, it is only because ontogenetic processes can be semantically informed by natural selection that individual organisms can survive and reproduce and contribute to the next generation of their populations. Thus, niche construction fuels the evolutionary process as a consequence of the interactions of individual organisms with their environments, while natural selection informs the evolutionary process by selecting for "fit" genotypes. The result is an intimate interplay between phylogenetic and ontogenetic processes in evolution. Neither process on its own suffices to account for either the evolution of populations or the development of individuals. Together they help to account for both.

CHAPTER 5

Niche Construction and Ecology

5.1 INTRODUCTION

In the previous chapter we established two fundamental points relevant to the ecological consequences of niche construction. The first is that if there are general qualitative characteristics of niche construction then it follows that the evolutionary process must have general and characteristic impacts on the local environments of evolving species. This raises the possibility that niche construction may have implications for ecosystem-level ecology and that a niche-construction perspective may shed light on problems traditionally considered within the domain of ecology. In a sense this much is already proven, as we have only to equate our concept of "niche construction" with Jones et al.'s. "ecosystem engineering," and point out some of their more interesting findings, to illustrate the utility of this perspective. In this chapter we will indeed draw heavily on insights from ecosystem engineering.

The second point is that niche construction may be understood as the expression of genetically encoded semantic information in local environments, semantic information that has previously been accrued through natural selection and that is about the history of selection in those environments. Even aspects of abiota can legitimately be regarded as the expression of semantic information if the abiota are transformed into altered states through niche construction. As natural selection is a process that results in the acquisition and updating of semantic information through the Darwinian algorithm, it is reasonable to envisage semantic information flowing through entire ecosystems. Niche construction connects the prior expression of semantic information to the subsequent acquisition of this information by modifying sources of natural selection in environments. This connection may occur via either biotic or abiotic ecosystem components. As a result the ecological consequences of natural selection are no longer local, as characterized by standard evolutionary theory, but involve chains of events in which information-guided niche construction generates modified natural selection pressures and, in the process, accrues further information that guides additional niche construction.

In this chapter we will attempt to demonstrate that once these two points are accepted several classic problems in ecology may begin to be resolved. These problems are issues to which standard evolutionary theory either has had little to contribute or else has itself been an obstacle to understanding, but which an extended evolutionary theory that includes niche construction as one of its central processes may illuminate.

One such problem concerns the relationship between evolution and ecosystem-level ecology. More specifically, when ecosystems either change or resist change over time are their dynamics partly governed by evolutionary processes? If so, how can evolutionary theory help describe these dynamics given that ecosystems include abiotic components that lack genes or any equivalent heritable information? At first sight the presence of abiota does appear to rule out the application of evolutionary theory to entire ecosystems. Yet if evolutionary theory really is irrelevant to ecosystems, then it should be possible to understand most of what we need to know about their dynamics exclusively in terms of nonevolutionary processes (Sterner 1995). To date, however, all such attempts to understand ecosystems in terms of bioenergetics, biogeochemistry, or stoichiometry have proved insufficient. Apparently something is left out (Reiners 1986). But if it is evolution, how the structures and dynamics of ecosystems may indeed depend in part on the evolution of their constituent organisms has not yet been resolved. In this chapter we will argue that evolution is an indispensable co-contributor to any complete description of ecosystem dynamics (Brown 1995; Holt 1995; Jones and Lawton 1995). An evolutionary theory that stresses niche construction can shed light on how ecosystems may be governed by engineering webs of connectivity.

Historically researchers have finessed this difficulty by partitioning the discipline of ecology into population-community ecology, which is able to use evolutionary theory by circumventing abiota, and process-functional ecology, which is able to include abiota by leaving out evolution (O'Neill et al. 1986). This creates a second problem, namely, that these two subdisciplines are difficult to integrate within a single framework in spite of repeated attempts to do so. Below we argue that an extended evolutionary theory with niche construction that takes account of how evolving organisms affect both biota and abiota can provide a rich, integrative evolutionary framework for ecology.

A third set of problems relates to the defining characteristics of ecosystems. Are ecosystems self-designing and self-regulating systems? Is there any scientifically meaningful, nonmystical sense in which, perhaps under restricted conditions, an ecosystem can be described as a complex adaptive system, or even a super-organism? If so, what are these conditions? In this chapter we open a discussion of these problems that is pursued further in chapter 8.

Our goal is to explicate in terms of niche construction how evolutionary theory can be applied to ecosystems, in spite of the presence of nonevolving abiota, and in a manner that clearly illuminates and advances ecosystem-level ecology. Initially we will seek only to demonstrate how, *in principle*, ecosystem-level ecology, including its abiota, can be integrated with population genetics and evolutionary theory. To do this we will first develop a simple and intuitive conceptual model to describe the structure of ecosystems and the relations between their constituent components. This allows us to define more precisely why an evolutionary theory that incorporates niche construction is more useful to ecologists than standard evolutionary theory. In later sections of this chapter, we describe how our proposed integrated approach to ecosystem-level ecology may lead to a better understanding of "engineering control webs" in ecosystems, and we will suggest how, because of the presence of niche-constructing organisms, ecosystems may be partly self-regulating. Lastly, in section 5.5, we consider some of the *practical* opportunities and problems that are introduced by our approach. These include problems of scale and questions about when evidence either of an ongoing evolutionary event or of prior evolution on ecosystem structures and functions is likely to be observed in the field. We return to these issues in chapter 8, where we also offer some empirical predictions.

5.2 A SIMPLE CONCEPTUAL MODEL OF AN ECOSYSTEM

For simplicity, we will assume that, regardless of scale, the structure and function of all ecosystems depends on only three factors: (i) the number and kind of constituent parts or components in the ecosystem; (ii) the number and kind of interactive links that connect these components together; and (iii) the different kinds of "currency," such as energy or nutrients, required to keep track of the state of ecosystem components when they interact. These assumptions, although simplifying, are nevertheless sufficiently realistic to leave the basic problem intact. We explain each in more detail.

The first of these factors refers to the many different species of organisms and biotic and abiotic resources in the ecosystem. For our purposes, it will be sufficient to reduce all of these diverse components to just two kinds, biotic components, which are alive, versus abiotic components, which are not (fig. 5.1a). Under this scheme we will then treat the awkward cases of dead organic matter, organism-generated detritus, and artifacts as not alive and therefore as abiotic.

The second factor refers to the different ways in which the biotic and abiotic components interact with each other to establish the dynamic links that constitute whole ecosystems. Here, we will assume that every ecosystem component, whether biotic or abiotic, receives at least one input from, and emits at least one output to, at least one other ecosystem component (fig. 5.1b). We will also assume that in every interaction one "source" component in the ecosystem serves as the origin of a physical input to a second "sink" component in the same ecosystem.[1] We will also assume that each interactive connection between each source-sink pair is unidirectional; whatever entity is conveyed in any particular interaction travels only from the source to the sink. A causal arrow (→) is associated with each separate source-sink interaction, and if a source component emits an output to a sink component, the source component will typically "cause"

[1] To avoid confusion we stress that we are using the terms "source" and "sink" in their most general sense, and not in the more restricted ecological sense, which refers solely to the demographics of either source or sink populations or habitats (e.g., Holt 1996). We note a nonstraightforward relationship between our use of "source" and "sink" and Leibold's (1995) "requirement" and "impact" niches (see chapter 2), but we will not develop it here.

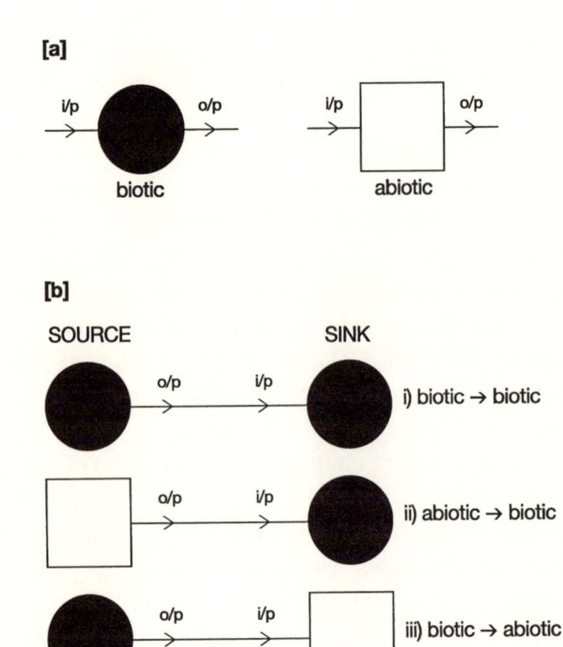
[a]
i/p
o/p
biotic
i/p
o/p
abiotic
[b]
SOURCE
SINK
o/p
i/p
i) biotic → biotic
o/p
i/p
ii) abiotic → biotic
o/p
i/p
iii) biotic → abiotic
o/p
i/p
iv) abiotic → abiotic

[c]

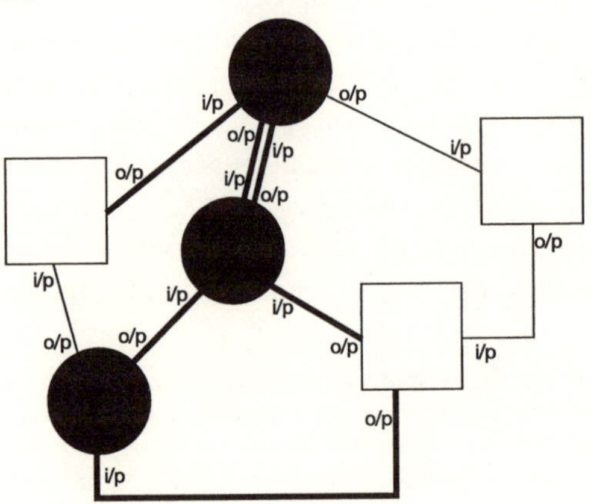
i/p
o/p
o/p
i/p
o/p
i/p
i/p
o/p
o/p
i/p
i/p
o/p
o/p
i/p
o/p
i/p
o/p
i/p

a change of state in the sink component. (A possible exception to this rule, which we note but will otherwise ignore, occurs if an intermediate ecosystem component serves as a catalyst for an interaction between two other ecosystem components, without itself being affected.) We use the term "cause" loosely to include cases in which the state of the source component affects the sink component in such a way as to generate a change in the sink component. We also use these links to signify both ecological and evolutionary interactions. When these two kinds of ecosystem components are combined with their possible interactions, they yield a set of four elementary interactive connections between different ecosystem components, which from here on we will call "links." These are (i) a biotic → biotic link (for example, a predator-prey or host-parasite interaction), (ii) an abiotic → biotic link (e.g., the uptake of water by plant roots), (iii) a biotic → abiotic link (e.g., the production of oxygen by photosynthetic bacteria), and (iv) an abiotic → abiotic link (e.g., a chemical reaction).

The third factor refers to the "currency" required to describe the inputs and outputs of source and sink components in ecosystems via these four different kinds of interactive links. Of course there are many different kinds of ecosystem inputs and outputs, but we will assume that each can be reduced to a measurable unit or currency, which ranges from solar energy to different kinds of nutrients and chemical elements. For any particular feature of an ecosystem, such as the nitrogen or hydrological cycle, it may be necessary to assign to each link its own specific currency, such as atoms of nitrogen or molecules of water (O'Neill et al. 1986; DeAngelis 1992). We will

---

FIGURE 5.1. Elementary components and links in our model ecosystem. (a) Two kinds of ecosystem components: biotic (circles) and abiotic (squares). Arrows indicate that each component is capable of receiving an input (i/p) from and emitting an output (o/p) to another ecosystem component. Black coloration indicates the expression of genes in the biota. (b) The four kinds of elementary link between ecosystem components. In each case a source component emits an output (o/p) that translates into an input (i/p) for a sink component. (c) A web of connections in an ecosystem made out of an assembly of elementary components and links. Thick lines represent evolutionary links. Thin lines represent nonevolutionary links.

TABLE 5.1. The Three Universal Ecosystem Currencies

	Currency	*Input from a Source*	*Output to a Sink*
Ecological	Energy	Influx	Efflux
	Matter	Inflow	Outflow
Evolutionary	Fitness	Acquisition of genetically encoded semantic information by populations subject to natural selection	Expression of semantic information by individual niche-constructing phenotypes

assume, however, that at least in principle all currencies can be reduced to the universal ecological currencies of energy and matter that apply to all energy fluxes and material flows in all ecosystems, and the universal evolutionary currency of *fitness* that applies to every organism.[2]

Table 5.1 lists these three universal ecosystem currencies, and their associated inputs from, and outputs to, source and sink ecosystem components. The two ecological currencies shown in this table are relatively straightforward because they are based on familiar physical concepts. Biotic and abiotic ecosystem components can serve as both sinks and sources of these ecological currencies. As we saw in the last chapter, however, fitness is anything but straightforward because it involves both the acquisition and the expression of information encoded in genes (Stuart 1985; Maynard Smith and Szathmary 2000). In line with our previous discussion we will treat fitness as a measure of the evolutionary utility of semantic information accrued in the gene pools of populations of organisms through natural selection and expressed by individual organisms in niche construction. We will treat the acquisition of semantic information by any biotic ecosystem component as an input, and the expression of semantic information by any biotic ecosystem component as an output.

[2] In suggesting that there is a *universal* currency of fitness we mean only to imply that all evolutionary analyses require fitness to be taken into account, and not that there is necessarily a single quantitative fitness measure by which all species in all environments can be compared.

Our final assumption is that all webs and chains of connection in ecosystems, on all possible scales, are made up of assemblies of the four kinds of elementary interactive links between biotic and abiotic ecosystem components (fig. 5.1c). Thus different webs of connectivity in different ecosystems refer either to different linkage patterns, based on different numbers of links between different numbers of ecosystem components, or to the different kinds of ecosystem currencies that can be carried by these different links. We assume that these chains of connection are largely responsible for the structures and dynamics of ecosystems, and that every ecosystem is ultimately an open system. We stress that this conceptual framework is not in itself an attempt to understand ecosystems but rather a simple working model that will allow us to illustrate a number of points concerning the role of evolution in ecosystems.

5.3 DIFFERENT APPROACHES TO ECOSYSTEMS IN BIOLOGY

To what extent do the webs and chains of connection that occur in ecosystems depend on evolutionary as well as on ecological processes? We will approach this question by first considering the ways in which different kinds of ecologists, working in different subdisciplines, treat the ecosystem factors we have just introduced in the light of standard evolutionary theory. Then we will consider how the same set of ecosystem factors might be handled differently with the inclusion of niche construction.

The two principal approaches to ecosystems in ecology are the process-functional approach and the population-community approach (O'Neill et al. 1986; DeAngelis 1992). Superficially, these two approaches may appear to be separated by little more than their choice of subject matter, by the different kinds of questions they ask, and by the different kinds of data they collect. However, for many years these two subdisciplines have proved remarkably difficult to integrate (O'Neill et al. 1986; Schoener 1986; Levin 1989; Roughgarden et al. 1989; Tilman 1989; Hagen 1989, 1992; Thompson 1994; Grimm 1995; Brown 1995; Jones and Lawton 1995), which suggests that there is probably something more fundamental than mere choice of subject holding them apart.

5.3.1 Process-Functional Ecology

Process-functional ecologists seek to understand how whole ecosystems work, including both their biotic and abiotic components (O'Neill et al. 1986; Odum 1971, 1989; Tilman 1989; Patten and Jorgensen 1995; Lawton and Jones 1995; Vitousek et al. 1997). Typically, they ask questions about the composition of ecosystems, their scale and their boundaries, their origin and their diversity, their structural and functional design, and their regulation. They also ask about the productivity, resilience (the rate at which things change), and the resistance of ecosystems to different kinds of perturbation (DeAngelis 1992), for example, perturbations caused by animal or plant invasions (Vitousek 1986). One of the principal goals of process-functional ecology is to understand how energy and matter flow through ecosystems, and, more particularly, how energy fluxes drive nutrient and material flows through both biotic and abiotic ecosystem compartments. For this reason, process-functional ecologists frequently grapple with complete biogeochemical cycles such as the hydrological, carbon, or nitrogen cycles (Hutchinson 1948; Odum 1989). The currencies they deal with are therefore energy and matter (table 5.1), and they look primarily to the physical laws of thermodynamics, to the availability of free energy, to the conservation of energy and matter, and to stoichiometry for guidance about the general principles underlying the organization and regulation of ecosystems (Reiners 1986).

The strengths and limitations of process-functional ecology have been summarized by O'Neill et al. (1986). The main strength lies in its capacity to model complete biogeochemical cycles, as well as the overall regulation and energetics of entire ecosystems, without having to decompose them into separate subsystems. For example, because process-functional ecologists deal with all four kinds of elementary interactive links in ecosystems, they do not separate biotic and abiotic ecosystem components from each other but can incorporate the contributions of both biota and abiota to the overall energy flows and nutrient cycles in entire ecosystems. Its main weakness is that it tends to lose sight of the organisms that are partaking in, and frequently driving, these ecosystem flows and cycles. One reason why whole organisms fall out of the picture is that different parts of the same organism, the leaves versus the roots of plants, for example, typically contribute to different functional processes in ecosystems and have to be assigned to different chemical cycles. Ironically, the conceptual integration of ecosystems seems to require the conceptual disaggregation of their constituent organisms. A second reason why organisms fall out of the picture is that process-functional ecologists do not incorporate the evolution of organisms in their models. Typically, they implicitly assume that organisms have evolved but are no longer evolving, or at least are not evolving on a time scale meaningful to ecological investigation. Surviving and reproducing organisms are assumed to be "fit" or well adapted, but neither the evolutionary process nor the fitness currency is directly included.

In sum, process-functional ecologists handle both kinds of ecosystem component, biotic and abiotic, and all four kinds of link between components, but they usually only deal with two of the three universal ecosystem currencies, ignoring fitness (table 5.2a).

TABLE 5.2. Process-Functional and Population-Community Ecology Compared in Terms of the Subset of Ecosystem Factors Each Handles

Components	*Links*	*Currencies*
(a) Process-Functional Ecology		
Biotic	(i) Biotic → biotic	Units of energy
	(ii) Abiotic → biotic	
Abiotic	(iii) Biotic → abiotic	Units of matter
	(iv) Abiotic → abiotic	
(b) Population-Community Ecology		
Biotic	(i) Biotic → biotic	Units of energy
	(ii) Abiotic → biotic (restricted)	Units of matter
Abiotic (restricted)	(iii) Abiotic → biotic (restricted)	Fitness
	(iv) _____	

5.3.2 Population-Community Ecology

In contrast, population-community ecologists make whole organisms, populations of organisms, and communities of different species of organisms their principal subject matter. For ecologists working in this subdiscipline: "The biota *are* the ecosystem. . . . The biota may interact with the abiotic environment, but the environment is largely viewed as the backdrop or context within which biotic interactions occur" (O'Neill et al. 1986, pp. 8–9, their italics).

The questions population-community ecologists ask are therefore strongly organism-oriented. Their subjects include the dynamics of population interactions, fluctuations in population sizes, the abundance of organisms, and the distribution of different species of organisms in communities. Like process-functional ecologists, population-community ecologists are interested in energy and matter flows, but only with respect to populations and communities of organisms; abiota are not usually included. Their primary focus is therefore on food webs, food chains, and patterns of trophic connections (the "who eats whom" question) in communities (DeAngelis 1992). When studying these problems, population-community ecologists refer to the same energy and matter currencies as the process-functional ecologists, but they apply them to biotic subsystems and to highly confined forms of biotic-abiotic interaction, instead of to entire ecosystems. One consequence of this restriction is that population-community ecologists are mainly concerned with only one of the two ecosystem components, namely, biota, and primarily with only three of the four links, the biotic → biotic, biotic → abiotic, and abiotic → biotic links. Moreover, population-community ecologists deal with the latter two links primarily by devising ecological rather than population-genetic models, by focusing on demographics rather than changes in genotype frequencies, and by concentrating on limited forms of change in abiota. Thus, population-community ecologists ignore or short-change abiotic → abiotic links, and use biotic → abiotic and abiotic → biotic links in a restricted manner (see table 5.2).

The strengths and limitations of the population-community approach have also been summarized by O'Neill et al. (1986). By focusing on biotic subsystems, population-community ecologists do keep whole organisms in sight, but lose sight of the whole ecosystem by placing many of the abiotic components in the organisms' environment outside the boundaries of the particular biotic subsystems they are studying. Particularly in population-genetic coevolutionary models, they circumvent abiota by assuming that biotic ecosystem components are directly linked to each other when in reality they may be linked indirectly via intermediate abiota. These various devices carry the penalty of making it all but impossible to keep track of complete biogeochemical cycles. Instead, what is followed is a series of incompletely connected, or even disconnected, entries and exits from environments to organisms, and from organisms back to their environments (O'Neill et al. 1986).

Population-community ecologists do, however, frequently incorporate evolution. Yet in the main they do so by restricting their study to biota, effectively "editing out" all those components of the ecosystem that do not evolve. This tactical simplification yields a workable research program as their biotic subsystems can then be explained not only in terms of physics and chemistry, but also in terms of evolution. There is active research that endeavors to link community dynamics and ecosystem processes (DeAngelis 1992; Abrams 2001), but the presence of abiotic → abiotic links and the restricted means of dealing with biotic → abiotic and abiotic → biotic links impose a fundamental constraint on this research.

Thus population-community ecologists can use explanations based on population genetics, natural selection, and adaptation, as well as explanations based on the laws of physics and chemistry, in their approaches to ecosystems. It is here, however, that population-community ecologists start to part company with process-functional ecologists. Evolutionary explanations introduce the third universal currency of fitness to ecosystem-level ecology (table 5.1), and as we saw in chapter 4, fitness is an information currency. In sum, population-community ecologists do sometimes utilize all three types of ecosystem currency, but usually only when they restrict their focus to biota, and primarily to the direct links between biota (table 5.2b).

5.3.3 The Relationship between Standard Evolutionary Theory and Ecosystems

Let us consider to what extent standard evolutionary theory addresses the four types of links between ecosystem components. Populations

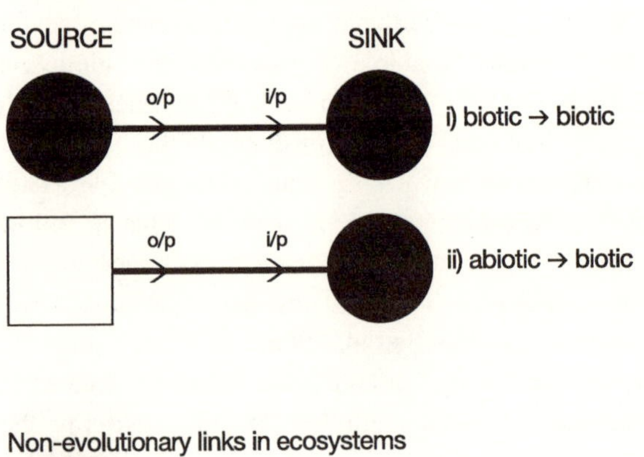

FIGURE 5.2. The two kinds of elementary ecosystem link that are incorporated by standard evolutionary theory, compared to the two kinds of elementary ecosystem link that are not. Thick lines represent evolutionary links. Thin lines represent nonevolutionary links. Circles indicate biota. Squares indicate abiota. Inputs and outputs are indicated by (i/p) and (o/p). Black coloration indicates the expression of genes in the biota. In this figure, genes are expressed only in biota.

of organisms respond to natural selection imposed by other biotic ecosystem components. Therefore, link (i) biotic → biotic interactions (fig. 5.2) are consistent with evolutionary theory. In biotic → biotic links, all three of the universal ecosystem currencies are implicated in both the outputs from the source ecosystem component and the inputs to the sink component. Thus for two coevolving populations, all three ecosystem currencies—energy, matter, and fitness—are required to describe whatever natural selection pressures the source population may impose on the sink population. This is because the energy and matter outputs of the source population expressed, for example, by the number of its organisms will be at least partly influenced by the semantic information in the source population's genes, for example, through fecundity or survival, and therefore by the fitness currency, as well by the energy and matter currencies. Similarly, if a sink population responds to a natural selection pressure from a source population, there will not only be altered genotypic frequencies due to fitness differences between individuals, but also changes in the population's energy and matter states, for instance, due to a change in the number of individuals in the current generation. There is, therefore, symmetry in biotic → biotic links with respect to the three ecosystem currencies, because all three are needed to describe both the biotic sources and the biotic sinks in these interactions.

For population-community ecologists this degree of symmetry is very useful. It means that biotic → biotic links are potentially reversible, so that what was previously a sink population can become the source of a new natural selection pressure for a previous source population via a second biotic → biotic link. This not only introduces an evolutionary feedback but also sets up coevolutionary scenarios, such as those frequently modeled in standard coevolutionary theory (Futuyma and Slatkin 1983; Thompson 1994). If, on the other hand, a sink population becomes the source of a new selection pressure for a third population via a new biotic → biotic link, then it may initiate a chain of interactions, involving many species, which may or may not ultimately feed back on itself to form a web of interacting and potentially coevolving populations as, for instance, in a food web. The same logic also applies to subsets of organisms in single populations in cases where one group of organisms acts as the source of a natural selection pressure for another group of organisms in the same population. This happens, for instance, in populations where there is either frequency-dependent or density-dependent selection.

Populations may also respond to natural selection pressures from abiotic ecosystem components; therefore abiotic → biotic links (fig. 5.2) are consistent with evolutionary theory. Here, however, there is no symmetry between the nonliving source and the living sink ecosystem components, because the abiotic source of natural selection has neither genes nor a fitness state and can express only two of the three ecosystem currencies. Yet all three currencies, energy, matter, and fitness, are still needed to describe the responses of sink populations to these abiotic sources of natural selection. It follows that abiotic → biotic links cannot be evolutionarily reversible.

In link (iii) biotic → abiotic interactions, the source population will partly depend on the semantic information encoded in its genes, and it will partly direct whatever energy and matter outputs it may transmit to the abiotic sink by semantic information via its phenotypes. However, because there are no genes in abiota, the abiotic sink can only respond to its interactions with this biotic source by changing its energy and matter states. So every time a biotic → abiotic link occurs in an ecosystem, it must apparently bring any potential flow of semantic information through that system to a halt. Even though biotic → abiotic links are clearly ecologically significant and can be explored using ecological models, from the perspective of standard evolutionary theory they appear to be evolutionary "dead ends."

The remaining abiotic → abiotic link also appears to be irrelevant for evolution. Here, neither the source nor the sink abiotic component can acquire or express the fitness currency, and hence it is impossible to include this ecosystem interaction in models of standard evolutionary theory. It would seem that abiotic → abiotic links refer only to energy and matter interactions that are governed exclusively by physics and chemistry.

We can now see why it has proved so difficult to integrate a complete ecosystem-level ecology with standard evolutionary theory, or to integrate process-functional ecology and population-community ecology. Standard evolutionary theory recognizes only two out of the four kinds of elementary interactive links that occur in ecosystems, links (i) biotic → biotic and (ii) abiotic → biotic, and regards both the other links, (iii) biotic → abiotic and (iv) abiotic → abiotic, as evolutionary irrelevancies (fig. 5.2). Moreover, neither process-functional ecology nor population-community ecology usually treats the complete set of components, links, and currencies that collectively make up ecosystems. Worse, each of these ecological subdisciplines is concerned with different subsets of factors. Process-functional ecologists need all four types of links between components, including nonevolutionary biotic → abiotic and abiotic → abiotic links. In contrast, population-community ecologists focus on evolutionary links, namely, biotic → biotic and abiotic → biotic links, but primarily by restricting their attention to the biotic components of ecosystems. Yet, if biotic → abiotic links are indeed evolutionary "dead ends," and abiotic → abiotic links are evolutionary irrelevancies, then their existence in ecosystems is sufficient a priori to rule out the possibility that semantic information could ever flow through an entire ecosystem. Semantic information can flow through the biotic subsystems that population-community ecologists study, but not through an entire ecosystem. So, on the basis of standard evolutionary theory, we would be forced to conclude that ecosystems cannot be governed by evolutionary processes.

5.4 THE ROLE OF NICHE CONSTRUCTION IN ECOSYSTEMS

We now reexamine the relationship between evolution and ecology in the light of our extended evolutionary theory, which includes niche construction.

5.4.1 With Niche Construction All Links Become Evolutionarily Significant

A reexamination of the same four links in the light of an evolutionary theory that explicitly includes niche construction reveals that all links now become evolutionarily significant. Biotic → abiotic and abiotic → abiotic links take on a new role in ecosystems through niche construction.

Biotic → abiotic links are analogous to the cartoon shown in figure 4.1b in chapter 4. Because niche-constructing organisms have the characteristics shown in the left-hand column of table 4.1 and the kind of Maxwell's demon properties illustrated in figure 4.1b, they can drive not only themselves but also some of the abiotic components in their local environments into nonequilibrium states. Moreover, since all organisms are governed by the second law of thermodynamics, it follows that they are compelled to modify their local environments by niche construction to some degree, because they must impose net material and thermodynamic costs on their environments that compensate for the benefits they receive in order to maintain their own far-from-equilibrium states. Abiotic ecosystem components could never reach some of the nonequilibrium states they achieve in the absence of niche-constructing organisms (Turner 2000). It follows that, to the extent that abiota do deviate from the physical states they would have occupied in the absence of niche construction, their deviations must reflect the expression of the semantic information that underpins this niche construction. Moreover, differential changes in abiota that are caused by the activities of niche-constructing organisms must reflect the differential fitness of organisms. With niche construction, biotic → abiotic links are no longer an evolutionary dead end.

In the case of biotic → biotic links, the organisms in the sink component are likely to have been naturally selected to resist, counteract, or possibly take advantage of the niche-constructing activities of the organisms in the source ecosystem component in complicated ways. In biotic → abiotic interactions, however, the abiotic component can only react to the niche-constructing activities of the organisms in the source ecosystem component by "blindly" obeying the laws of physics and chemistry. Hence, the expression of fitness by organisms in biotic → abiotic interactions should take a simpler and far more direct form than in biotic → biotic interactions. Paradoxically, it should actually be easier to observe the physical consequences of the expression of fitness differences by niche-constructing organisms in abiota than it is in biota.

In principle it should be possible to partition any abiotic ecosystem component into two subcomponents after it has been acted on by niche construction. One subcomponent should reflect the physical state the abiotic sink ecosystem component would have occupied if it had not been driven into a new state by organisms. We could describe this either as a "null" state, or as a "dead-planet" state (the state prior to niche construction). The second subcomponent would then reflect the abiotic component's actual deviation from its "null" or "dead-planet" state as a consequence of the activities performed on it by niche-constructing organisms. This partitioning of abiotic sink components into two subcomponents is illustrated in figure 5.3.

Beyond this generality, there is the additional and empirically useful point that in any particular case of a biotic → abiotic link, the

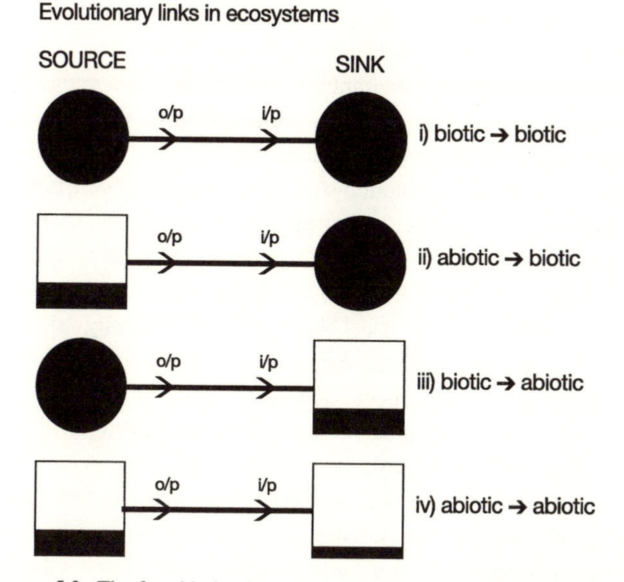

FIGURE 5.3. The four kinds of elementary ecosystem link that are incorporated by extended evolutionary theory. Thick lines indicate that all four links are now evolutionary. Circles indicate biota. Squares indicate abiota; inputs and outputs are indicated by (i/p) and (o/p). Black coloration indicates the expression of genes in both biota and abiota, in the latter case through niche construction. The expression of genes in abiota occurs when a biotic source component (niche-constructing organisms) drives an abiotic sink into a new physical state that it could not have occupied if it had not been driven there by the activities of organisms. The resulting deviation of the abiotic sink from its "dead-planet" state thus indirectly expresses the genes influencing the niche-constructing activities of the organisms that constitute the biotic source. Here, the partitioning of the abiotic components into shaded and unshaded parts indicates that the abiotic ecosystem component in links (ii), (iii), and (iv) has previously been modified by niche construction.

semantic information that is expressed by the organisms when they niche-construct will be clade-typical, species-typical, and individual-specific. To the extent that an abiotic ecosystem component is driven out of equilibrium and into a new physical state by its interactions with a biotic ecosystem component, it should reflect the specific information that is carried in the genes of the niche-constructing organisms. It is also true that if an abiotic ecosystem component, for instance the soil,[3] is driven into a new physical state by more than one biotic → abiotic link, for example, if the soil is acted upon by several different species of organisms simultaneously, then because of the specificity of each species' interactions with the soil, it might be possible to partition the abiotic ecosystem component still further by separating out the contributions of each different species to the soil's current deviant physical state, even potentially on a gene-by-gene basis. Indeed, soil scientists already undertake very similar studies.

Even abiota → abiota links become evolutionarily significant when niche construction is introduced into evolutionary theory. Provided neither component in the abiota → abiota link is in its dead-planet state, evolution-dependent ecosystem-engineering effects may occur in the ecosystem. This is because if the link occurs in the context of a web or chain of interactions in an ecosystem, and the initial abiotic source has previously been modified by niche-constructing organisms in a prior biota → abiota link, then a subcomponent of the abiotic source should reflect these niche-constructing activities. The resulting biotically induced change in this abiotic source should then have a "knock-on" effect on the abiotic sink in the subsequent link. This point is also illustrated in figure 5.3. If the sink abiota in this abiota → abiota link acts as the source of a natural selection pressure for another biotic ecosystem component, then in the context of a wider ecosystem chain or web, there should be an indirect evolutionary connection in which the selection pressures acting on the sink population will result indirectly from the prior niche-constructing activity of other organisms. Thus even abiota → abiota links may facilitate coevolutionary scenarios via specific environmentally mediated associations between populations in ecosystems.

For the field ecologist, the observation that with niche construction all links become evolutionarily significant may be regarded as something of a mixed blessing. After all, the simplifications adopted by practicing ecosystem and population-community ecologists have the advantage that they render a dauntingly complicated system em-

[3] Although, strictly speaking, soil is not exclusively abiotic, our general point holds.

pirically tractable. The explicit recognition of further links between ecosystem components would be of little value if it made progress more difficult for ecologists. However, in the sections below, and later in chapter 8, we suggest that due recognition of the role that niche construction plays in ecosystems is likely to be of value to empirically minded ecologists.

5.4.2 Ecosystem Engineering

The expression of semantic information by niche-constructing organisms in abiota introduces ecological consequences that go beyond present standard ecological models, or standard evolutionary theory. These consequences are seldom formally recognized as evolutionarily driven probably because, as we have seen, it is difficult to model them on the basis of standard evolutionary theory. They do, however, correspond closely to the phenomena that have begun to be described theoretically (Gurney and Lawton 1996) and empirically by ecosystem ecologists (Chapin et al. 1997; Jones et al. 1994, 1997) as *ecosystem engineering*. We will illustrate these ecological consequences with some examples of biota → abiota links in which niche construction has a significant impact. For this, we draw heavily on the important insights of ecosystem-engineering researchers.

The interaction of beavers with rivers makes a good example because it is well known and has been used both by Dawkins (1982) to illustrate his extended-phenotype concept and by Jones et al. (1994, 1997) to illustrate ecosystem engineering. When beavers build dams and lodges in rivers, they cause deviations in river flows and change many other characteristics of the river (Naiman 1988; Naiman et al. 1988). Jones et al. (1994) describe the building of dams by beavers as an example of "allogenic ecosystem engineering," which they define as any change caused by organisms in either their own environment or the environments of other species, "by transforming living or non living materials from one physical state to another via mechanical or other means" (p. 374). Jones et al. contrast allogenic engineering with "autogenic engineering," which they restrict to cases where organisms change their environments simply through their own physical structures as, for instance, when submerged macrophytes in freshwater lakes or rivers create weed beds through their own growth (Carpenter and Lodge 1986). Jones et al. describe many of the consequences of the beavers' allogenic engineering, over and above the changes in the flow and level of water in the rivers themselves. For example, when beavers cut down trees and use them to construct their dams, they create ponds which can have profound effects on a whole series of resource flows used by other organisms, including flows of nutrients, the buildup of sediments, and the availability of the river water itself. They may also create wetlands that can persist for centuries. They may modify the structure and dynamics of the riparian zone, and they can influence the composition and diversity of plant and animal communities.

Jones et al. point out the extraordinary degree to which some species of ecosystem engineers exercise control over the energy and matter flows in their ecosystems by their niche-constructing activities (e.g., Shachak et al. 1987; Jones and Shachak 1990). Sometimes this control is established by organisms that are not even participants in the energy and matter flows they are controlling, beavers being an example:

> The elemental ratios of the materials in a beaver dam, or of the organisms in the pond, bear no relationship to the elemental stoichiometry of the beaver. Perhaps the fundamental reason why energy, mass, and stoichiometry appear to be of little value in understanding engineering is that engineers do not have to be a part of the energy and material flows among the trophically connected organisms they affect. They are controllers of these flows, not participants in the flows (Jones et al. 1997, p. 1952).

If a niche-constructing or engineering species does exercise some control over the energy and matter flow of an ecosystem, the control itself must ultimately stem from the specific semantic information that is carried by that species. In other words, ecosystem engineering depends on adaptations. Moreover, when niche-constructing organisms drive abiotic ecosystem components into nonequilibrium states, they have the potential to drive them in either thermodynamic direction, toward greater local organization or toward greater local dissipation. The dams and lodges built by beavers are examples of the former. Both are beaver-"designed" artifacts that are neither completely abiotic nor biotic. They are nonliving objects, yet they could not exist without the evolved niche-constructing activity performed on them by living beavers. But beavers can only build dams and lodges by accelerating the dissipation of resources elsewhere in their environment. For instance, beavers have to import sticks to the sites where they build their dams from other places, sometimes from considerable distances.

Artifacts make special demands on both organisms and their environments, as we have already seen. As "designed" systems they are especially prone to dissipation because they are too far out of equilibrium to persist for long without maintenance. Generally speaking, construction and maintenance of artifacts are therefore likely to demand extra work from organisms. However, because organisms have evolved the behavior that enables them to carry out their construction, artifacts are unusually potent modifications of environments, to the benefit of their builders and other organisms (Hansell 1984). Dams are such important modifications of beavers' environments that Jones et al. (1994, 1997) describe beavers as "keystone" or dominant species in ecosystems.

Artifacts are costly, requiring extensive work to construct and maintain, and this has important repercussions for both ecosystem structure and the evolution of the engineers. There is evidence that, because the costs of building and burrowing behavior are high, individuals are reluctant to leave their nests (e.g., social insects), are willing to wait to inherit nests (e.g., cooperatively breeding birds), and are prone to recycle nest materials (e.g., reuse of wax by honey bees), to reuse abandoned nests (e.g., many solitary bees), and to usurp other individuals' nests (e.g., egg dumping in birds) (Hansell 1993). Moreover, because artifacts are costly to produce, the presence of nest builders may significantly contribute to species richness (Hansell 1993). For instance, barn swallows often attach their nests to the old mud nests of wasps, while many small birds use spider or caterpillar silk in the construction of their nests. Niche construction may effectively promote further niche construction in other species.

The ecological consequences of niche construction are characteristic not only of highly intelligent animals such as beavers. For example, recall the contribution of the two species of snails that we first met in chapter 1, *Euchondrus albulus* and *Euchondrus desertorum*, to the weathering of limestone rocks in the Negev desert by eating endolithic lichens that grow under the surfaces of the rocks (Shachak et al. 1987, 1995; Jones and Shachak 1990). Another example is provided by a plant species, the black rush *Juncus gerardii*, which facilitates the amelioration of soil conditions in salt marshes by shading the soil surfaces in the marsh (which has the effect of reducing water evaporation, thereby reducing the accumulation of salt in the soil), and by delivering oxygen from the aboveground parts of the plant to its belowground arenchyma tissue (which has the beneficial side effect of keeping the soil in the marsh from becoming anoxic) (Hacker and Bertness 1995; Bertness and Leonard 1997; Hacker and Gaines 1997).

Jones et al. (1994, 1997) discuss several factors that scale up the ecological consequences of niche construction in ecosystems. One is the lifetime per capita activity of individual organisms. Different species of organisms vary hugely in this respect, some being far more metabolically "expensive" than others in terms of both the resources they take from their environments and the detritus they return to them. Boreal forest trees are an example of organisms that have large per capita effects on both hydrology and climate. "To a degree . . . the boreal forest 'makes its own weather' and the animals living therein are exposed to more moderate and slower variation in temperature and moisture than they would otherwise be" (Holling 1992). However, Jones et al. point out that even organisms whose individual impacts on their environments are very small may nevertheless cause huge ecological effects if they occur in sufficiently high densities, over large enough areas, and for sufficient periods of time. The examples they quote include bog-forming *Sphagnum* mosses, which produce peat that can persist for hundreds or thousands of years after the death of the living moss.

Another factor is the length of time a population remains in the same place. Here, the potency of niche construction is most likely to vary with its cumulative effects over many generations. For example, if the niche-constructing activities of a population are directed by particular genes, and if the same genes are inherited by organisms in the same population in each succeeding generation, then each generation should invest in the same niche-constructing activities as its ancestors. Therefore, after a time, the sheer repetition of these same niche-constructing activities might act like a persistent unidirectional "pump" and eventually cause a very considerable change in some abiotic ecosystem component, for example, the scaled-up contribution of individual photosynthesizing microorganisms to the earth's aerobic atmosphere.

Jones et al. also stress the durability of the constructs or artifacts that organisms leave behind after they are dead or have moved on. Here their examples include beaver dams again, which can continue to have considerable ecological consequences long after the beavers have left. Termite nests are a similar example. Finally, Jones et al. emphasize the number and types of resource flows that are modulated by the niche-constructing activities of a species and the number of other species that depend on the same flows. Here they point out that the impact of organisms is greatest when the resource flows or abiotic ecosystem components that they modulate are utilized by many other species. It follows that some of the most significant consequences of niche construction in ecosystems are found in soils, sediments, rocks, in hydrology, and in fire ecology, and even in wind resistance. One example of the latter is from Puerto Rico where *Dacryodes excelsa* trees are able to withstand hurricane-force winds because their extensive roots and root grafts bind and stabilize bedrock and superficial rocks (Basnet et al. 1992). In this case the trees can literally hold the mountainside together.

5.4.3 Environmentally Mediated Genotypic Associations

When a biotic → abiotic link is not isolated, but occurs in the context of a wider ecosystem web, chain, or network made up of different kinds of links, then it is unlikely to be an evolutionary dead end. In particular, if it is immediately followed by an abiotic → biotic link that either feeds back to the same biota or feeds forward to a new biotic sink, then the initial link may become evolutionarily significant via its influence on the sink organisms. This should happen whenever the niche-constructing activities of the organisms in the original biotic source result in sufficient qualitative and quantitative changes in the intermediary abiotic component to modify a natural selection pressure stemming from the abiota and generate selection on whatever organisms are in the biotic sink. In this case the original link becomes part of a three-component biotic → abiotic → biotic "bridge" that indirectly connects two sets of evolving biota. This bridge works by connecting the *expression* of semantic information by niche-constructing organisms in an initial biotic source to the *acquisition* of semantic information in a subsequent biotic sink as a consequence of the modification of natural selection pressures generated through the intermediate abiotic component.

While such bridges are a common feature of ecological models (e.g., competitive exclusion models) they are not a common characteristic of population-genetic models, for the reasons given above. However, the local consequences of a flow of semantic information across such a bridge may be both evolutionary and ecological. There will be evolutionary consequences if the modification of an intermediate abiotic selection pressure by the niche-constructing organisms in a source population selects for different alleles in a sink population, and thereby affects further evolution in the sink population. The evolutionary outcomes should then resemble those demonstrated by the population-genetic models we presented in chapter 3. Recall that in our models we considered the coevolution of two diallelic genes **E** and **A**, in a single population, via an intermediate resource in the external environment, called **R**. In these models we assumed that the interaction between **E** and **A** is based on the expression of niche construction by the **E** locus alleles, the modification of **R** by this niche construction, and selection at the **A** locus by modified natural selection pressures that stem from **R**. All we need to do is equate the three (**E**, **R**, and **A**) variables in our models with the three components of a biotic → abiotic → biotic chain, by defining the initial biotic source as the population that carries the **E** locus alleles and therefore expresses the niche construction, allowing the environmental component **R** to be abiotic, and defining the sink biotic component to be the population that carries the **A** locus alleles that are subject to selection by a modified natural selection pressure from **R**. The results we have previously obtained can then be taken as examples of the kind of evolutionary consequence we might expect from abiotic bridges in ecosystems. In this respect it makes little difference that in these models the **E**-locus and the **A**-locus alleles are in the same population, which reduces the bridge to a feedback loop from and to the niche-constructing population itself. Linkage aside, the logic is broadly the same even when the **A** locus is in a second population. Then the modification of the intermediate abiotic **R** by the niche-constructing organisms in the first population feeds forward to a second population, instead of feeding back to itself. More realistically, the state of the abiotic component in a bridge is likely to be only partly controlled by the prior niche construction of a source population, because it will usually also be under the influence of processes independent of niche construction. This is captured by model 2 in chapter 3 where the amount of **R** is partly dependent on the impact of the niche construction influenced by alleles at the **E** locus, and partly on other processes of renewal and depletion affecting **R**.

However, the modification of an intermediate abiotic natural selection pressure by a source population's niche construction may also cause the local extinction or exclusion of the current sink population and subsequently create the opportunity for a replacement sink population to invade (Armstrong 1979; Tilman 1982; DeAngelis 1992; Holt et al. 1994; Abrams 2000). This is the kind of phenomenon that is observed in an ecological succession. Dramatic changes in species composition and ecosystem behavior can occur in a few generations by ecological processes that facilitate population replacement, such as competitive exclusion. But even here evolution is likely to be relevant. For example, instead of evolving in situ in response to a natural selection pressure that has been modified by a source population's niche construction, an incumbent sink population may relocate or become locally extinct. The original incumbent may then be replaced by another population that already expresses genes that are better suited to the modified environment. For this to happen, genotypes in the replacement sink population should already have a higher fitness than those in the original sink population because of prior evolution, at least relative to the selection pressure that has been modified by the niche construction.

When this occurs the replacement population will most likely belong to a different species from the original incumbent. It may be a species that belongs to the same guild as the original sink population, or it may even belong to a different guild. What matters is that the replacement population will eventually be better adapted to the changed selection pressures introduced by the source population's niche construction than was the previous sink population. The replacement population must also be reasonably well adapted to all the other selection pressures that are present at that particular time and place in the ecosystem. Of course, the replacement population could eventually become extinct as well. This scenario implies that after the modification of a natural selection pressure by a niche-constructing source population, the favored genes in the sink population are likely to be highly specific, whereas the species that express those genes need not be. The selection pressure modified by the source population's niche construction may be indifferent to which species is carrying the genes that are now favored. In an extreme case, for example, any photosynthesizing sink species might substitute for another photosynthesizing sink species, provided it can cope better with the changed selection pressure caused by a source population's niche construction. Moreover, much of the semantic information that is expressed in niche construction is unlikely to be species-specific, but rather to be characteristic of a higher-level taxonomic group or a guild.

This ecological consequence of niche construction has implications for species invasions and species replacements in ecosystems (see section 5.5 below, and chapter 8). The immediate point is that, regardless of whether the consequences of the niche-constructing activities of source populations in ecosystems are evolutionary, ecological, or both, they are always partly governed by flows of semantic information through ecosystems, generated by the Darwinian algorithm.

If there is an evolutionary response by the incumbent sink population, the flow of semantic information across a biotic → abiotic → biotic bridge connects the *prior* expression of semantic information by the genes of a niche-constructing source population to the *subsequent* acquisition of new semantic information by the sink population via the intermediate abiota. Here the involvement of evolution is clear, as there is genetic evolution of the sink population in response to a selection pressure that has previously been modified by the source population's niche construction. However, the evolutionary influence is still present in the case where there is local extinction and replacement. In this case a flow of semantic information across a biotic → abiotic → biotic bridge should now connect the *prior* expression of semantic information by the genes of a niche-constructing source population, to the *prior* acquisition and current expression of old semantic information by a replacement sink population. Even when the niche-constructing activities of a source population only promote ecological consequences in ecosystems, any attempt to understand these consequences in terms of ecology's energy and matter currencies only is unlikely to be sufficient. A comprehensive understanding will probably demand some reference to evolution's semantic information currency as well (table 5.1).

Henceforth, we will call any indirect but specific connection between distinct genes that occurs via an ecosystem component in the external environment an *environmentally mediated genotypic association*, or EMGA, and we will do this without regard to whether the intermediate environmental component on which a particular genotypic association depends is biotic or abiotic. Although we have so far described semantic information flows across biotic → abiotic → biotic bridges, they could also occur across biotic → biotic → biotic bridges in which the intermediate ecosystem component is another evolving population. We will therefore allow EMGAs to refer to cases where both genes are in the same population (as they are in our theoretical models, in which biotic → biotic and biotic → abiotic links are present, and in examples of "indirect genetic effects"), or in two different populations (as they are in both biotic → abiotic → biotic bridges, and in biotic → biotic → biotic chains in ecosystems). We also anticipate that some EMGAs will be mediated by one or more abiotic → abiotic links. Provided that the first such link is preceded by a biotic → abiotic link, and the last is followed by an abiotic → biotic link, an EMGA may still exist. Thus even abiotic → abiotic links are potentially no barrier to the flow of semantic information through ecosystems.[4]

[4] EMGAs may also be relevant to the development of organisms, as, for example, when the prey selection and pigment utilization genes in flamingos interact via the crustaceans in the flamingos' environment to generate a pink plumage. See also the literature on "indirect genetic effects" (Wolf et al. 1998, and chapter 10). However, here we focus primarily on the ecological and evolutionary consequences of EMGAs.

However, we do suggest that the evolutionary dynamics of EMGAs are likely to be very different when they are mediated by intermediate biotic, as opposed to abiotic, ecosystem components. Where the EMGA is mediated by an intermediate population of organisms, the genes underlying niche construction in the initial source population will indirectly generate modified selection acting on the genes in the ultimate sink population that depends on the biological characteristics of the intermediate population. Such biological characteristics include the size of the intermediate population, its population structure, its generation time, and whether the organisms in the intermediate population are themselves under selection that may have resulted in niche-constructing activities counteracting or enhancing the niche construction of the source population. However, if the intermediary component in the EMGA is abiotic, for example, a resource such as a sediment, soil, or water, none of this is relevant, and the impact of alleles in the initial source population will depend largely on physical and chemical reactions in the abiota. These reactions are likely to demonstrate dynamics that depend on the characteristics of the component, as for example, different chemical elements, or different nutrients that have dramatically different residence times or return times in an ecosystem (DeAngelis 1992).

EMGAs potentially provide a new tool for dissecting out the overall influence of evolving organisms in an ecosystem on any particular chemical element or molecule, in any particular biogeochemical cycle, such as the carbon or nitrogen cycle. By initially focusing on the abiotic component in the middle of each biotic → abiotic → biotic EMGA, and by choosing the particular chemical element or molecule of interest, it should be possible to work backward to the particular niche-constructing activities in the initial source population that are responsible for modifying the physical state of this chemical element. In principle, it might even be possible to identify the gene or genes in the source population that are expressed in this niche construction. Similarly, it might also be possible to work forward to the gene or genes in the sink population that are selected by the modified natural selection pressure stemming from the biotically induced changes in the intermediate abiota. The same step could then be repeated each time an EMGA occurs in the biogeochemical cycle, until the contributions of evolving populations to the whole cycle are established.

Note, however, that for any EMGA in a biogeochemical cycle, there is no simple chain of genes underpinning the cycle, because the genes in the sink population that are selected by a modified abiotic component are not necessarily genes that will influence the niche construction of the sink population in the next link in the biogeochemical cycle, although other genes in the same population may well do so. Although we have described this procedure in terms of single source and sink populations, in many instances it may be more appropriate to regard the source or sink as a guild rather than a species, and then it will not matter if the various genes expressed in the guild's niche construction are not identical by descent, so long as they are functionally equivalent. The same applies to any guild subject to selection resulting from niche construction.

Of course, the task of tracing the possible consequences of a single act of niche construction will rarely be straightforward, because that act may be broken up by the intermediate abiotic components into several different selective consequences, each with different dynamics. For instance, if a gene in an initial niche-constructing population of plants influences the timing of litter fall, the same gene could conceivably play a role in several different EMGAs by being associated with other genes in other populations, via different intermediate chemical elements in the soil, perhaps involving long chains of intermediate components, such as earthworms, fungi, and bacteria. While that may make it difficult to pin down the full set of evolutionary and ecosystem-engineering consequences of a single act of niche construction, it is nevertheless likely that, in some cases, the major effects will be quite visible and accessible to analysis.

Other consequences of EMGAs stem from bridging abiotic components that are modified over time scales greater than a generation of the source population, in which case a transformed state *is* an ecological inheritance to the sink population. It is important to note that major alterations in the topography and biochemistry of habitats can be brought about through continuous niche construction by a particular population over long periods of time, sometimes hundreds or even thousands of years (Hansell 1993). For instance, termite mounds and badger sets are commonly reported to be several hundred years old (Hansell 1993). The effect of such an ecological inheritance in EMGAs raises the prospect of some bizarre indirect evolutionary events. For instance, in an extreme case an initial niche-constructing population, such as the *Sphagnum* moss referred to earlier, might change an abiotic ecosystem component and then become either locally or globally extinct. The extinct population might, nevertheless, still influence the evolution of a later population by its earlier niche construction if the niche construction happened to modify the action of natural selection on a later population through the mediation and continuing presence of a modified intermediate abiotic component. The result would be a form of indirect and time-lagged evolution or coevolution, involving both a contemporary population and the "ghost" of a now extinct population. In this manner an extinct population could continue to exercise an influence over its ecosystem long after its disappearance.

5.4.4 An Integrated Approach to Ecosystem-Level Ecology

When niche construction is treated as a fundamental process in evolution it is easier to see how, *in principle*, the process-functional and population-community approaches to ecosystems may be integrated, and how process-functional ecology may incorporate evolutionary theory.

Thus, when evolutionary theory is extended by adding niche construction, all the links that can occur in ecosystems become theoretically capable of connecting the expression of semantic information by niche-constructing organisms with the acquisition of semantic information by naturally selected organisms. There should therefore be no dead ends to prevent the full application of evolutionary theory to ecosystems. The one exception to this claim is the possible presence in an ecosystem of dead-planet or biologically uncontaminated abiotic → abiotic links, which have never previously been affected by any interactions with biota. However, after four billion years of life on earth, it is doubtful whether any dead-planet abiotic → abiotic links exist anywhere in the biosphere (Jones and Lawton 1995). Similarly, it is equally unlikely that any dead-planet abiotic → biotic links do exist. This means that the four links in figure 5.1, only two of which are evolutionarily significant, have been replaced by the four links in figure 5.3, all of which are evolutionarily consequential.

However, if all four links can connect the expression of semantic information by niche-constructing populations to the acquisition of semantic information in naturally selected populations, then there should be no remaining barriers between process-functional ecology and population-community ecology. For example, if we now return to table 5.2, which illustrates the ecosystem factors process-functional and population-community ecology could incorporate in the light of standard evolutionary theory, niche construction can fill in all the gaps. The fitness currency becomes relevant to process-functional ecology because it is not possible to understand ecosystem structures, or biogeochemical cycles, without taking into account the fitness of the niche-constructing organisms that, in part, control the flows of energy and matter around the cycles and establish engineering control webs. Moreover, the links that were previously out of bounds to population-community ecology can now be replaced by EMGAs between two or more evolving populations.

Therefore, an extended evolutionary theory that incorporates niche construction should be generally applicable to entire ecosystems. This is illustrated by comparing figure 5.1c with figure 5.4. Figure 5.1c shows which interactions in any arbitrary ecosystem web either can or cannot be of evolutionary significance according to standard evolutionary theory. Nonevolutionary links act as barriers to the flow of semantic information through ecosystems. In contrast, figure 5.4 illustrates that with our extended evolutionary theory no such barriers exist, as abiota are bridged by EMGAs.

5.5 APPLICATION TO UNDERSTANDING ECOSYSTEMS

In this section we use our extended evolutionary perspective to build on the analysis of the consequences of ecosystem engineering by Jones et al. (1994, 1997). We consider and illustrate some of the practical problems and opportunities that are likely to be associated with our approach, for example, when, and on what scale, ecologists may expect to see an ongoing evolutionary response to niche construction occurring in the field, and when they are likely to witness only an ecological response, albeit one guided by prior evolution.

5.5.1 Controlling Webs in Ecosystems

Jones et al. (1997) point out that a major ecological consequence of the physical engineering by organisms is that it establishes "engineer-

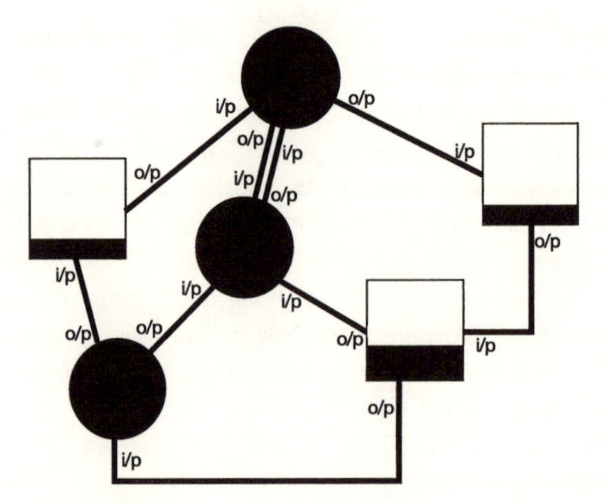

FIGURE 5.4. The same web of connections in the same ecosystem as that shown in figure 5.1c. Circles indicate biota; squares indicate abiota; inputs and outputs are indicated (i/p) and (o/p). Black coloration indicates the expression of genes in both biota and abiota, in the latter case through niche construction. In contrast to figure 5.1c all the links in this web are now evolutionary as well as ecological, as is indicated by the thick lines in this figure.

ing webs," or control webs, in both communities and ecosystems. Such webs are not well explained by conventional ecological theory, largely because ecosystem engineers are not necessarily part of the flows or cycles they control. To give an example, it is generally assumed that trophic relations must conform to the principles of mass flow and conservation of energy (i.e., the mass consumed minus the wastes produced times the growth efficiency equals the mass gained by the consumer). But the amount of mass or energy put into a beaver (minus its wastes and the energy it uses to build its dam) does not equal the mass of the dam or the water it holds, nor the magnitude of the many ecosystem effects that flow from dam construction (Jones et al. 1997). Moreover, it is also difficult to predict which engineers are likely to have the biggest effect on an ecosystem merely from knowing about energy and matter flows.

Control webs are difficult to understand exclusively in terms of these energy and matter flows because niche construction (or ecosystem engineering) depends on the adaptations, or possibly exaptations, of organisms. For example, the beaver's dam depends on the expression of specific behavioral and morphological adaptations by beavers.

Therefore, insofar as the adaptations of organisms are involved in the control of energy and matter flows in ecosystems by niche-constructing engineers, so is evolution. In this respect EMGAs are likely to play a key role in establishing control webs in ecosystems.

When organisms utilize niche-constructing adaptations, they are likely to modify natural selection pressures that act either on themselves or on other populations in their ecosystems. If the prior expression of semantic information by a source population is connected to the subsequent selection of semantic information in a sink population in this manner, then niche construction should activate an EMGA. The activated EMGA should then promote either an evolutionary response in the sink population, or an ecological response, for instance, a local extinction. Any natural selection pressures that have previously been modified by a source population's niche construction may select for different alleles, leading to the acquisition of some new or revised semantic information in the sink population, and thence to its further evolution. Alternatively, the modified selection may lead to the local exclusion of the sink population from the ecosystem, and perhaps to its replacement by an alternative population. If so, the replacement sink population could be one that is already better adapted to the modified selection pressures, because of the different semantic information it carries as a consequence of its own prior evolution.

Either way, the flow of semantic information carried by an EMGA between any source and any sink population in an ecosystem should grant the source population a degree of control over the sink population's status and presence in the ecosystem, via whatever natural selection pressure it has modified in the ecosystem. Hence, EMGAs introduce an additional and relatively novel type of control to ecosystems, one based on the evolution of populations and on evolution-generated flows of semantic information between populations, rather than exclusively on ecosystem ecology's classic energy and matter flows.

5.5.2 A Case Study: Desert Isopods and Soil Erosion

These points are illustrated by a case study taken from Shachak and Jones (1995) concerning the relationship between a desert isopod, the soil, and the local hydrology, which we use to draw attention to the points in this system where EMGAs might be important.

[a]

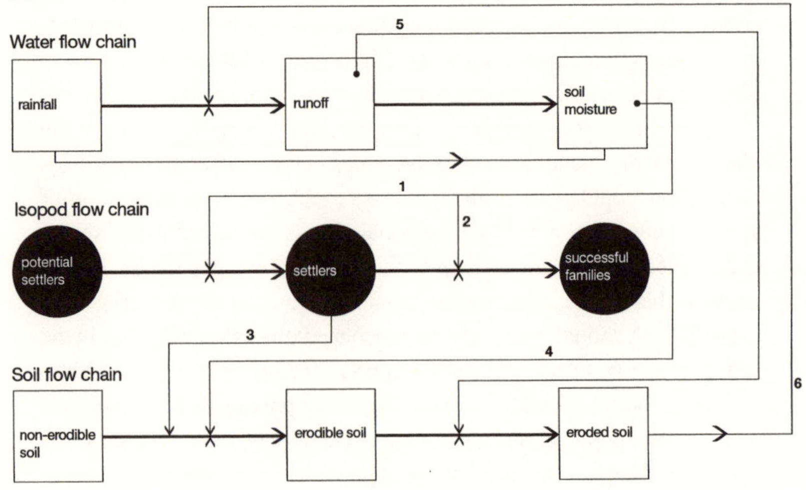

[b]

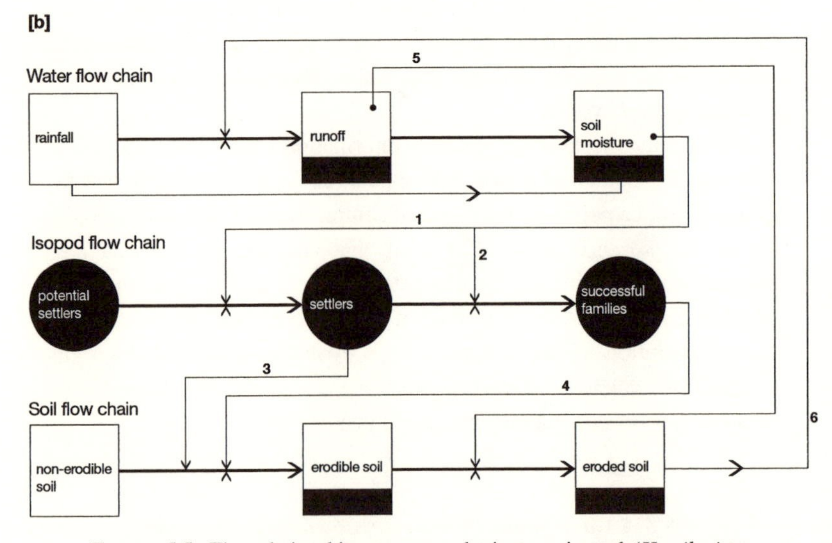

FIGURE 5.5. The relationships among a detrivorous isopod (*Hemilepistus reamuri*), soil erosion, and hydrology in an ecological system in the Negev desert. Isopod population dynamics, soil erosion, and hydrology are represented as three ecological flow chains. There are six major interconnections among the three flow chains: (1) soil moisture controls the number

Shachak and Jones describe the annual life cycle of a detrivorous arthropod (*Hemilepistus reamuri*), the pertinent ecosystem engineer of its ecosystem in the Negev desert. Every February adult isopods leave the burrows in which they hatched and seek new sites in which to settle, in the process dispersing distances of up to 1 km. Only 10% survive this hazardous journey. Females select a suitable patch of soil, dig a fresh burrow, and after a brief courtship give birth around May to 80–120 offspring. An appropriate location for the burrow is critical, as only those families that dig burrows at sites with high moisture content will survive (approximately half of the surviving immigrants, and 5% of the original dispersing population). The amount of moisture in the soil depends partly on the amount of rainfall and partly on runoff water, and because of this runoff, moisture is highest in areas with a high rock-to-soil ratio. Parents and offspring continue to live in the burrow until the following year, consuming large quantities of soil, and depositing fecal pellets around the entrances to their burrows. While the undisturbed surface is relatively nonerodible because it is colonized by a microphytic crust of cyanobacteria, algae, mosses, and lichens, the isopods' activity creates erodible soil that is washed away periodically when there is high rainfall. The isopods thus increase or maintain a high rock-to-soil ratio, which increases soil-moisture content and feeds back to create a favorable environment for themselves and their offspring (see Shachak and Jones 1995, pp. 281–283 for details).

Shachak and Jones illustrate how none of the three ecological flow

---

of potential isopod settlers and (2) the number of settlers that raise families. The settlers (3) and the successful families (4) control the transformation of nonerodible to erodible soil. Rain runoff (5) transforms erodible to eroded soil, and (6) eroded soil controls the flow of rain to runoff. (a) From the perspective of standard evolutionary theory, only two of these interconnections (1) and (2) are potentially evolutionarily significant. (b) In contrast, the niche-construction perspective renders all six of the interconnections between flow chains potentially evolutionary significant. Engineering organisms directly or indirectly modify the state of abiotic components in the system, in the process changing selection pressures. Further details are given in the text. (From Jones and Lawton, eds. [1995], fig. 27-1.)

chains they introduce to describe state changes in the water, isopod, and soil can be regarded as independent of the others (see fig. 5.5). They identify six major interconnections between the three component systems: (1) soil moisture controls the number of potential isopod settlers and (2) the number that raise families; (3) the settlers and (4) the successful families control the transformation of nonerodible to erodible soil; (5) runoff transforms erodible to eroded soil; and (6) eroded soil controls the flow of rain to runoff. It is immediately apparent that interconnections 1 and 2 correspond to abiotic → biotic links, (3) and (4) are biotic → abiotic links, while (5) and (6) are abiotic → abiotic links.

From the perspective of standard evolutionary theory, only two of these interconnections (1 and 2) are potentially evolutionarily significant (fig. 5.5a), although other links may perhaps be taken for granted. Variability in soil moisture potentially generates selection pressures favoring isopods that can choose, survive, and reproduce in appropriate sites. As soil moisture is an abiotic environmental component, and as standard evolutionary theory would ignore the possibility that isopods may exert control over soil moisture, a traditional evolutionary perspective would contribute little toward an understanding of this ecosystem.

In contrast, a niche-construction perspective reveals that interconnections 1 and 2 are not the only links that can be influenced by evolution (fig. 5.5b). The isopods experience feedback in the selection acting on them as a consequence of prior generations' soil-perturbing niche construction, which creates an ecological inheritance of modified soil states and consequent modified selection pressures. The isopods' activities could counteract, exacerbate, cancel out, or modify the intensity and direction of selection acting on them and their descendants, favoring alternative adaptations for habitat choice, burrowing behavior, fecal-pellet removal, or other traits affecting survival and reproduction. In the process they may also maintain stasis or generate change in soil-moisture content and rock-to-soil ratio. In figure 5.5b this feedback depends on the isopods exerting a direct controlling influence via interconnections 3 and 4 and, as a consequence, an indirect controlling influence via interconnections 5 and 6. There is, therefore, a flow of semantic information around this ecological system, reflecting the expression of the isopods' genes in niche construction. For instance, there should be EMGAs connecting those isopod genes expressed in burrowing behavior and those expressed in habitat selection. Note that there should be EMGAs operating in both directions, with the perturbational niche construction of burrowing potentially (indirectly) affecting selection of isopod genes that underpin habitat choice, and the relocational niche construction of habitat selection potentially affecting selection on burrowing. Note also that EMGAs are also likely to connect the genes related to the engineering activity of the isopods with the genes expressed by the cyanobacteria, algae, mosses, and lichens that form the microphytic crust on the nonerodible soil, driving indirect ecological and coevolutionary scenarios via abiotic soil and water components. None of these feedback loops would be captured by standard evolutionary theory, but they are captured by an evolutionary theory that recognizes niche construction.

5.5.3 Ecological and Evolutionary Scales

The last issue we need to consider is when and how ecologists might reasonably expect to observe evolutionary as well as ecological responses to the EMGAs that are the result of niche construction in ecosystems, and when they should only expect to see ecological consequences. This issue primarily concerns the relationship between the temporal and spatial scales at which ecologists can work and observe in the field, and the rates and scales at which evolution can be detected.

We can use Shachak and Jones's (1995) Negev desert study again as an example of the kind of scales at which ecosystem ecologists work in the field. Shachak and Jones point out that the three ecological flow chains shown in figure 5.5 represent different types of flow that operate across alternative spatial and temporal boundaries. Ecological scalar properties can be added to the functional properties of each flow chain by including the spatial range and temporal period over which change in each of the component states takes place. For instance, the smallest spatial scale pertinent to the isopod flow chain is the size of their burrows (between 10^{-4} and 10^{-3} m^2) and the largest is the distance they disperse (up to 10^3 m). Similarly, temporal scales pertinent to the water flow chain range from the minutes it takes for runoff to develop after rain to the years that moisture can remain in the soil. The scalar properties of the entire ecological system correspond to the collective boundaries that are set by the three ecological flow chains and their interconnections (Shachak and Jones 1995). There is considerable overlap of the flow chains over spatial ranges, which means that large-scale events can influence small-scale events, and vice versa. There is also a degree of temporal overlap, although this is more limited.

Insofar as these time scales are indicative of the rates of change at which ecosystem-level ecologists typically work, the comparatively slow rate at which evolution is commonly deemed to occur may at first sight appear to rule out the possibility that the ongoing evolution of a population in response to prior niche construction would be detectable. For instance, ecologists might be able to detect the rate at which natural selection pressures are being modified by a source population's niche construction, particularly when this generates an accumulating ecological inheritance for another population, possibly over a large span of time. In this case it might be possible to measure the rate at which a niche-constructed resource accumulates, but it might still not be possible to detect any evolutionary response to this ecological inheritance in any sink population. This may be the case when the sink populations are long-lived species for which the rate of evolution is slow relative to the life span of a human observer.

Ecological responses to niche construction in ecosystems may be detectable, however, because ecological change is frequently faster than evolutionary change, sometimes much faster. For example, it might take many generations for a population to evolve detectably in response to a selection pressure modified by another population's niche construction. However, the same population might be excluded quite rapidly from an ecosystem by a competitor population that is already better adapted to the same modified selection pressure, perhaps in less than a single generation. In general, therefore, ecologists may be more likely to detect ecological responses to niche construction, for example, successional phenomena, species replacements, or invasions in ecosystems, than they are to detect ongoing evolutionary response in an incumbent population.

That might appear to render our evolutionary approach to ecosystems irrelevant to the fieldworker, and to most ecologists. But, as we saw before, it is not just evolutionary responses, but also ecological responses to niche construction that are controlled, or partly controlled, by EMGAs, and therefore by evolution-generated semantic information flows in ecosystems. Ecologists should, therefore, still be able to detect the operation of engineering or control webs in ecosystems as a consequence of the semantic information that is already carried by populations because of their *prior* evolution, and the semantic information that is *currently* expressed by those same populations when they engage in niche construction. The niche construction expressed by incumbent populations in an ecosystem should partly determine which populations can and cannot replace some other population in the ecosystem. The second reason why our evolutionary approach should not be irrelevant in practice is because there are some circumstances where ongoing evolutionary episodes in ecosystems may be observed. The most obvious example concerns microorganisms (and other fast-evolving species). Whenever microorganisms are involved, in a detrital food chain for instance, detectable evolution in response to niche construction may be observed in the short term. The dramatic rate at which bacteria evolve resistance to human-produced antibiotics is such an example (Ewald 1994; Palumbi 2001a,b). To return to Shachak and Jones's example, microbes could evolve rapidly on the salty erodible soil that is produced by the isopod digging (Jones, personal communication).

Some multicellular organisms are also capable of rapid and observable evolution (Endler 1986; Kingsolver et al. 2001). Observations of natural selection in the wild have led to the conclusion that biological evolution may be extremely fast, with significant genetic and phenotypic change sometimes observed in just a handful of generations (Endler 1986; Dwyer et al. 1990; Thompson 1998). The best-known examples include the rapid evolutionary responses of many species to pesticides (Roush and McKenzie 1987; Palumbi 2001a,b) and of cactus finches and guppies to changed selection pressures (Grant and Grant 1995a,b; Reznick et al. 1997). Similarly, artificial selection experiments also show the potential for fast rates of evolutionary change in multicellular organisms (Hill and Caballero 1990). Rates of genetic change vary enormously according to the nature of the gene and its effect on the selected phenotype, but most evolutionary biologists would agree that biological evolution does not always require millions of years. The distinction between ecological and evolutionary time scales may therefore sometimes be blurred (Thompson 1998).

This last point is illustrated by the theoretical study of Kerr et al. (1999). They demonstrated that the evolution of anatomical and reproductive traits that allow plants such as eucalyptus and chaparral species to persist in fire-prone environments may indirectly be driven by the existence in the same plants of traits that enhance their flammability, such as accumulating oils or litter. The pertinent point here is that niche construction on the part of the plants creates the tinderbox environment that favors fires, and Kerr et al. demonstrate how the frequency and extent of fires can plausibly change as a function of the changing frequency of flammability alleles in the evolving plant population. The prevalence of fire feeds back to generate a selection pressure favoring persistence in a fire-prone environment. Here ecological and evolutionary events are operating within a similar time frame.

Clearly, introducing evolution to Shachak and Jones's ecological system will increase the range of temporal scales over which the system is perceived to operate. Nevertheless, there is ample evidence that the time scales of evolution can be sufficiently short to warrant its inclusion in many circumstances. Indeed, in a recent analysis, Kingsolver and colleagues (2001) reviewed 63 studies that measured the strength of natural selection in 62 species, including over 2,500 estimates of selection. They concluded that the median selection gradient was 0.16, which would cause a quantitative trait to change by one standard deviation in just 25 generations. Such rates of change are certainly detectable. Evolution is generally regarded as a gradual process when it is viewed over long periods of time. Gingerich (1983) notes that at short time scales organic evolution looks fast, whereas at long time scales it appears slow. There are several possible explanations for this discrepancy; for instance, the relationship between evolutionary rate and the time interval over which it is measured may be an artifact, or selection may fluctuate over longer periods of time. However, for the field ecologist, the rate of evolution over long time scales is essentially irrelevant. What matters is the rate over short time scales, and that rate may sometimes be sufficiently fast for biologists to detect evolutionary change.

Moreover, evolutionary responses to niche construction might be expected in situations where ecological responses are not possible. For instance, population replacements, invasions, and competitive exclusions all require that other species are in the vicinity, and that they are able to survive and reproduce more effectively in those environments modified by the niche construction of the source population. This will not always be the case. For instance, in depauperate environments, such as a desert, there may be no alternative populations waiting to replace or exclude an incumbent population. In this case the incumbent population will typically either evolve or become locally extinct. In some circumstances, this evolutionary response may be observable. In other situations there may be species present that are potentially capable of exploiting the modified selection pressures, but they may be unable to invade because they are outcompeted by the incumbent. Conceivably, some species may be too specialized to "upstream" niche construction in their ecosystem to be easily replaced by less specialized alternatives. For instance, in the case of some mutualisms the two species have coevolved to such an extent that is unlikely that either could independently be replaced by invading species. From a global perspective, many coevolving species can be regarded as only weakly coupled, as a given species may have hundreds of potential competitors, predators, or pathogens across its entire range. However, for a specific ecosystem the coupling between coevolving populations may be much tighter, and each species may not be locally replaceable. For all the above reasons, our evolutionary perspective may be relevant to the ecology of multicellular organisms even in the short term, provided sufficient genetic variation on which selection can act is present in such populations.

This leads us to the view that it may sometimes be useful to include evolutionary scales into ecological analyses. More important, the apparent differences between ecological and evolutionary time scales do not prevent engineering webs from also being evolutionary webs, since at whatever rate evolutionary processes work we can expect ecological and evolutionary responses to niche construction to be partly controlled by EMGAs.

5.6 LINKING EVOLUTION TO ECOSYSTEMS

Reiners (1986) argued that there must be a third complementary approach to ecosystems, in addition to the approaches based on energy and matter flows, that is missing from ecosystem-level ecology. He suggests that this missing element could be based on population interactions and resultant causal networks. We believe that Reiners is right, and that what has been left out is evolution. However, what is required here is the kind of evolutionary theory that can fully connect abiotic and biotic components of ecosystems, through the flow of semantic information encoded in organisms' genes. Such an evolutionary perspective is provided by niche construction.

As we have noted, Jones et al. (1994, 1997) already introduced several ecosystem phenomena that cannot be understood in terms of energy and matter flows only. The creation of physical structures and other modifications of their environments by organisms partly controls the distribution of resources for other species. Furthermore, Gurney and Lawton (1996) demonstrated theoretically that the efficacy with which an engineering population degrades (by negative niche construction) a virgin habitat determines not only whether there will be no engineers, a stable population of engineers, or population cycles in the frequency of engineering, but also the extent of virgin and degraded habitat.

We suggest that evolutionary phenomena associated with niche construction compliment and add to Jones et al.'s observations of the ecological repercussions of engineering. First, when they engineer, niche-constructing organisms may influence their own evolution by modifying their own selective environments via EMGAs that include abiotic components or chains of such components. Second, niche-constructing organisms also influence the evolution of other populations, perhaps indirectly, via EMGAs with intermediary abiotic components. Such interaction may occur between particular species or across entire guilds. Third, through their inceptive niche construction, organisms may create new niches for themselves, possibly through technological innovation, or they may relocate to a novel environment, which again will influence the dynamics of their ecosystems (Vitousek 1986). For example, the effect of improved paper technology would appear to have had a massive effect on the geographic distribution, colony size, and social complexity of Polistinae wasps (Hansell 1993). Fourth, it is important to recognize that the theoretical tradition in population genetics of finessing intermediary abiotic components in biota → abiota → biota coevolutionary events, by assuming that the intensity of selection on the biotic sink population is proportional to the frequency of genotypes in the biotic source population, is a matter of mathematical convenience. While we accept the cardinal importance of simplification in theoretical analyses, there are likely to be occasions when such idealization of the processes will obscure rather than elucidate ecosystem dynamics. Genetic coevolutionary models analogous to the **E-R-A** models in chapter 3 are required here. Fifth, it is also important to recognize that, while energy and matter typically flow through ecosystems in the same direction, semantic information can sometimes travel in the opposite direction. For instance, when a predator repeatedly captures and consumes prey, energy and matter flow from the prey to the predator, but the predator imposes a selection pressure on the prey, ultimately revising semantic information in the genes of the prey population to reflect the foraging methods of the predator. Sixth, evolutionary and coevolutionary events can operate on time scales within the range of ecological temporal scales, which means that the dynamics of abiotic components may reflect evolution due to genotype frequency changes in engineering species. Finally, seventh, there may be predictable patterns associated with ecological responses to prior niche construction, such as replacements, competitive exclusions, and invasions, as a function of the characteristics of the niche construction expressed by the incumbent populations in an ecosystem.

These observations can be subsumed under two general points. First, all three of the universal currencies shown in table 5.1, energy, matter, and fitness, and all three of the ecosystem flows they give rise to, namely, energy, matter, and semantic information, are needed for a comprehensive understanding of ecosystems. Without any one of them the analysis of ecosystems is inevitably left incomplete, as Reiners (1986) has indicated. The second is that the ecological flows of energy and matter and the evolutionary flow of semantic information depend on each other, probably strongly, in most if not all ecosystems. The details of how energy and matter flow through ecosystems cannot be understood without reference to the control imposed by engineering webs, in terms of the semantic information that accrues in populations through natural selection because this kind of control depends on the niche-constructing adaptations (and exaptations) of organisms that cause changes to energy and matter flows in ecosystems. Conversely, the semantic information that underlies evolution's fitness currency depends on the ability of organisms to tap into ecology's energy and matter flows in such ways that they do manage to survive and reproduce. In this case it is ecology's energy and matter flows that put the "semantics" into evolution's semantic information, since these ecological flows are the ultimate sources of the "meaning" in semantic information flows. Again, it is impossible to understand any one of these flows without reference to the others because they depend on each other and partly control each other. Again Reiners (1986) was right.

Obviously, the control that organisms do exert on ecosystems is rarely absolute, as ecosystem components are also subject to some physical and chemical processes that are independent, or partially independent, of their evolving engineers. Ecosystems are subject to control webs but they are not completely specified by them. As a result, at this stage we have no reason to believe that ecosystems will ever prove to be fully integrated, self-regulating, and self-designing systems. Nevertheless, an advantage of the conceptual framework that we have developed in this chapter is that it potentially facilitates exploration of questions such as to what degree ecosystems can legitimately be described as complex adaptive systems (see chapter 8). We anticipate that to the extent that ecosystems do demonstrate self-regulating and self-designing qualities, it will be precisely because niche-constructing organisms can exert sufficient control over their local environments to establish engineering control webs.

CHAPTER 6

Human Niche Construction, Learning, and Cultural Processes

6.1 INTRODUCTION

The conventional view of evolution is that, through the action of natural selection, organisms have come to exhibit those characteristics that best enable them to survive and reproduce in their environments. The fact that, in standard evolutionary theory, it is always changes in organisms rather than changes in environments that are held responsible for generating the organism-environment "matching" relationship is made explicit by the term used to describe the *process* of evolutionary change itself, "adaptation." Organisms are assumed to adapt to their environments, but environments are not assumed to "adapt" to their organisms. The same term, however, is used to describe the ever-changing *products* of natural selection, the "adaptations" that organisms exhibit. This double usage of "adaptation" provides the clearest possible indication that the process of adaptation, whereby organisms respond to their environments, is usually regarded as the only process thought to be capable of generating complementarity between organism and environment in evolution.

One consequence is that, hitherto, the most common evolutionary approaches to the study of humans have been "adaptationist" in nature, in the sense that they have placed sole emphasis on the process of selection and on purported adaptations that underlie human behavior (Laland and Brown 2002). Sociobiologists and evolutionary psychologists provide explanations for the characteristics of human behavior, human relationships, and human institutions in terms of natural selection's furnishing our ancestors with functional solutions to problems posed by ancestral environments. Hence, any match that is observed between the characters or features that humans possess and the factors in the environments that their ancestors experienced is assumed to have come about either by chance or exclusively through natural selection.[1] Yet, in chapter 2, we saw that niche construction provides a second evolutionary route to the dynamic match between organism and environment. Such matches need no longer be treated as products of a one-way process, exclusively involving the responses of organisms to environmentally imposed problems: instead they should be thought of as the dynamical products of a two-way process involving organisms both responding to their environments, and changing their environments through niche construction (fig. 2.1).

In this chapter we investigate some of the principal repercussions of this niche-construction perspective for the human social sciences. A focus on niche construction has important implications for the relationship between genetic evolution and cultural processes. One implication is that niche-constructing organisms can no longer be treated as merely "vehicles" for their genes because they also modify selection pressures in their own and in other species' environments, and in the process they can introduce feedback. A second is that there is no requirement for niche construction to result directly from genetic variation in order for it to modify natural selection. Humans can and do modify their environments mainly through cultural processes. The human dependence on cultural processes, however, does not make human niche construction unique. As we saw in chapter 2, niche construction is a general process exhibited by all organisms, and species do not require advanced intellect or sophisticated technology to change their world. Hence, the general replacement of a single role for phenotypes in evolution by a dual role removes from cultural processes any claim to a unique status with respect to their capacity to modify natural selection. Nonetheless, as we shall see, cultural processes provide a particularly powerful engine for human niche

[1] In contrast, human-behavioral ecologists and researchers studying gene-culture coevolution tend to place much greater emphasis on human *adaptability*, especially that resulting from ontogenetic and cultural processes. Nonetheless, virtually all of the work carried out in these traditions is squarely within the framework of standard evolutionary theory.

construction. Moreover, this dual role for phenotypes in evolution does imply that a complete understanding of the relationship between human genes and cultural processes must not only acknowledge genetic inheritance and cultural inheritance, but also take account of the legacy of modified selection pressures in environments, or ecological inheritance. Again, it is readily apparent that contemporary humans are born into a massively constructed world, with an ecological inheritance that includes a legacy of houses, cities, cars, farms, nations, e-commerce, and global warming. Niche construction and ecological inheritance are thus likely to have been particularly consequential in human evolution.

6.2 THE RELATIONSHIP BETWEEN EVOLUTIONARY AND CULTURAL PROCESSES

The relationship between genetic evolution and cultural processes raises two causal issues. The first concerns the extent to which contemporary human behavior and society are constrained or directed by our biological evolutionary heritage. The second concerns whether hominid genetic evolution has itself been influenced by cultural activities.

Evolutionary biology has often been invoked to provide biological explanations for human behavior and social institutions. These explanations have spawned human sociobiology, human behavioral ecology, evolutionary psychology, as well as evolutionary approaches to archaeology, economics, medicine, and other subjects. They have also been expressed outside of academia, through applications such as social Darwinism (Kuper 1988). However, evolutionary approaches to human behavior have typically provoked strong opposition, and the relevance of biological evolution to the human sciences remains widely disputed (Sahlins 1976; Segerstrale 2000; Laland and Brown 2002).

Less familiar, but equally deserving of attention, are empirical data and theoretical arguments suggesting that human cultural activities have influenced human genetic evolution by modifying sources of natural selection and altering genotype frequencies in some human populations (Bodmer and Cavalli-Sforza 1976; Feldman and Cavalli-

Sforza 1976; Boyd and Richerson 1985; Durham 1991; Feldman and Laland 1996). Cultural information, expressed in the use of tools, weapons, fire, cooking, symbols, language, agriculture, and trade, may also have played an important role in driving hominid evolution in general, and the evolution of the human brain in particular (Holloway 1981; Byrne and Whiten 1988; Dunbar 1993; Aiello and Wheeler 1995; Diamond 1998; Klein 1999; Wrangham et al. 1999). There is evidence that some cultural practices in contemporary human societies continue to affect ongoing human genetic evolutionary processes (Cronk et al. 2000; Smith 2000).

We contend that these issues are inextricably linked: the significance of evolutionary theory to the human sciences cannot be fully appreciated without a more complete understanding of how organisms, and human beings in particular, modify significant sources of natural selection in their environments, thereby codirecting subsequent biological evolution. Our principal goal in this chapter is to explore the two-way interactions between biological evolution and cultural change. We consider both the influence of human genetics on human cultural processes, and the influence of cultural processes on human genetic evolution, through the cultural modification of natural selection in culturally changed environments.

Figure 6.1 depicts how three independent approaches, human sociobiology, contemporary gene-culture coevolutionary theory, and our own proposed extension of gene-culture coevolutionary theory, model the interactions between biological evolution and cultural processes (based on Laland et al. 2000b). We will work through each.

6.2.1 Human Sociobiology, Human Behavioral Ecology, and Evolutionary Psychology

Historically, evolutionary theories have suggested two means by which feedback from an organism's activities could affect its evolution: the Lamarckian process of the inheritance of acquired characteristics and the Darwinian process of natural selection. For our own species, these correspond to the ideas that either our activities may directly change the genes that humans pass on to their descendants by generating mutations that are guided by experience, or they merely change the probabilities that different humans survive and reproduce. The first alternative was ruled out by the failure of Lamarckism. The so-called Weismann barrier effectively stops genes from being affected by any of the acquired characteristics of phenotypes, including the culturally acquired characteristics of human beings (Mayr 1982). True, modern molecular biologists do interfere with genes directly on the basis of their acquired scientific experiences, but this innovation is too recent to have had any impact on human genetic evolution. The failure of this Lamarckian route therefore left only the second Darwinian alternative, which has encouraged sociobiology's claims that phenotypes of all species, including humans, reduce to "survival machines" or "vehicles" for their genes (Dawkins 1989) and that the only role of phenotypes in evolution is to survive and reproduce differentially in response to natural selection and chance. This subordinate status for phenotypes does not cut off human cultural processes from human genetic evolution entirely, as it still allows cultural processes to contribute to human adaptations and adaptive behavior (Alexander 1979), and hence to genotypic fitness. However, cultural processes are regarded as simply another component of the human phenotype, a distinctive component perhaps, but no different from, say, our naked skin or bipedal manner of locomotion.

This perspective is described by figure 6.1a, which represents much of human sociobiology (Wilson 1975; Alexander 1979; Trivers 1985) and related disciplines such as evolutionary psychology (Barkow et al. 1992) and human behavioral ecology (Cronk et al. 2000).

These approaches encompass an array of different views of cultural processes. In most cases, human cultural processes are regarded as little different from any other aspect of the human phenotype. Yet for some sociobiologists and human-behavioral ecologists, cultural processes are part of the unusually broad and flexible evolved behaviors that are elicited by ecological conditions (Alexander 1979; Smith 2000; Cronk et al. 2000). Humans are assumed to be predisposed to learn that which maximizes their inclusive fitness by satisfying various proximal goals. For other sociobiologists and evolutionary psychologists (e.g., Tooby and Cosmides 1990; Barkow et al. 1992), human behavior is composed largely of cultural universals closely tied to our biological nature. Most evolutionary psychologists believe

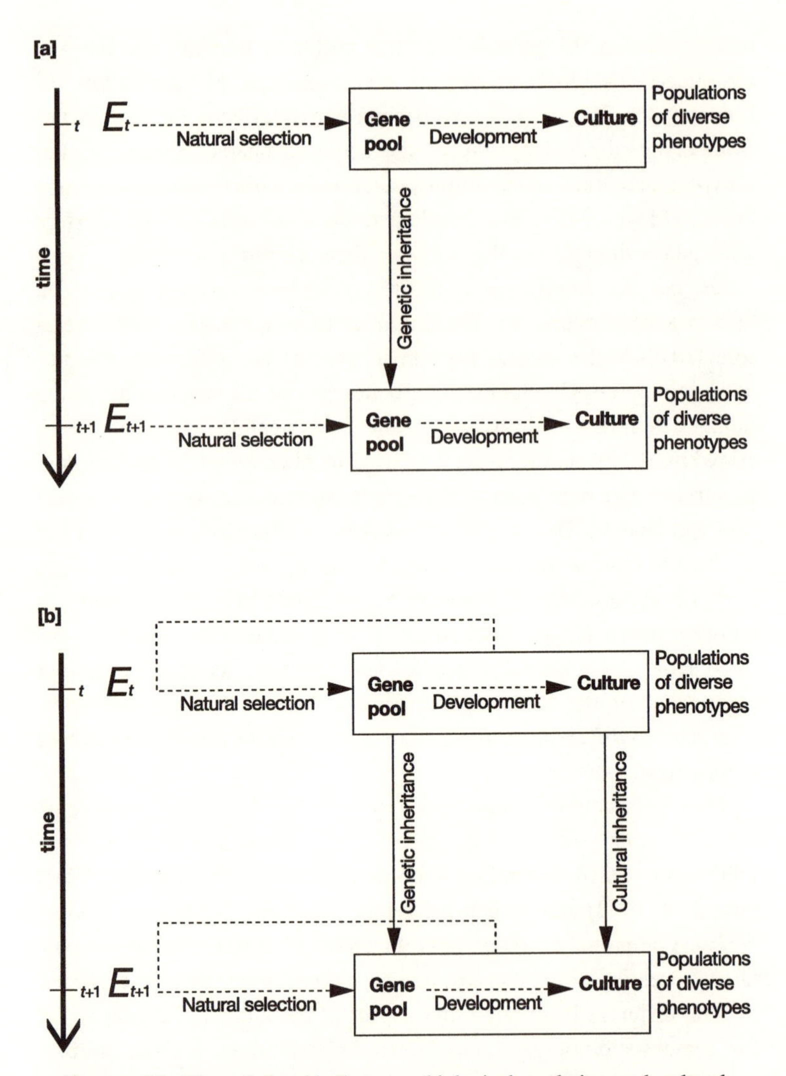

FIGURE 6.1. The relationship between biological evolution and cultural change. (a) *Sociobiology and related approaches*: Culture is treated like any other feature of the phenotype. Each generation, populations of diverse phenotypes in their environment ($\mathbf{E}_t$) are subject to natural selection that influences which genes are passed to the next generation. These genes may be expressed throughout development and may affect the characteristics of the population's culture. Cultural inheritance is regarded as inconsequential, and is not included. (b) *Gene-culture coevolutionary theory*: As before, except culture is treated as shared ideational phenomena (ideas, be-

that human minds are organized into complex, evolved information-processing structures that channel learning aptitudes toward that which was beneficial in ancestral environments (Tooby and Cosmides 1990). In all cases, cultural diversity is generally assumed to reflect variation in the environments in which humans develop, although some human sociobiologists have also suggested that genetic variation underlies behavioral differences between societies (Wilson 1975; Lumsden and Wilson 1981).

These approaches to cultural processes are built upon the standard evolutionary viewpoint portrayed in figure 1.2a in chapter 1. Here the potential interactions between biological evolution and cultural change are comparatively straightforward. The only way in which cultural processes can affect genetic evolution is by influencing the fitness of individual organisms, and hence the probability that different individ-

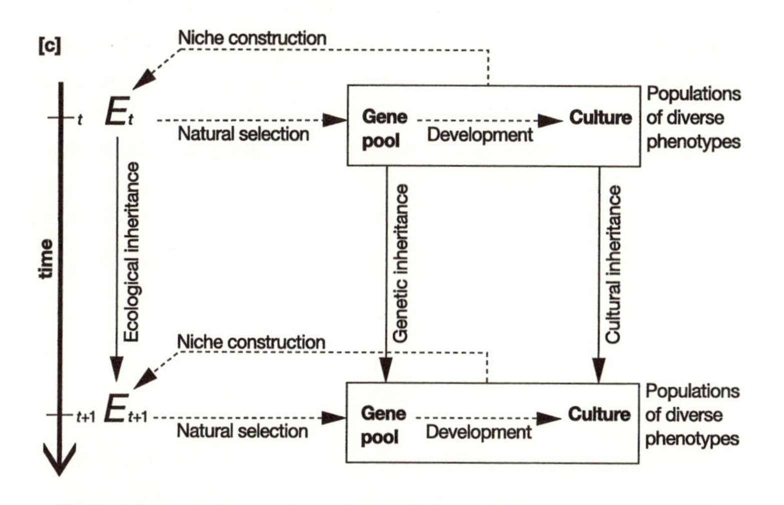

liefs, knowledge) that are learned and socially transmitted between individuals as a cultural inheritance. Cultural activities may modify some natural selection pressures in human environments and thereby affect the frequencies of some selected human genes. (c) *Extended gene-culture coevolutionary framework*: As before, except niche construction from all ontogenetic processes modifies human selective environments, generating a legacy of modified natural selection pressures that are bequeathed by human ancestors to their descendants. (From Laland et al. [2000b], fig. 2. Reprinted with the permission of Cambridge University Press.)

uals in a population will survive and reproduce to pass on their genes to the next generation. However, this initial conceptual model leaves us with a rather restricted understanding of how human genetic evolution interacts with human cultural change. For instance, as admitted by Wilson (1994), these perspectives largely neglect cultural inheritance and ignore the fact that cultural activities can modify human environments and feed back to affect selection on the population.

The scheme portrayed in figure 1a has also fostered the contrary view, maintained by many of the critics of sociobiology (Sahlins 1976; Montagu 1980), that, at least in modern humans, cultural inheritance is so powerful that in many cases it no longer interacts with genetic inheritance at all, but overrules it. This position fails to explain data indicating that, to varying degrees, human cultural processes are only possible because of human genetic aptitudes. Cultural processes need a priori knowledge in the form of evolved, genetically encoded information to get started (Daly and Wilson 1983; Durham 1991; Barkow et al. 1992; Smith 2000; Cronk et al. 2000). For example, there is now considerable evidence that the capacity for language is the result of biological adaptations (Pinker 1994; Deacon 1997). In this sense, the existence of cultural processes cannot always be meaningfully decoupled from some basic biology that is specific to *Homo sapiens*. This is not to say that there should be a connection between cultural and genetic variation in humans, although this may sometimes exist.

6.2.2 Gene-Culture Coevolution

According to gene-culture coevolutionary theory as portrayed in figure 6.1b (Cavalli-Sforza and Feldman 1973a,b, 1981; Feldman and Cavalli-Sforza 1976, 1979, 1984; Boyd and Richerson 1985; Durham 1991; Smith 2000), cultural activities are assumed to affect the evolutionary process by modifying selection pressures. Here, mathematical and conceptual models describe not only how human genetic evolution influences cultural evolution, but also how human cultural change can drive or codirect genetic evolution in human populations (Cavalli-Sforza and Feldman 1981; Boyd and Richerson 1985; Durham 1991). These models include investigations of the evolution of language (Aoki and Feldman 1987, 1989) and of handedness (Laland et al. 1995), the emergence of incest taboos (Aoki and Feldman 1997), changes in the genetic sex ratio in the face of sex-biased parental investment (Kumm et al. 1994), the coevolution of hereditary deafness and sign language (Aoki and Feldman 1991), sexual selection with a culturally transmitted mating preference (Laland 1994), and cultural group selection (Boyd and Richerson 1985). The common element among all these models, and part of what distinguishes gene-culture coevolutionary theory from classical sociobiology, human behavioral ecology, and evolutionary psychology, is that cultural processes modify the selective environment to which humans are exposed and may affect which genetic variants are transmitted. There is no assumption about the direction of causality, and genetic and cultural variation are viewed symmetrically.

Gene-culture coevolution is still based on standard evolutionary theory, except that here the interactions between human genetic evolution and cultural processes become richer. Human populations share information (ideas, beliefs, values) that is learned, expressed in cultural activities, and transmitted socially between individuals in the form of a cultural inheritance. This concept of culture abandons more diffuse and all-encompassing notions of culture that in the past have been common in the human sciences (e.g., Tylor 1871), while building on ideational and cognitive perspectives, in an attempt to operationalize the units of cultural transmission (Durham 1991). The novelty of the gene-culture approach is that it assumes that some human cultural activities may feed back to modify some natural selection pressures in human environments, and thus cultural transmission may affect the fate of some human genes (Feldman and Cavalli-Sforza 1976; Cavalli-Sforza and Feldman 1981; Boyd and Richerson 1985; Durham 1991; Feldman and Laland 1996), or even those of other species (Palumbi 2001a,b). In this theory, genetic- and cultural-transmission systems are not treated as independent; what an individual learns may depend on its genotype, or on the cultural traits of its parents, or both, and the selection acting on genes may be generated or modified by the spread of a cultural practice. Thus, in figure 6.1b, the relevant aspect of human selective environments is defined as cultural. This selection arises from the impact of cultural activities on human environments and is sufficient to affect human evolution.

Cultural-transmission systems (Cavalli-Sforza and Feldman 1981; Boyd and Richerson 1985) provide humans with a second, nongenetic "information-carrying" inheritance system. If the cultural inheritance of an environment-modifying human activity persists for enough generations that it mediates a stable selection pressure, it may be able to codirect human genetic evolution. The culturally inherited traditions of pastoralism provide a case in point. Apparently, the persistent domestication of cattle, and the associated dairying activities, did alter the selective environments of some human populations for sufficient generations to select for genes that today confer greater adult lactose tolerance (Feldman and Cavalli-Sforza, 1989; Durham 1991).

Gene-culture coevolutionary theory explicitly recognizes two distinct processes via which information that is assumed to be fully integrated throughout development (Smith 2000) is transmitted between generations. The dual inheritance is a way to include interactions between nature and nurture in a tractable system. Classical sociobiology, human behavioral ecology, and evolutionary psychology all assume that, given environmental variability, genetic inheritance and biological evolution are sufficient processes to explain cultural diversity, and that the formal modeling of transmitted and inherited cultural information is an unnecessary complication. However, several gene-culture coevolutionary analyses have provided evidence that conventional models of biological inheritance do not explain data as well as gene-culture models (Otto et al. 1995; Laland et al. 1995) and can lead to erroneous conclusions (Kumm et al. 1994; Feldman and Cavalli-Sforza 1989). Moreover, these interactions between genetic and cultural transmission can change the evolutionary process, for instance, by generating a new form of group selection (Boyd and Richerson 1985).

It is frequently argued that genetic evolution is too slow, and cultural change too fast, for the latter to drive the former (e.g., Adenzato 2000). However, as we saw in chapter 5, selection experiments and observations of natural selection in the wild have led to the conclusion that biological evolution may be fast, with significant genetic and phenotypic change sometimes observed in just a handful of generations (e.g., Dwyer et al. 1990; Grant and Grant 1995a,b; Reznick et al. 1997; Thompson 1998; Ehrlich 2000; Kingsolver et al. 2001). Furthermore, observations of hominid stone-tool technologies reveal that cultural change may at times have been extremely slow (Klein 1999). Cultural institutions such as labor markets can also be extremely persistent, albeit on a shorter time scale (Bowles 2000). Furthermore, theoretical analyses have revealed that cultural transmission may change selection pressures to generate unusually fast genetic responses to selection in humans (Feldman and Laland 1996). Thus genetic and cultural evolution could conceivably operate at similar rates.

Cultural niche construction can, of course, cause rates of environmental change that really are too fast for human genetic evolution to track (see below). In the last 25 to 40 thousand years the dominant mode of human evolution has probably been purely cultural. However, that does not mean there has been no evolutionary feedback from niche construction; it merely switches the evolutionary responses to the cultural domain. Under such circumstances, cultural niche construction should have favored further cultural transmission, or coevolution between subsets of culturally transmitted information, sometimes called memes (Aunger 2000a).

The conceptual model in figure 6.1b extends figure 6.1a, yet it still oversimplifies the causal pathways connecting genes and cultural processes, because it requires that cultural inheritance should affect the fate of some human genes directly, in the absence of any other mediating process. In most cases where gene-culture coevolutionary theory has been applied, this assumption is reasonable. Cultural processes may bias human mating patterns, they may bias other human interactions, such as trade or warfare, or they may bias the choice of which infants are selected for infanticide (Boyd and Richerson 1985; Laland 1994; Kumm et al. 1994). The assumption that human cultural inheritance can directly bias human genetic inheritance may also be reasonable even when the source of the natural selection pressure that is modified by cultural activities is no longer human, provided the relationship between whatever cultural information is being expressed and whatever natural selection pressure it is modifying is sufficiently direct. For example, the information that affected human genetic evolution in the lactose-tolerance case was that expressed in milk usage (Durham 1991). Here, gene-culture theory is again applicable, because the link between milk usage and its genetic consequences is sufficiently simple to allow it to be modeled without bringing in any intermediate variables (Feldman and Cavalli-Sforza 1989). However, in the next section we describe circumstances where gene-culture coevolutionary analyses are not adequate.

The gene-culture coevolutionary approach is explicitly species-specific. Although other species of animals have their "protocultures" (Galef 1988), it has generally been assumed that *Homo sapiens* is the only extant species with cultural transmission stable enough to co-direct genetic evolution (Boyd and Richerson 1985). If this were really the case, "cultural" processes could be used to explain little in primate evolution that happened prior to the appearance of powerful, accumulatory cultural inheritance. We think this particular human-centered perspective is misleading. Humans may be unique in their extraordinary capacity for cultural processes, but they are not unique in their capacity to modify their environments and hence the way in which they may be affected by natural selection. As chapter 2 demonstrates, countless other species do the same. We suggest that a deeper understanding of potential evolutionary relationships between genes and cultural processes can be facilitated by accepting that humans are far from unique in being able to change their own selective environments. Human cultural processes may allow humans to modify and construct their niches with spectacular social and ecological consequences, but in evolutionary terms they represent just one more set of processes in one particular species that results in niche construction.

6.2.3 Gene-Culture Coevolution plus Niche Construction

The gene-culture coevolutionary approach, while an advance on sociobiology, is still a simplification of the relationship between biology and cultural processes. Take, for example, the case of Kwa-speaking yam cultivators in West Africa, among whom the frequency of a hemoglobin allele that causes sickle-cell anemia increased as a result of the indirect effects of yam, or possibly cassava, cultivation

(Jackson 1996). These people traditionally cut clearings in the rain forest to grow their crops, creating more standing water and increasing the breeding grounds for malaria-carrying mosquitoes. This, in turn, intensifies selection for the sickle-cell allele, because of the protection offered by this allele against malaria in the heterozygous condition (Durham 1991). Here the causal chain is so long that simply plotting the cultural practice of yam cultivation against the frequency of the sickle-cell allele would be insufficient to yield a clear relationship between the cultural practice and allele frequencies (Durham 1991). The crucial variable is probably the amount of standing water in the environment caused by the yam cultivation, but standing water is an ecological variable, not a cultural variable, and it partly depends on factors (i.e., rainfall) that are beyond the control of the population. So here the simplifying assumption of a direct link between cultural and genetic inheritance distorts reality too much to allow their interaction to be modeled in the standard way. In this case the two human inheritance systems can only interact via an intermediate, abiotic, ecological variable, which should be included to complete the model (Laland et al. 2000b). This example is unlikely to be an exception, both because human modification of water conditions affects the transmission of other parasitic diseases (schistosomiasis, river blindness, etc.) that may exert selective pressures on the genes coding for functions of the immune system (Combes 1995; Klein 1999) and because there are countless other ways in which humans have modified their selective environments.

This leads us to propose an extended, or "triple-inheritance," version of gene-culture coevolutionary theory, a conceptual account of which is shown in figure 6.1c. The novelty here is the replacement of the genetic-inheritance scheme, described by standard evolutionary theory, as the proper basis of gene-culture coevolution, by the extended evolutionary scheme incorporating niche construction, portrayed in figure 1.2b. Here, niche construction from all ontogenetic and cultural processes modifies human selective environments (see below). Culturally modified natural selection pressures are now regarded not as unique, but as just a part of a more general legacy of modified selection pressures that are bequeathed by niche-constructing human ancestors to their descendants. Hence, instead of being exclusively responsible for allowing us to codirect our own evolution, cultural processes now become merely the principal way in which we humans do the same thing as other species.

This triple-inheritance version of human gene-culture coevolution differs from the earlier dual-inheritance versions in several respects. Two of the inheritance systems in figure 6.1c, genetic and cultural inheritance, are the same as in figure 6.1b. Now, however, genetic inheritance is directed by natural selection stemming from every kind of niche construction, and not just cultural niche construction. However, the third inheritance system in figure 6.1c, ecological inheritance, extends gene-culture coevolutionary theory. Ecological inheritance is explicitly directed by niche construction, and it potentially includes human artifacts (Aunger 2000a). It follows that human cultural inheritance may influence human genetic inheritance in two ways instead of one: first, directly, by influencing differential survival and reproduction, as already assumed by sociobiology, human behavioral ecology, and evolutionary psychology (fig. 6.1a), and second, indirectly, by contributing to cultural niche construction, and thence to a human ecological inheritance that includes culturally modified natural selection pressures. These modified selection pressures may then feed back to select for different human genes. Finally, the explicit inclusion of the mechanisms of niche construction in figure 6.1c may have another practical consequence. Smith (2000) criticized the dual-inheritance versions of gene-culture coevolution for being "theoretically rich" but "empirically impoverished" (p. 32). It is a valid criticism. Unless the mechanisms of niche construction are included, it is frequently difficult to pin down the empirical implications of gene-culture coevolutionary theory. Adding niche construction should make empirical research easier in future (see chapter 9).

6.3 MULTIPLE PROCESSES IN EVOLUTION

In this section we take a closer look at the set of processes by which populations of complex organisms, such as humans, acquire adaptive information, and how this information is expressed in niche construction. It is now widely recognized that several of the major evolutionary transitions involved changes in the way information is acquired,

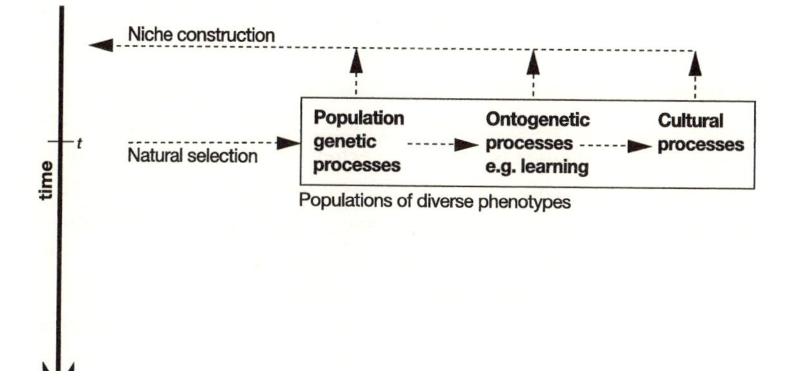

FIGURE 6.2. This figure zooms in on one of the boxes labeled "populations of diverse phenotypes" in figure 6.1. In each generation, populations of phenotypes are subject to natural selection. Genetic information is expressed throughout development and may affect each individual's ontogeny and possibly components of culture. Human behavior results from information-acquiring processes at three levels. Niche construction in humans depends on population-genetic processes, information-acquiring ontogenetic processes, and cultural processes, all of which can influence the selective environment. (From Laland et al. [2000b], fig. 3. Reprinted with the permission of Cambridge University Press.)

stored, and transmitted (Szathmary and Maynard Smith 1995). This is reflected in figure 6.2, which shows how populations of complex organisms can acquire relevant *semantic information* through a set of information-acquiring processes (J. Holland 1992) operating at three different levels (population genetic, ontogenetic, cultural), and with the information gained, interacting with niche-constructed environments at each level. In various combinations, these are the processes that supply all organisms with the knowledge that organizes their adaptations. Every species is informed by naturally selected genes. Many are also informed by complex, information-acquiring ontogenetic processes such as learning or the immune system, while hominids, and perhaps a few other species, are also informed by cultural processes. Any comprehensive treatment of the gene-culture relationship should include all three sets of processes because the links between genetic and cultural evolution cannot be understood without some reference to the intermediate ontogenetic processes, such as individual learning, that connect them (Plotkin and Odling-Smee 1981; Boyd and Richerson 1985; Durham 1991; Feldman and Laland 1996).

Genetic processes, ontogenetic processes, and cultural processes operate at three distinct but interconnected levels. Each level interacts with but is not completely determined by the others: that is, learning is informed, but only loosely, by genetic information, and cultural transmission may be informed, but not completely specified, by both genetic and developmental processes. Genes may affect information gain at the ontogenetic level, which in turn influences information acquisition at the cultural level. In addition, ontogenetic processes, particularly learning, may be affected by cultural processes, while population-genetic processes may be affected by both ontogenetic processes and cultural processes when humans modify their selection pressures. We spell out the relationships between these levels in more detail below.

6.3.1 Population-Genetic Processes

Darwinian evolution, based as it is on descent with modification and variation in the ability to survive and reproduce, comprises a natural algorithm for the acquisition of semantic information by organisms (Campbell 1960, 1974; Odling-Smee 1983; Dennett 1995; Hull et al. 2001; and see chapter 4). The genes that are inherited by each generation must necessarily reflect whatever selective biases were imposed on their ancestors by the blind sorting decisions of natural selection, according to the characteristics of the ancestors' environments. This process of genetic evolution is the most phylogenetically ancient process responsible for niche construction. As a consequence of the differential survival and reproduction of individuals with different genotypes, genetic evolution results in the acquisition, inheritance, and expression of genetically encoded information by individuals in populations. At this level, and this level alone, the variants that are subject to selection, that is, the genetic mutations, are entirely random (or, at least, random with respect to the selection process). Mutations with phenotypic effects that are beneficial will commonly be retained in the knowledge store of the gene pool and transmitted to the next generation, while those with negative repercussions will typically be weeded out. Each individual inherits this genetic information from its ancestors, and this information translates into developmental processes, expressing different phenotypes in different environments, the so-called "norm of reaction." This genetic information is the most fundamental source of information that underpins niche construction.

6.3.2 Information-Gaining Ontogenetic Processes

Many species have also evolved a set of more complicated ontogenetic processes that allow individual organisms to acquire other kinds of information. These more complicated processes are themselves products of genetic evolution, but they are unusual products because, rather than expressing adaptations, they function to accumulate further semantic information relative to the local environments of individuals. For example, specialized information-acquiring subsystems, such as the immune system in vertebrates or brain-based learning in animals, are capable of additional, individual-based, information acquisition. These secondary sources of information are complementary to the first; for instance, learning allows individual organisms to fine-tune their behavior to the idiosyncracies of their local circumstances in a manner that would have been impossible on the basis of inherited genetic knowledge alone.

Some factors in the environment can potentially change many times within the typical life span of the animal concerned. The fundamental Darwinian process of the natural selection of genetic variation in populations cannot furnish individual organisms with specific adaptations for each of these environmental contingencies. What it can do, however, is select for supplementary processes that permit characteristics of the phenotype to adjust on a within-lifetime basis. For instance, the vertebrate immune system is effective precisely because it is *not* clear which pathogen will confront individual animals during their lifetimes, if only because viruses and bacteria can evolve so much faster than vertebrates. Nevertheless, the vertebrate immune system's information-acquiring capability allows it to track rapidly changing pathogens and to determine which antibodies the individual animal should produce. Similarly, and analogously to the way in which military engineers have designed "smart" missiles with on-board guidance systems that "learn" about their target and adjust the missile's trajectory while in flight, natural selection has furnished organisms with onboard guidance systems that allow them to learn about their environment and, within some constraints, to adjust their behavior accordingly during their lives. These onboard guidance systems work through built-in processes such as the "law of effect" that guides learning in animals. This law states that actions that are followed by a positive outcome are likely to be repeated, while those followed by a negative outcome will be eliminated. Thus, natural selection is more likely to confer on animals a *feel-good* sensation when they act in a manner that enhances survival and reproduction, and a *feel-bad* sensation when they act in a way that is detrimental to survival and reproduction. As a consequence, individual animals can fine-tune their behavior by learning in a broadly adaptive manner, as they do things that elicit pleasure and avoid actions that elicit pain.

Ontogenetic processes such as learning and the immune response can also be regarded as operating in a manner loosely analogous to the Darwinian algorithm. In each case, variants (behavior patterns or antibodies) are produced, their utility is evaluated (e.g., their performance at generating pleasure or avoiding pain, or binding to antigens, is assessed by some kind of system that natural selection has previously selected), and those variants that are most effective are retained while the others are selected out. As a result these processes acquire and store information about the behavioral patterns or antibodies that are most effective in dealing with the pathogens in the local environment, and through a process frequently described as "trial-and-error" learning, individuals repeatedly generate behavioral variants, test them, eliminate the behavioral errors that produce pain or no positive feelings, and regenerate the successes that elicited more positive sensations. As a result of this loosely Darwinian process, relevant information pertaining to survival and reproduction is acquired and stored, while the behavior of individual animals is shaped to be functional and adaptive.

There are, however, at least three respects in which these ontogenetic information-gaining processes are not strictly Darwinian. The first is that they rely on evolved aptitudes that generate positive and negative sensations for behavior patterns that are broadly adaptive and maladaptive, respectively. These aptitudes are themselves adaptations resulting from the action of natural selection operating at the population-genetic level. Those associations and patterns of behavior that animals do learn critically depend on which stimuli are reinforcing under the influence of species-typical motivational and perceptual processes that are informed by genes (Hinde and Stevenson Hinde 1973; Plotkin and Odling-Smee 1981).

Second, while the variants that occur during genetic evolution (i.e., mutations) are random (or, at least, blind relative to natural selection), those acquired through ontogenetic processes are not. We have previously described such directed variation as *smart variants* (Laland et al. 2000b). For example, animals may be genetically predisposed to respond to both specific internal cues (e.g., hunger) and environmental contexts (e.g., sensory cues indicating food is nearby) by generating appropriate behavior patterns from their repertoires. Hence, during learning, animals typically demonstrate inherited a priori biases in their associations and patterns of behavior that are likely to be adaptive (Garcia et al. 1966; Seligman 1970; Bolles 1970). These biases influence the behavior of each individual animal, the associations it forms, the antibodies it generates, or the developmental pathways it takes, usually in the direction of being functional and adaptive.

Third, because the information-acquiring entity for these ontogenetic processes is no longer an evolving population, but is each individual organism in a population, the adaptive knowledge acquired cannot be inherited by successive generations. All the knowledge gained by individuals during their lives is erased when they die. However, processes such as learning can still be of considerable importance to subsequent generations because learned knowledge can guide niche construction, the consequences of which can be subject to ecological inheritance (fig. 1.3.b). In this respect, learning provides a second source of semantic information that can be expressed in niche construction (fig. 6.2).

This highlights one of the major differences that niche construction makes to the evolutionary process: acquired characteristics can play a role in evolution through their influence on the selective environment, in other words, through niche construction. The Galápagos woodpecker finch's use of tools, which we introduced in chapter 1, illustrates this point (Alcock 1972; Tebbich et al. 2001). These birds create a woodpecker-like niche by learning to use a tool such as a cactus spine to peck for insects under bark. One can think of this behavior as the consequence of an internal selective process operating at the ontogenetic level, in individuals, rather than at the genetic level, in populations. Like countless other species, this finch has apparently exploited the more general and flexible adaptation of learning to develop the skills it needs to grub in environments that reliably contain cactus spines and similar implements. Its learning certainly opens up resources in the bird's environment that would be unavailable to it otherwise and is therefore an example of niche construction. By modifying natural selection pressures, the same behavior probably also created a stable selection pressure favoring a bill able to manipulate tools rather than the sharp, pointed bill and long tongue characteristic of more typical woodpeckers, and it may have created selection pressures favoring enhanced learning capabilities as well.

6.3.3 Cultural Processes

A few species, including some vertebrates, have evolved a capacity to learn from other individuals. In humans this ability is facilitated by a further set of processes (e.g., language) that collectively underlie cultural processes. Cultural processes add a second knowledge inheritance system to the evolutionary process through which socially learned information is accrued, stored, and transmitted between individuals both within and between generations. Here, the information-acquiring entity is again a group of interacting organisms rather than an individual. Although all cultural knowledge is traceable to innovation and learning by particular individuals (with ontogenetic processes the ultimate source), major cultural changes may occur through learning from neighboring groups or immigrants and may be associated with ideological or organizational requirements.

Cultural change can also be regarded as loosely Darwinian in character, in the sense that cultural variants are generated by individuals, and as a result of social learning are culturally "selected" through their differential adoption. Cultural processes, like biological evolution, may accumulate functional responses to the environment.

Thus an analogue of the Darwinian algorithm can be regarded as operating at the cultural level, in addition to the population-genetic and individual-development levels. Most of the time, cultural processes can be regarded as a shortcut to acquiring adaptive information, as individuals rapidly learn, or are shown, what to eat, where to live, or how to avoid danger by doing what other more knowledgeable individuals do. Experienced others such as parents are a reservoir of "smart variants," allowing naive individuals to shortcut the many iterations of ontogenetic selection necessary to learn for themselves behavior patterns appropriate to their environment, and leapfrog to the functional and already tested solutions established by others.

Within a population, individuals share at least some of their learned knowledge with others, within and between generations. This information sharing can depend on several kinds of cultural inheritance, including vertical (from parents), horizontal (from peers), oblique (from unrelated older individuals), indirect (e.g., from key individuals), and frequency-dependent (e.g., from the majority) cultural transmission (Cavalli-Sforza and Feldman 1981; Boyd and Richerson 1985). Cultural inheritance therefore requires nongenetic channels of communication among organisms through which learned knowledge can be shared and spread. It probably also requires that organisms can decompose their store of cultural knowledge into discrete transmittable "chunks" or packages of information (Feldman and Cavalli-Sforza 1976; Dawkins 1989; Aunger 2000a), perhaps equivalent to a psychologist's "schemata," in either simple or compound form (J. Holland et al. 1986; Plotkin 1996). However, cultural processes may in some cases exhibit blending inheritance (Boyd and Richerson 1985), leaving the fidelity of cultural "units" open to question (Sperber 2000).

Cultural and biological evolution differ in other respects too, and many of the same caveats that pertain to the ontogenetic level apply again at the cultural level. For instance, human beings do not create or adopt cultural variants at random, since past phylogenetic and developmental aptitudes, including past asocial and social learning, inform these creative and selective processes. Thus, cultural information is not expressed in random behavior but in the production of smart behavioral variants. This behavior is also directed by evolved predispositions and motivational biases (what Durham [1991] calls "primary values"), as well as by individual reinforcement histories and past associations. Therefore, with some caveats that we detail below, we expect that the ideas, values, and acquired knowledge that underlie human behavior and cultural traditions will usually, but not always, be adaptive (Cavalli Sforza and Feldman 1981).

However, the adoption of cultural variants is also affected by socially transmitted cultural values, or "secondary values" (Durham 1991), arising through collective experience and social history, such as rules of thumb, proverbs, conventions, moral or ethical principles, and other information accrued through prior social learning. In addition, cultural evolution often involves social learning processes that are rare or unheard of at the genetic level. For instance, much human (and animal) social learning is characterized by a positive frequency dependence or conformity (Boyd and Richerson 1985), in which individuals bias their adoption of cultural information toward that expressed by the majority. In fact, a theoretical analysis by Boyd and Richerson (1985) found that most of the conditions under which natural selection favors social learning also favor the evolution of conformity. This "when in Rome do as the Romans" principle can result in conventions that only loosely track environmental change and, at least in the short term, may generate maladaptive traditions. In addition, members of a group may be particularly prone to adopting cultural variants exhibited by particularly authoritative or charismatic individuals, a process Boyd and Richerson (1985) describe as "indirect bias." Theoretical models have demonstrated that cultural processes can lead to the transmission of information that results in a fitness cost relative to alternatives (Cavalli-Sforza and Feldman 1981; Boyd and Richerson 1985), and strong cultural evolutionary processes will frequently be independent of genetic control. While socially learned smart behavioral variants will subsequently be tested by the individuals that adopt them, even nonreinforcing or maladaptive behavior may be expressed again if it is socially sanctioned, or if individuals are locked into conventions that penalize nonconformists. As a result, some cultural information may be propagated even when it is detrimental to individual fitness.

Much of human niche construction is guided by socially learned knowledge and cultural inheritance, but the transmission and acquisition of this knowledge is itself dependent on preexisting information acquired through genetic evolution, complex ontogenetic processes, or prior social learning. As a result, niche construction that is based on either learned or culturally transmitted information may be expressed "intentionally" relative to a specific goal. Some aspects of human cultural processes, for instance, the scientific process, clearly reflect this goal-directed, intentional quality. Individual scientists build on established knowledge stores in the scientific community and derive better explanations and superior solutions to problems in a highly directed manner. The extraordinary advances of human technology are another manifestation of goal-directed niche construction. Animal protocultures typically lack this directed and accumulatory quality. Other components of cultural processes, such as fads and fashions, are less clearly directed and may be subject to so many complex and frequency-dependent selective processes that their evolution is unpredictable and more difficult to describe quantitatively.

Cultural niche construction modifies selection not only at the genetic level, but also at the ontogenetic and cultural levels as well. By modifying the environment, niche construction creates artifacts and other ecologically inherited resources that not only act as sources of biological selection, but also facilitate learning and perhaps mediate cultural traditions (Aunger 2000a, 2002). For instance, the construction of villages, towns, and cities creates new health hazards associated with large-scale human aggregation, such as the spread of epidemics (Diamond 1998). Humans may respond to this novel selection pressure either through cultural evolution, for instance, by constructing hospitals, medicines, and vaccines, or at the ontogenetic level, by developing antibodies that confer some immunity, or through biological evolution, with the selection of resistant genotypes. As cultural niche construction typically offers a more immediate solution to new challenges, we anticipate that cultural niche construction will usually favor further counteractive cultural niche construction, rather than genetic change. However, where a culturally transmitted response is not possible, perhaps because the population lacks the requisite knowledge or technology, then a genetic response may occur. This is illustrated in figure 6.3. Consider the example of the Kwa-speaking yam cultivators of West Africa whose niche construction, in the form of agricultural activities, created breeding grounds for malaria-carrying

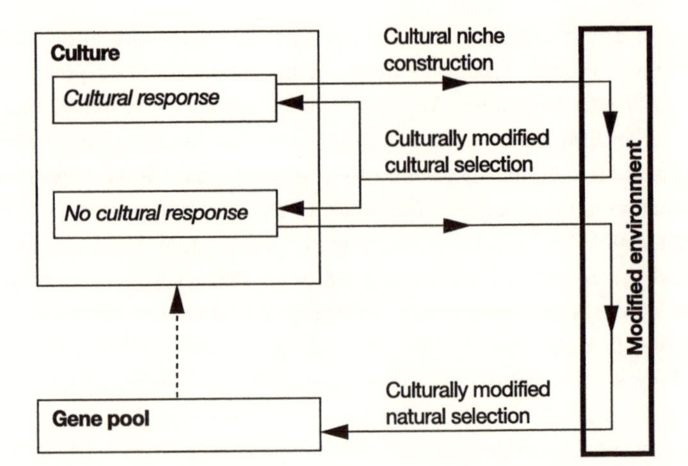

FIGURE 6.3. Culturally modified niche construction may result in a modification of the environment. This changed environment may lead to cultural selection; that is, individuals may adopt different cultural practices that are functional under the modified conditions. If the culture cannot adapt rapidly enough to the culturally modified environment, genetic change may become important. Here, to simplify this figure, processes operating at the ontogenetic level have been omitted.

mosquitoes, which generated a culturally modified selection pressure. If the population had understood the relationship between yam farming and malaria and had adapted its cultural processes accordingly, say with antimalarial drugs, there would have been no selective response at the genetic level. However, because the population failed to adapt to the new challenge at the cultural level, its modified selection pressures apparently persisted for long enough to increase the frequency of the sickle-cell allele, which, in the heterozygous condition, offers protection against malaria. A significant genetic response was the outcome.

It is apparent that the relationship between human cultural and biological evolution is considerably richer in the scheme based on niche construction (fig. 6.1c) than those based on standard evolutionary theory (figs. 6.1a and 6.1b), largely because of the proliferation of interactions between niche-constructing processes, biological evolution, and cultural processes. This proliferation is summarized in table 6.1, which illustrates each interaction with an example, organized in terms of the sources and consequences of niche construction.

TABLE 6.1. Sources of and Feedback from Niche Construction (examples)

Source of Niche Construction	*Feedback to Biological Evolution*	*Feedback to Cultural Change*
Population-genetic processes	Web spiders marking web or building dummy spiders (Edmunds 1974)	Sex differences in human mating behavior (Daly and Wilson 1983; Barkow et al. 1992)
Information-acquiring ontogenetic processes	Woodpecker finch, by learning to grub with a tool, alleviates selection for a woodpecker's bill (Grant 1986; Tebbich et al. 2001)	Learning and experience influence the adoption of cultural traits (Durham 1991)
Cultural processes	Dairy farming selects for lactose tolerance (Feldman and Cavalli-Sforza 1989)	Invention of writing leads to other innovations such as printing, libraries, e-mail

In 1871 Tylor defined culture as "that complex whole which includes knowledge, belief, art, morals, custom and any other capabilities and habits acquired by man as a member of society." In anthropological circles Tylor's rather cumbersome definition has been superseded (Aunger 2000a), but it still captures the intuitive notion of "culture" held by the layperson, and it presents a challenge for the evolutionary biologist. How could such an inextricably interwoven complex of ideas, behavior, and artifacts evolve? We believe that the framework presented in figures 6.1c and 6.3 is a step toward addressing that challenge. One advantage of this framework over the accounts developed by sociobiology and evolutionary psychology is that it explicitly incorporates culturally transmitted information (ideas), niche construction (behavior), and ecological inheritance (artifacts). As anthropologist Robert Aunger (2000a, p. 228) puts it: "Having three forms of inheritance (genes, memes and artefacts) is a means by which a sophisticated theory of mutual constraint relations between individual, societal and cultural replicator levels can be constructed within an explicitly evolutionary framework."

Our goal here is to provide a more useful and more acceptable evolutionary framework for the social and other human-oriented sciences than standard evolutionary accounts, but it is important that theoretical frameworks are sufficiently simple to be put to work. The utility of a conceptual model is usually assessed by its capacity to stimulate further theoretical advances or useful empirical work. The next section contains an illustrative example of how aspects of this new evolutionary framework can be translated into formal models. Later, in chapter 9, we present a number of examples of ways in which the framework developed here can be, and is being, tested empirically.

6.4 CULTURAL NICHE CONSTRUCTION AND HUMAN EVOLUTION

As described in the preceding section, most previous gene-culture coevolutionary models entail that the natural selection on genes that is generated by cultural activities is a function of the frequencies of the cultural variants in the population. For example, it may be assumed that the intensity of natural selection is directly proportional to the frequency of the cultural practice. While this is generally a reasonable assumption for those systems to which the models have been applied, these models cannot be used to explore such examples as how the farming practices of the Kwa of West Africa affect the frequency of a gene for sickle cell. In such cases, the human genetic and cultural inheritance systems interact via an intermediate, abiotic, ecological variable, which is required to complete the model (Laland et al. 2000b).

Given that cultural inheritance can exhibit modes of transgenerational transmission that differ from genetic inheritance, and may operate on different time scales, it is pertinent to ask how culturally generated niche construction may affect genetic evolution. Cultural processes have almost certainly been enhancing the human capacity for niche construction for many thousands, perhaps even millions, of years and have probably exposed human populations to changing environments at a faster rate than gene-based niche construction. Thus cultural processes that precipitate niche construction might be expected to have played a critical role in human evolution. Here we present the results of two gene-culture coevolutionary models that explore the evolutionary consequences of culturally transmitted niche construction (based on Laland et al. 2001). The two models presented here (models 3 and 4) are gene-culture analogues of the simple genetic model (model 1a) and the more complex ecological model (model 2) presented in chapter 3. As in the models in chapter 3, we explore the interaction between a niche-constructing trait, which affects the amount of a resource in the environment, and a recipient trait exposed to selection pressures that depend on the frequency of the resource. In models 3 and 4, however, the niche-constructing trait is a socially learned cultural trait, rather than a trait that results largely from genetic variation, while the recipient trait is under the influence of genes.

6.4.1 Models 3 and 4: Gene-Culture Models of Niche Construction

Consider a population of humans or human ancestors capable of the cultural transmission of information from one generation to the next. Individuals express one of two variants (E and e) of a package of culturally transmitted information (**E**). For simplicity we assume a one-to-one correspondence between the information that individuals acquire from others and its expression in behavior, although we recognize that this correspondence will not always be tight (Cronk 1995). As a convenient shorthand, we refer to the alternative behavior patterns E and e that result from socially transmitted information as "cultural states." Hence, E and e correspond to the presence or absence (or greater or lesser impact) of a socially learned niche-constructing behavior. We assume that the population's capacity for cultural niche construction is a function of the frequencies of individuals expressing states E or e. This cultural niche construction affects the amount R of a resource **R** in the environment. Thus, R (where $0 < R < 1$) is a function of the amount of cultural niche construction over n generations (that is, the frequency of cultural state E), as well as other independent processes of resource renewal or resource depletion. Genotypes at locus **A** (with alleles A and a) have fitnesses that are functions of the frequency with which the resource **R** is encountered by organisms in their environment. The three genotypes and two cultural states can be found in six possible combinations, or

TABLE 6.2. Fitnesses of Individuals with Cultural States E or e and Genotypes AA, Aa, and aa

	E (α_1)	e (α_2)
AA (η_1)	$w_{11} = \alpha_1\eta_1 + \varepsilon R$	$w_{12} = \alpha_2\eta_1 + \varepsilon R$
Aa (1)	$w_{21} = \alpha_1 + \varepsilon\sqrt{R(1 - R)}$	$w_{22} = \alpha_2 + \varepsilon\sqrt{R(1 - R)}$
aa (η_2)	$w_{31} = \alpha_1\eta_2 + \varepsilon(1 - R)$	$w_{32} = \alpha_2\eta_2 + \varepsilon(1 - R)$

phenogenotypes,[2] namely, *AAE*, *AAe*, *AaE*, *Aae*, *aaE*, *aae*, which have fitnesses w_{ij} (given in table 6.2). We shall use p to denote the frequency of allele A, and x for that of state E.

As in our earlier treatments, fitnesses of phenogenotypes are assumed to be functions of a fixed viability component and a frequency-dependent viability component. The fixed components (given by the α_i and η_i terms) represent selection from the external environment, independent of cultural niche construction. These are analogous to the fitnesses of genotypes in the standard two-locus multiplicative viability model. The frequency-dependent components of the contribution to fitness of genotypes AA, Aa, and aa are again functions of R, $\sqrt{R(1 - R)}$, and $1 - R$, respectively, chosen so that allele A will be favored by this component of selection when the resource is common, and allele a when it is rare. The coefficient of proportionality (ε) determines the strength of the frequency-dependent component of selection relative to the fixed fitness component, with $-1 < \varepsilon < 1$. The value of ε is positive if an increase in the amount of resource results in an increment in the fitness of genotypes containing allele A, and negative if an increase in R favors a. Again, positive and negative niche construction, respectively, refer to processes that increase and decrease R. Model 3 considers the simple case in which the amount of the resource is proportional to the amount of cultural niche construction over the previous n generations, as in model 1a. In model 4, R is also a function of independent processes of resource renewal and depletion, as in model 2. The mathematical structure of the resulting models is outlined in appendix 4.

[2] A more detailed discussion of the phenogenotype concept can be found in chapter 9.

TABLE 6.3. Cultural-Transmission Parameters

Matings	*E Offspring*	*e Offspring*
$E \times E$	b_3	$1 - b_3$
$E \times e$	b_2	$1 - b_2$
$e \times E$	b_1	$1 - b_1$
$e \times e$	b_0	$1 - b_0$

Vertical cultural transmission occurs according to standard rules (Cavalli-Sforza and Feldman 1981) given in table 6.3. In the following analyses we focus on three special cases of cultural transmission. The first is *unbiased transmission* (given by $b_3 = 1$, $b_2 = b_1 = 0.5$, $b_0 = 0$) where offspring adopt states in direct proportion to their parents' cultural state, and where transmission would not by itself change the frequency of the states in the population. The second is *biased transmission* (given by $b_3 = 1$, $b_2 = b_1 = b$, $b_0 = 0$, with $b \neq 0.5$), where the offspring of mixed or heterocultural matings (that is, father and mother exhibit different cultural states) preferentially adopt one of the states over the other. The third we call *incomplete transmission* (given by $b_3 = 1 - \delta$, $b_2 = b_1 = b$, $b_0 = \delta$, with $\delta > 0$), where only some of the offspring of parents with the same cultural state adopt that state.

The entries in tables 6.2 and 6.3 give rise to the recursions given by equations A4.1 in appendix 4.

As in chapter 3, we first describe the behavior of models 3 and 4 when there is no external source of selection, just selection resulting from the cultural niche construction. Then we consider externally imposed selection that favors one of the alleles, or one of the cultural states, and finally focus on the effects of cultural niche construction when the external selection generates overdominance at the **A** locus. The two models generate consistent findings, and consequently in the following section it can be assumed that the reported results apply to both models unless otherwise stated.

6.4.1.1 NO EXTERNAL SELECTION: $\alpha_1 = \eta_1 = \alpha_2 = \eta_2 = 1$

(i) Unbiased Cultural Transmission ($b_3 = 1$, $b_2 = b_1 = 1/2$, $b_0 = 0$). With unbiased cultural transmission, model 3 behaves exactly

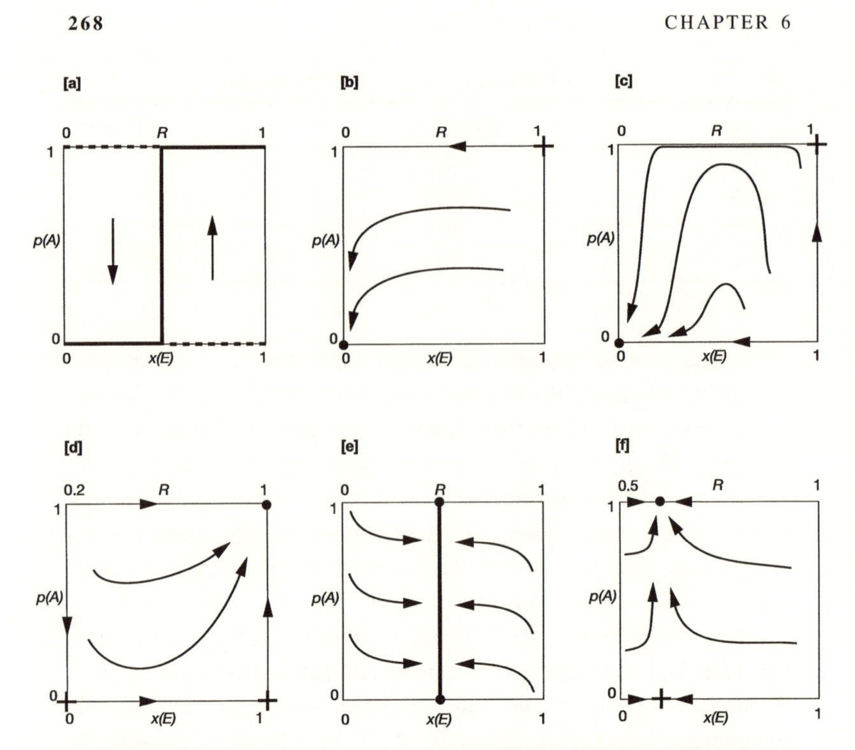

FIGURE 6.4. Dynamics (arrows), stable equilibria (thick lines, filled circles), and unstable equilibria (dashed thick lines, pluses) for systems with no external selection. Model 3 with (a) unbiased cultural transmission ($b_3 = 1$, $b_2 = b_1 = 0.5$, $b_0 = 0$), (b) strongly biased cultural transmission ($b_3 = 1$, $b_2 = b_1 = 0.25$, $b_0 = 0$), and (c) weakly biased cultural transmission ($b_3 = 1$, $b_2 = b_1 = 0.49$, $b_0 = 0$). (d) Model 4 with biased cultural transmission ($R_t = \lambda_1 R_{t-1} + \lambda_2 x_t + \lambda_3$, $\lambda_1 = 0.5$, $\lambda_2 = 0.4$, $\lambda_3 = 0.1$, $\gamma = 0$, $b_3 = 1$, $b_2 = b_1 = 0.55$, $b_0 = 0$). (e) Model 3 with incomplete transmission ($b_3 = 0.9$, $b_2 = b_1 = 0.5$, $b_0 = 0.1$). (f) Model 4 with incomplete transmission ($R_t = \lambda_1 R_{t-1} + \lambda_2 x_t + \lambda_3$, $\lambda_1 = 0.8$, $\lambda_2 = 0.1$, $\lambda_3 = 0.1$, $\gamma = 0$, $b_3 = 0.9$, $b_2 = b_1 = 0.25$, $b_0 = 0.1$). Note, as R depends on x, on the top axis of each figure we present the value of R that corresponds to the value of x given on the bottom axis, where x ranges from 0 to 1. In all cases $\alpha_1 = \eta_1 = \alpha_2 = \eta_2 = 1$, $\varepsilon = 0.3$, $n = 1$, $f = 1$, and $r = 0.5$.

like its genetic counterpart, model 1, with the same characteristic curve of E-frequency equilibria (fig. 6.4a), the formal details of which are given in appendix 1. Provided there is no gene-culture linkage disequilibrium, the amount of the resource is constant and given by the frequency of E (see appendix 4 for details). The cultural niche construction generates selection that changes the frequency of alleles at locus **A**. If niche construction increases the amount of the resource (ε is positive), then in the selective environment characterized by high frequencies of E ($x > 1/2$) allele A will be favored, while in the selective environment provided by low frequencies of E ($x < 1/2$) allele a will be favored. The reverse is true if niche construction decreases the amount of the resource R (if ε is negative).

Model 4 also behaves like the equivalent genetic model, model 2, generating a curve of polymorphic equilibria at $\hat{R} = 1/2$. With only positive niche construction ($R_t = \lambda_1 R_{t-1} + \lambda_2 x_t + \lambda_3$, $\gamma = 0$), for frequencies of state E from $x = 0$ to $x = 1$, the corresponding values of R range between 0 and 1, depending on λ_1, λ_2, λ_3. An expression for the equilibrium resource value $\hat{R}$ is given by equation A3.2 in appendix 3. With only negative niche construction [$R_t = \lambda_1 R_{t-1}(1 - \gamma x_t) + \lambda_3$ and $\lambda_2 = 0$], for values of x increasing from 0 to 1, R decreases in an interval of values contained in [0,1], depending on the values of λ_1, λ_3, and γ. The equilibrium value of R for any given value of $\hat{x}$ is given by equation A3.3 in appendix 3. The direction of selection on the gene generated by the resource, and in part influenced by cultural niche construction, switches at $R = 1/2$. For positive values of ε, selection favors a when $R < 1/2$, A when $R > 1/2$, with no selection on A when $R = 1/2$. For negative values of ε the reverse is true.

(ii) Biased Transmission ($b_3 = 1$, $b_2 = b_1 = b$, $b_0 = 0$). Biased transmission results in an increase in the frequency of cultural state E if $b > 1/2$ and in e if $b < 1/2$. If $b < 1/2$, so that cultural state e is favored, and (with the exception of populations on the $x = 1$ boundary) cultural state E is lost. If ε is positive, there is fixation on ae if $R < 0.5$ at $x = 0$ (e.g., figs. 6.4b and 6.4c), on Ae if $R > 0.5$ at $x = 0$, and there is a neutrally stable curve of values of p at $x = 0$ for $R = 0.5$. With strong transmission bias in favor of e, populations rapidly converge on the $x = 0$ boundary as a result of cultural transmission, and then the selection from the cultural niche construction fixes A or a (fig. 6.4b). With weak transmission bias in favor of e and an initial value of $R > 0.5$, p initially increases (if ε positive), before cultural transmission can take populations into the range ($R < 0.5$) where the selection generated by the cultural niche construction favors a (fig. 6.4c). In cases like figure 6.4c, while the frequency of A is transiently large, in populations of small size the a allele might be lost by random drift before it becomes fixed by selection. Similarly, if $b > 0.5$, cultural transmission bias favors the E cultural state, and (with the exception of the $x = 0$ boundary) the e cultural state is lost. If ε is positive, there is fixation on aE if $R < 0.5$ at $x = 1$, and on AE if $R > 0.5$ at $x = 1$ (fig. 6.4d). All values of p are possible on the $x = 1$ boundary if $R = 0.5$. The reverse is true if ε is negative. These findings apply to both models.

(iii) Incomplete Transmission ($b_3 = 1 - \delta$, $b_2 = b_1 = b$, $b_0 = \delta$). For model 3, if there is no association between the cultural state and the alleles ($D = 0$), if $\delta > 0$ and $b \neq 1/2$, all populations remain polymorphic for the cultural state (Cavalli-Sforza and Feldman 1981), and there is convergence on a line of gene-frequency equilibria with the frequency of E given by equation A4.3 in appendix 4. When $b = 1/2$, this neutral curve is at $\hat{x} = 1/2$ (fig. 6.4e). If $b \neq 1/2$, the value of R determines the stability of the equilibria, with (for positive ε) A fixed for $R > 1/2$ at $\hat{x}$ (see fig. 6.4f), a fixed for $R < 1/2$ at $\hat{x}$, and all values of $\hat{p}$ possible for $R = 1/2$ at $\hat{x}$ (fig. 6.4e). Again, we may substitute allele A for a if ε is negative. Note that if oscillations occur in the frequency of the cultural state, they could generate cycles of selection that alternately favor A or a, depending on whether the maximum and minimum values of R are greater or less than 1/2, respectively. For example, this could happen if $\delta > 1/2$, which causes the frequency of E to cycle (Cavalli-Sforza and Feldman 1981).

6.4.1.2 External Selection at the A Locus Only: $\alpha_1 = \alpha_2 = 1$, $\eta_1 \neq 1$, $\eta_2 \neq 1$

(i) Unbiased Cultural Transmission ($b_3 = 1$, $b_2 = b_1 = 0.5$, $b_0 = 0$). From equation A4.2 in appendix 4, if $D = 0$ at equilibrium then x remains constant. Here there is no selection on the cultural state, so in both models $\hat{x}$ is given by the initial value of x, and the system behaves in an identical manner to the genetic model (see appendix 4 for details). An illustrative example showing the dynamics for p is given in figure 6.5a.

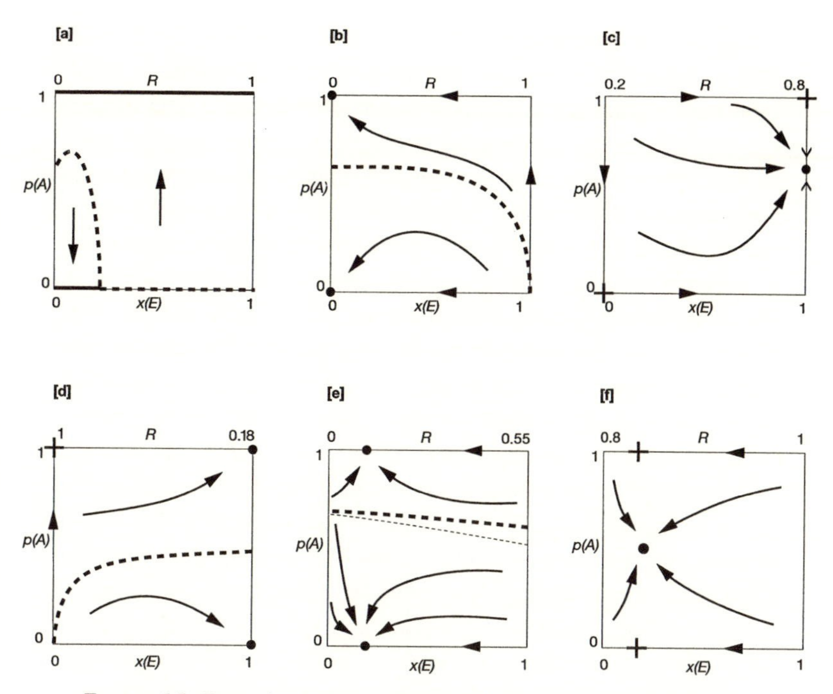

FIGURE 6.5. Dynamics (arrows), stable equilibria (thick lines, filled circles), and unstable equilibria (dashed thick lines, pluses) for systems with external selection at the **A** locus. Model 3 with (a) unbiased cultural transmission with ε positive ($b_3 = 1, b_2 = b_1 = 0.5, b_0 = 0, \varepsilon = 0.3$) and (b) biased transmission with ε positive ($b_3 = 1, b_2 = b_1 = 0.25, b_0 = 0, \varepsilon = 0.3$). Model 4 with biased cultural transmission ($b_3 = 1, b_2 = b_1 = 0.75, b_0 = 0$) and (c) positive niche construction with ε negative ($R_t = \lambda_1 R_{t-1} + \lambda_2 x_t + \lambda_3$, $\lambda_1 = 0.5$, $\lambda_2 = 0.3$, $\lambda_3 = 0.1$, $\gamma = 0$, $\varepsilon = -0.3$), and (d) negative niche construction with ε positive [$R_t = \lambda_1 R_{t-1}(1 - \gamma x_t) + \lambda_3$, $\lambda_1 = 0.9$, $\lambda_2 = 0$, $\lambda_3 = 0.1$, $\gamma = 0.5$, $\varepsilon = 0.3$]. Model 4 with incomplete transmission ($b_3 = 0.9, b_2 = b_1 = 0.25, b_0 = 0.1$) and positive niche construction, with (e) ε positive ($R_t = \lambda_1 R_{t-1} + \lambda_2 x_t + \lambda_3$, $\lambda_1 = 0.5$, $\lambda_2 = 0.275$, $\lambda_3 = 0$, $\gamma = 0$, $\varepsilon = 0.3$), and (f) ε negative ($R_t = \lambda_1 R_{t-1} + \lambda_2 x_t + \lambda_3$, $\lambda_1 = 0.75$, $\lambda_2 = 0.05$, $\lambda_3 = 0.2$, $\gamma = 0$, $\varepsilon = -0.3$). In (e) the thick dashed line represents the separatrix for $n = 1$ and the thin dashed line that for $n = 25$. On the top axis of each figure we present the value of R that corresponds to the value of x given on the bottom axis. In all cases $\alpha_1 = \alpha_2 = 1$, $\eta_1 = 1.1$, and $\eta_2 = 0.9$.

(ii) Biased Transmission ($b_3 = 1$, $b_2 = b_1 = b$, $b_0 = 0$). With bias in cultural transmission there are no curves of equilibria. If cultural transmission favors cultural state E ($b > 0.5$), the final equilibrium for all initially polymorphic populations depends upon the value of R at $\hat{x} = 1$. For $\varepsilon > 0$, since positive niche construction increases the amount of the resource, cultural transmission favoring E reduces the likelihood that R will be sufficiently small for niche construction to generate selection that counters the external selection favoring A. However, cultural transmission favoring e ($b < 0.5$) increases this likelihood, and the stronger the bias in cultural transmission (the smaller the value of b) the broader the parameter space over which ae eventually becomes fixed (cf. figs. 6.5a and 6.5b). For ε negative, there are three possible equilibria at $\hat{x} = 1$, with AE fixed, aE fixed, or an equilibrium polymorphic for A and a (fig. 6.5c), at which $\hat{p}$ is given by equation A1.25 in appendix 1 (see appendix 4 for further details). When ε is negative, high values of $\hat{x}$ are required for populations to reach equilibria polymorphic for A and a, so cultural transmission favoring E increases the parameter space over which populations converge on such polymorphic equilibria, while transmission favoring e reduces it.

For ε positive, negative niche construction decreases the amount of the resource, so cultural transmission favoring E increases the likelihood that R will be sufficiently small for cultural niche construction to generate counterselection to the a fixation boundary (fig. 6.5d). As a result, the parameter space in which this counterselection is generated is considerably enlarged, relative to unbiased transmission. The effects of cultural transmission favoring e depend similarly on the value of R at $\hat{x} = 0$. For ε negative, there are three possible equilibria at $\hat{x} = 0$, with Ae fixed, ae fixed, or an equilibrium polymorphic for A and a, at which $\hat{p}$ is again given by equation A1.25 in appendix 1.

(iii) Incomplete Transmission ($b_3 = 1 - \delta$, $b_2 = b_1 = b$, $b_0 = \delta$). If $\delta > 0$, all populations remain polymorphic for the cultural state and converge to equilibria at which $\hat{x}$ is given by equation A4.3 in appendix 4. Here, the value of R only partly determines the stability of the equilibria, which is also dependent on the external selection. If ε is positive, there are only two possible stable equilibria, with A or a fixed at $\hat{x}$ (fig. 6.5e). Depending on initial conditions, low values of R generate counterselection that takes populations below the separatrix to the a fixation equilibrium. The position of the separatrix is affected by n, the number of generations of cultural niche construction influencing the amount of the resource (see legend of fig. 6.5e). With ε negative, fully polymorphic equilibria are possible when R is large, with the line of equilibria characteristic of unbiased transmission collapsing to a single equilibrium point at $\hat{x}$ (fig. 6.5f), with $\hat{p}$ given by equation A1.25 in appendix 1.

6.4.1.3 Selection of the Cultural State: $\alpha_1 \neq 1, \alpha_2 \neq 1, \eta_1 = \eta_2 = 1$

(i) Unbiased Cultural Transmission ($b_3 = 1, b_2 = b_1 = 0.5, b_0 = 0$). Here both models (3 and 4) behave exactly like the equivalent genetic models (1 and 2, respectively).

(ii) Biased Transmission ($b_3 = 1, b_2 = b_1 = b, b_0 = 0$). With bias in cultural transmission and selection on E, there are two processes influencing the frequency of the cultural state. When E is favored either by natural selection ($\alpha_1 > 1 > \alpha_2$) or by cultural transmission ($b > 0.5$), positive values of ε result in fixation of AE, while negative values cause the fixation of aE. When e is favored either by natural selection ($\alpha_1 < 1 < \alpha_2$) or by cultural transmission ($b < 0.5$), positive values of ε result in fixation of ae, while negative values cause the fixation of Ae. If these two processes reinforce each other in favoring E or e, the same fixation points are reached as when one or the other acts alone, although the approach to equilibrium is faster. When they are in conflict then the final fixation point depends on the relative strength of the two processes. In some cases cultural transmission may take E or e to fixation in the face of strong natural selection favoring the other state (e.g., $\alpha_1 = 1.1, \alpha_2 = 0.9, b = 0.25$).

(iii) Incomplete Transmission ($b_3 = 1 - \delta, b_2 = b_1 = b, b_0 = \delta$). If $\delta > 0$, there will be a polymorphism in the cultural state, with either allele A or a fixed, and if there is no natural selection, $\hat{x}$ is given by equation A4.3 in appendix 4. If there is natural selection, the frequency of the cultural state at allele-fixation equilibria is given by a solution to equation A4.4 in appendix 4. The frequency of the cultural state typically differs at the two allelic fixation states, although this difference is small for realistic parameter values (see appendix 4 for an explanation).

As in the genetic models, for $n > 1$, we see time lags in the response to selection at the **A** locus, following the spread of a niche-constructing state. Typically, the time lags are shorter than in the case of the purely genetic system, principally because the cultural state reaches equilibrium faster than an analogous gene. This is first because there are only two cultural states (analogous to a haploid genetic system) compared to three genotypes, so disadvantageous cultural states are not shielded from selection as deleterious alleles are in heterozygotes, and consequently rare cultural states take less time than rare alleles to be eliminated by selection. Second, cultural transmission is typically faster than the natural selection of genes (except possibly in microorganisms). Only if there is no selection and a weak transmission bias do we see time lags of the order seen in the genetic model. With a weak bias, and strong selection generated by the cultural niche construction, the A allele may reach fixation before E. Moreover, with a weak bias, allele A may reach very low frequencies (i.e., $p < 10^{-7}$) before it is selected toward fixation, and in small populations it might be lost. If $\delta > 0$, E does not go to fixation but, provided $b > 0.5$, A will eventually become fixed. In general, a cultural niche-constructing state has to spread through the population only enough to increase the frequency of the resource R above 0.5, before it generates selection that will fix A.

6.4.1.4 Overdominance at the **A** Locus: $\eta_1 < 1$, $\eta_2 < 1$

(i) Unbiased Cultural Transmission ($b_3 = 1$, $b_2 = b_1 = 0.5$, $b_0 = 0$). With unbiased cultural transmission the models behave differently from our earlier genetic models, since in this case there is no selection on the cultural state but there is external overdominant selection on the gene. Polymorphic equilibria at the **A** locus are possible provided the selection generated by the resource does not overcome the external source of selection to negate the heterozygote advantage. Provided R is not so small that a may become fixed, or so large that A may become fixed, there may be a line of fully polymorphic cultural equilibria. At equilibria polymorphic for **A**, the frequency of A is given by equation A1.25 in appendix 1, while the values of R for stability of $\hat{p} = 1$ are given by inequalities A3.6 and the values of R for stability of $\hat{p} = 0$ are given by inequalities A3.7 in appendix 3. As with the genetic models, the curve giving equilibrium values of A can have some interesting features. For instance, with $\alpha_1 = \alpha_2 = 1$, $\eta_1 = 0.95$, $\eta_2 = 0.9$, $\varepsilon = -0.3$, as x increases the frequency of A increases, then decreases, and then increases again.

(ii) Biased Transmission *($b_3 = 1$, $b_2 = b_1 = b$, $b_0 = 0$)*. With bias in cultural transmission the curves of equilibria (for example, as seen in fig. 6.4a) are no longer stable, and initially polymorphic populations converge on a single stable equilibrium. The bias in cultural transmission takes E or e to fixation, with the frequency of A given by equation A1.25 in appendix 1.

(iii) Incomplete Transmission *($b_3 = 1 - \delta$, $b_2 = b_1 = b$, $b_0 = \delta$)*. With incomplete transmission, if $\delta > 0$ and $D = 0$ there is convergence on a single equilibrium where $\hat{x}$ is given by equation A4.3 in appendix 4 and $\hat{p}$ by equation A1.25 in appendix 1. As with the genetic model, the selection generated by the resource can shift the position of polymorphic equilibria. The direction of the shift is in favor of a when ε is positive and $R < 0.5$ and also when ε is negative and $R > 0.5$, and in favor of A when ε is positive and $R > 0.5$ and also when ε is negative and $R < 0.5$. The selection generated by the resource is strongest when R is close to 0 or 1, and weakest at $R = 0.5$. By influencing the amount of the resource, cultural niche construction can significantly change the position of the equilibrium and may change the direction of selection resulting from R.

6.5 DISCUSSION

With the exception of the case of overdominance at the **A** locus, the gene-culture niche-construction models with unbiased cultural transmission simplify to the equivalent purely genetic niche-construction models. The exception occurs because, under our framework, there is no pattern of frequency-independent natural selection that can maintain polymorphism in a two-state cultural-transmission system (unless cultural transmission is incomplete). Here, the case of unbiased cultural transmission with overdominance at the **A** locus has some interesting features: for example, curves of stable fully polymorphic equilibria are found that exhibit complex relationships between the frequencies of the cultural state and the alleles at the **A** locus, similar to those found with the purely genetic models when selection operates at the **A** locus (Laland et al. 1999). Such curves might represent situations like that of the effect of yam cultivation (the cultural niche-constructing state, or E) on the frequency of the sickle-cell allele (the allele maintained through overdominance, or A) and illustrate the sensitivity of allele frequencies to cultural niche construction.

Biased cultural transmission can increase the range of parameter space over which niche construction has an impact. For instance, in the face of external selection favoring allele A, cultural transmission may generate counterselection that increases the likelihood of fixation on a, if cultural transmission favors state e and a decrease in the amount of the resource results in a decrement in the fitness of genotypes containing A (ε is positive). Similarly, cultural niche construction will increase the chance of convergence to equilibria polymorphic for A and a, if cultural transmission favors E when an increase in the amount of the resource results in a decrement in the fitness of genotypes containing A (ε is negative). In both cases cultural niche construction is driving genetic evolution. These situations are manifestations of the scenarios depicted in figure 6.3. Biased cultural transmission can also reduce the range of parameter space over which niche construction has an impact, if cultural transmission favors E when ε is positive, or e when ε is negative. This means that if there is external selection at the **A** locus, the effect of biased transmission is that cultural niche construction is likely to have a more extreme effect (either much smaller or much bigger) than that resulting from either unbiased transmission or purely gene-based niche construction. Because cultural processes typically operate on a faster timetable than natural selection, biased cultural transmission is likely to have a much greater influence on the consequences of niche construction than if **E** were a gene. These findings illustrate processes by which cultural niche construction may have played an instrumental and active role in hominid evolution, initiating novel evolutionary events through the creation of novel selection pressures and changing the direction of evolution by modifying established selection pressures. Moreover, they support the hypothesis that the hominid capacity for niche construction was enhanced by, and coevolved with, a capacity for cultural transmission.

There are at least two reasons why biased cultural transmission might be expected to sweep E or e to fixation more rapidly than natural selection. First, there are only two cultural states, compared with three genotypes, so rare cultural states take less time to be eliminated by selection than rare alleles. Second, cultural transmission is typically faster than the natural selection of human genes, with b much more likely to deviate substantially from 0.5 (indicative of a significant bias in cultural transmission) than α_1 and α_2 are to deviate substantially from 1 (indicative of strong natural selection). This means that niche construction resulting from cultural processes is more likely to cause dramatic changes in the frequency of the key resource **R** than niche construction resulting from genes. If niche construction has its greatest impact when R is large or small, by taking E or e to fixation, biased transmission will strengthen selection generated by cultural niche construction.

Weak transmission biases favoring a cultural niche-constructing behavior can also generate interesting evolutionary scenarios. For instance, if transmission bias results in a change in frequency of cultural niche-constructing states, then selection at the **A** locus may be modified or even reversed, as R may have increased or decreased beyond the $R = 1/2$ switch point. In the case of weak biases, there may be many more generations of selection favoring one of the alleles at the **A** locus than would be the case for strong biases, before selection switches to favor the other allele, and as a consequence one or the other allele may reach a very low frequency before increasing in frequency again. In reality, we anticipate that small populations that follow this trajectory because they are exposed to a weak cultural-transmission bias may lose genetic variation at the **A** locus before selection could favor the allele that had previously been selected against. This type of process could easily create and maintain genetic differences between semi-isolated populations, and in hominids may have played a role in biological speciation events.

If cultural transmission and natural selection on **E** conflict, there are circumstances under which cultural transmission can overwhelm selection. If the two processes act in concert, cultural transmission accelerates the rate at which the cultural state spreads.

Incomplete cultural transmission maintains variation in the cultural state, even when there is directional natural selection on **E**. This means that oscillations in the frequency of E could lead to corresponding fluctuations in the frequency of A. It also means that a single fully polymorphic equilibrium is more likely to be found when there is overdominance at the **A** locus. Under such circumstances, the gene-culture models with incomplete transmission most closely resemble the purely genetic models with overdominance.

When the amount of the resource is a function of more than one generation of niche construction ($n > 1$), we find time lags at the **A** locus, in response to a change in selection pressures acting on **E**. Typically, the time lags are shorter than in the case of the purely genetic systems, principally because the cultural state reaches equilibrium faster than an analogous gene. It is only if there is no selection and weak transmission bias that we see time lags of the order seen in the genetic models. With incomplete transmission, neither E nor e goes to fixation, but, provided a cultural-transmission bias favors state E ($b > 0.5$), A eventually becomes fixed. Thus a cultural niche-constructing state only has to spread through the population enough to increase the frequency of the resource R above 0.5, before it can generate selection that will fix A.

As with gene-based niche construction (Laland et al. 1996, 1999), these models demonstrate that cultural niche construction can generate counterselection that compensates for or counteracts a natural selection pressure in the environment. In chapter 2 we suggested numerous examples of organisms that seem to have evolved counteractive niche-constructing behaviors allowing them to regulate the environment so as to buffer out particular natural selection pressures. A reasonable inference from such findings would be that competent niche constructors should be more resistant to genetic evolution in response to autonomously changing environments than less able niche constructors. As cultural processes enhance the capacity of humans to alter their niches, it seems plausible that hominid niche construction in general has been more flexible than that of other mammals. In subsequent chapters we take this further in developing a number of predictions about animal (chapter 7) and human evolution (chapter 9).

6.6 CONCLUDING REMARKS

In this chapter we have explored the repercussions for the human sciences of an extended evolutionary theory that incorporates niche construction. The evolutionary framework that we have developed should be regarded as a broad conceptual model, designed to act as a hypothesis-generating framework around which human scientists can structure evolutionary approaches to their disciplines. Despite the complexity of the processes involved in human evolution it is possible to extract from this conceptual model particular subprocesses, or derive specific hypotheses, that are subject to empirical test, or can be developed into formal models. Researchers may choose the levels or processes in the model that they feel are most appropriate, but they will have the advantage of working within the context of an overarching framework.

Section 6.4 illustrates how aspects of this new evolutionary framework can be translated into formal models. In extending gene-culture coevolutionary models along the lines suggested by figure 6.1c, models 3 and 4 indicate that cultural niche construction may have been important in driving human genetic evolution throughout the last two million years. Cultural processes can operate on time scales necessary for gene-culture coevolution to occur. All of the results characteristic of gene-based niche construction, described in chapter 3, are also found for cultural niche construction, although frequently cultural niche construction is even more potent. Any bias in cultural transmission (differences in the rate of copying alternative acquired information) can increase the impact of niche construction over and above that resulting from genes. Where cultural transmission and natural selection conflict, there is a broad range of circumstances under which cultural transmission can overwhelm selection. This may be one reason why maladaptive behavior is possible among humans.

A cursory glance at figures 6.4 and 6.5 reveals that there is no simple function that relates the frequency of the cultural niche-constructing activity (x) to the frequency of the population's genes (p). Gene-culture models that do not treat the resource as an independent variable (illustrated in fig. 6.1b) usually assume that a simple relationship exists between the incidence of cultural activity in the present generation and the resulting selection on genes. In those situations referred to earlier, where the selection pressures acting on a human or hominid population depend partly on niche construction and partly on independent processes (as for model 4), or where niche construction generates an ecological inheritance with the activities of past generations affecting selection on contemporary populations (as in models 3 and 4), treating the resource as a variable generates a much wider range of evolutionary possibilities.

The models presented in this chapter suggest the possibility of making estimates of gene frequencies given knowledge of cultural activity, or vice versa, which may enhance the accuracy of population-genetic and demographic censuses. For instance, in the case of the yam-cultivating peoples who have modified selection pressures by increasing the amount of standing water, it might be possible to use information about levels of rainfall or water surplus in different regions, together with the level of slash-and-burn agriculture in that region, to predict the frequency of the sickle-cell allele among these peoples. These ideas are developed further in chapter 9.

In chapter 1 we argued that there have been two principal reasons why many human scientists have, to date, found it difficult to make use of evolutionary theory. We can now return to them. One is that the theory appears to have too little to offer them. Human scientists are predominantly interested in phenotypes, namely, human beings. Chiefly, they are concerned with what humans do, with economic and social activities, with the inheritance of property, and with other aspects of cultural inheritance. They are far less interested in genes and genetic inheritance, except with respect to individual human development. Although the origins of evolutionary theory reside in the study of phenotypes and their variation, much of modern theory and practice has focused on genes and their inheritance rather than phenotypes. In fact, in recent years, conventional evolutionary theory has, if anything, been further downplaying the role of phenotypes, sometimes reducing phenotypes to mere "vehicles" for their genes, and genes to mere molecules. These biological priorities are probably unattractive to the majority of human scientists because they do not appear to offer the human sciences any useful point of contact with evolutionary theory.

One reason why our framework may be more appealing to human scientists is that, in adding niche construction to evolutionary theory, we are proposing an additional role for phenotypes in evolution. Humans are not just passive vehicles for genes; they also actively modify sources of natural selection in environments. As it is abundantly obvious that humans do construct niches, any evolutionary theory that formally acknowledges this fact is likely to be more in tune with most human scientists' thinking. Also, because the niche construction of complex organisms such as humans depends so heavily on ontogenetic processes, learning, and cultural processes, which are the subject matters of the human sciences, it follows that human scientists may not only derive benefits from placing their own subjects in an evolutionary context, but they may also make real contributions to our understanding of human evolution.

The second reason why human scientists have difficulty with evolution is the oversimplicity of most evolutionary models. Adaptationist approaches to conventional evolutionary theory enjoy the advantage of being relatively simple, and their simplicity is regarded as a virtue since it allows shortcuts to the development of hypotheses and the drawing of conclusions. However, many human scientists regard the theory as too simple to explain their observations, and for that reason it repels them (Heyes and Huber 2000; Fodor 2000). Adding niche construction inevitably makes evolutionary theory more complicated, and any extra complexity must earn its keep; otherwise, adding it would be pointless. Niche construction will be seen to earn its keep as soon as human scientists start using it in new ways. In fact, the fertile nature of the niche-construction perspective is beautifully illustrated in some recent innovative analyses (Aunger 2000b, 2002; Barkow 2000; Bowles 2000; J. Lanata, personal communication; Robson Brown 2000; Simonton 2000; Thompson 2000; Griffiths and Gray 2001; Sterelny, in press; Fragaszy and Perry, in press). In chapter 9 we describe some of these ideas in more detail. We also delineate a number of ways in which our framework can be applied in the human sciences, and make a number of predictions that can be tested with empirical research.

CHAPTER 7

Testing Niche Construction 1

Empirical Methods and Predictions for Evolutionary Biology

7.1 INTRODUCTION

Chapters 3–6 have used formal and narrative arguments for an extended evolutionary theory that incorporates niche construction and that may produce fundamentally different conclusions from standard evolutionary theory. There should therefore be a set of empirical predictions that would generate data more consistent with such an extended evolutionary theory, and less consistent with the more conventional evolutionary perspective. If niche construction is going to be a valuable extension to standard evolutionary theory, it must stimulate useful empirical work.

In this chapter we describe how our hypotheses concerning the evolutionary role of niche construction might be tested through laboratory and field experiments and theoretical studies. We outline a series of methods to detect significant evolutionary consequences of niche construction and propose a number of testable predictions. The methods include experiments that investigate the consequences of canceling or enhancing a population's capacity for niche construction, ways to detect the evolutionary consequences of niche construction in the wild, and direct tests of the predictions of our theoretical models. We suggest that there are rich possibilities for testing the evolutionary credentials of niche construction and that the hypotheses generated by our extended evolutionary theory are likely to expose new lines of empirical research in evolutionary biology.

Before introducing these hypotheses and methods, however, we engage in the rhetorical exercise of playing devil's advocate by once again asking in which ways our perspective might be questioned. Our arguments can be attacked at a number of different levels, and each level requires a particular kind of empirical evidence in order to be refuted. As we noted in chapter 1, at the simplest level the cautious skeptic may simply maintain, as did George Simpson (1949) and Theodore Dobzhansky (1955), that organisms do not construct or regulate their niches to any significant degree, or that the impact of their niche construction is too weak, too transient, or too capricious to significantly change selection pressures. To a large extent this is an empirical issue, and this criticism is countered by evidence for niche construction and its consequences in tables 2.2–2.7 of chapter 2. Other comparable data also support our argument (Hansell 1984; Jones et al. 1997; Lewontin 2000; Turner 2000). To us, this is the least tenable of all skeptical positions, since it is now well established that a vast array of organisms, from bacteria to humans, repeatedly and consistently alter their environments and those of their descendants (Mousseau and Fox 1998b), with impacts ranging from the extremely local to the global. We therefore claim that the ubiquity and broad impact of niche construction are no longer open to question.

It is less clear that niche construction is likely to have generated consistent selection pressures that have been the source of further evolutionary change. We argue that environmental impacts are likely to be evolutionarily consequential, because to modify natural selection pressures organisms need only alter their local environments. Nevertheless, the empirical data described in section 2.3 of chapter 2, while strongly suggestive that niche construction has often had significant evolutionary consequences, cannot be regarded as proof that niche construction is an important evolutionary process. However, we do not have to rely on such data, and in the next section we describe several methods for detecting the evolutionary impact of niche construction.

A second class of skepticism would accept that on some occasions organisms modify their own selection pressures, but would question whether this changes the nature of the evolutionary process. This position might, for instance, maintain that it is irrelevant whether selection pressures originate from the niche construction by the organism or from some independent process, in which case it would not matter whether environment-altering traits coevolve with traits whose fitness depends on alterable environmental sources of natural selection or whether they evolve in sequence, since the end result, the *product* of evolution, will be the same. Such arguments neglect the fact that niche construction changes the *process* of evolution by generating forms of feedback in evolution that are generally ignored. In the preceding chapters we have described how niche construction may frequently generate an ecological inheritance that allows organisms to modify the selection acting on many subsequent generations. Our population-genetic analyses imply that this ecological inheritance may generate evolutionary time lags, with momentum or inertia effects. Related theoretical work on cultural inheritance, maternal inheritance, and indirect genetic effects support these conclusions (Feldman and Cavalli-Sforza 1976; Kirkpatrick and Lande 1989; Moore et al. 1997; Mousseau and Fox 1998a,b; Wolf et al. 1998, 2000). These predictions represent a departure from those of conventional population-genetic models. We have also argued that acquired characters, such as learned information, can play a non-Lamarckian role in the evolutionary process by altering selection pressures.

Whether or not niche construction changes the evolutionary process can be assessed directly. In section 7.2 we sketch a suite of empirical methods designed to investigate how the process of evolution is modified by niche construction. These methods range from experiments designed to establish whether ecological inheritance is present and playing an evolutionary role, to tests of theoretical predictions of time lags in the evolutionary dynamic, to development of a modified version of Hamilton's rule.

A third class of skepticism would accept that on some occasions organisms do modify their own selection pressures, and that when they do, this may change aspects of the evolutionary process, but would question whether such phenomena occur sufficiently often to merit our proposed extension of evolutionary theory. Given the evidence that we have presented in this book, this is perhaps the only reasonable skeptical position. If most evolution can be characterized as adaptation through the natural selection of organisms in preestablished environments, with only a few cases being better represented by the interaction of natural selection and niche construction, then perhaps the added complexity of extended evolutionary theory is not justified. Again, this is largely an empirical issue, and reservations concerning the utility of our extended evolutionary theory will be overcome to the extent that the niche-construction perspective proves useful to evolutionary biologists.

At this early stage, we see two uses for treating niche construction as an evolutionary process. The first, which we have already touched on, is that researchers engaged in the study of evolutionary biology will be given a suite of fresh hypotheses to test, and a battery of methods to employ, which are likely to provide them with a stimulating new program for research. The second is that there are a number of topics central to evolutionary biology that have not yet been investigated from a niche-construction perspective but for which there is reason to believe such a perspective might offer new insights. These include speciation, cooperation, and rates and patterns of evolutionary change, and in section 7.3 we sketch our reasons for suggesting that niche construction might prove helpful. An extended evolutionary theory may help to make sense of some of the stubborn quantitative anomalies in the field.

7.2 METHODS FOR DETECTING THE EVOLUTIONARY CONSEQUENCES OF NICHE CONSTRUCTION

7.2.1 Experiments That Investigate the Consequences of Canceling or Enhancing a Population's Capacity for Niche Construction

One fairly straightforward experimental design that allows researchers to assess the evolutionary consequences of niche construction is to contrast, over several generations, experimental populations of niche-constructing organisms with otherwise identical populations for which the impact of their niche construction is experimentally blocked. Control conditions, for instance, without the subject organism, may be required in order to establish the nature and magnitude of the niche-constructed changes in the experimental environment. This information can then be exploited in the experimental condition in order to cancel out the effects of niche construction. A different condition could enhance, instead of negate, the effects of niche construction on the environment. Jones et al. (1997) propose a virtually identical method to investigate the effects of ecosystem engineering experimentally using virgin, engineered, and degraded habitats. In fact, researchers may choose to focus on the ecological impacts of niche construction, as Jones et al. suggest, or on its evolutionary consequences in organisms, either in phenotypes (how have target traits changed with or without niche construction?) or at a genetic level (how have target alleles changed in frequency in populations with or without niche construction?). Obviously, the feasibility of evolutionary experimentation depends very much on the generation time of the organism concerned, and experiments that focus on changes in allele frequencies may be restricted to comparatively short-lived organisms such as microbes. One possibility here would be to speed up the turnover of generations in longer-lived organisms by creating artificial generations. This could be achieved by removing and then replacing successive batches of niche-constructing organisms in an environment in which the consequences of each "generation's" niche construction were allowed to accumulate. That would expose each new "generation" to a changed selection regime as a consequence of the niche-constructing activities of the prior generations (J. Lawton, personal communication).

7.2.2 Comparative Analyses That Explore the Phylogeny of Trait Evolution across Related Species

7.2.2.1 Using the Comparative Method to Detect the Dependence of the Evolution of Recipient Traits on the Prior Evolution of Niche-Constructing Traits

For any clade of organisms, it might be possible to determine, using established comparative methods (Harvey and Pagel 1991), those phenotypic characters (recipient traits) that might have been selected as a consequence of feedback from prior niche-constructing traits. Pertinent characters could be measured in closely related organisms that do and do not exhibit this niche construction. It would then be possible to use statistical methods to test whether the recipient character changes correlate with niche-constructing activity, and whether the characters are derived, thus allowing inference as to whether the niche-constructing trait preceded the evolution of the recipient trait and to what extent the niche-constructing trait is necessary for the evolution of the recipient trait.

McLennan, Brooks, and McPhail (1988), for instance, carried out a phylogenetic analysis of nest building and mating behavior of several stickleback species (*Gasterosteus* spp.). They found that simple nest building preceded the evolution of nest-show displays, nest maintenance, and fry-retrieval behaviors. This is consistent with our argument that niche construction (here the building of a nest) sets up the selection pressures for many other features of stickleback nest building, mating, and parental behavior. The generality of this finding might be assessed with a more extensive comparative analysis, for instance, regressing nest-building behavior on aspects of courtship and parental behavior across entire classes or orders of fishes. This analysis would utilize methods such as independent contrasts or equivalent procedures that control for the phylogenetic relationship between species and would require that degrees of freedom be adjusted in order to take account of their nonindependence (Harvey and Pagel 1991).

Hansell (1987, 1993, 1996) has proposed that building and burrowing activity have played a particularly significant role in the evolution of the social behavior of animals. Nests are valuable resources, constituting not just a rich concentration of free energy, but also a structure that is potentially valuable to a broad range of usurpers and squatters. Nest building can therefore initiate a coevolutionary arms race between nest constructors and nonconstructing exploiters (Hansell 1996). Moreover, Hansell (1996) argues that the creation of a nest has consequences that positively reinforce the evolution of social complexity. Other researchers have suggested that nest building may have been instrumental in the evolution of eusociality in the social insects (Bourke 1997). These hypotheses can be tested in the manner described above and, if confirmed, would constitute evidence that niche construction has been important in the evolution of both cooperation and parasitism. It is also conceivable that niche construction has been important in several of the major evolutionary transitions. For instance, the evolution of burrowing behavior coincides with the Cambrian explosion, and there may be a general relationship between burrowing and speciosity. Whether such a relationship exists could also be investigated using comparative methods.

Another example is food storage in birds. Comparative studies of memory in food-storing versus non-food-storing birds have found that species that hide food in a large number of caches in their environments, and later retrieve it, typically have better spatial memories, and larger relative hippocampal volumes, than do other closely related species that do not store food (Sherry et al. 1989; Healy and Krebs 1992, 1996; but see Bolhuis and MacPhail 2001). The relevance of the hippocampus is that it is a part of the brain that is involved in spatial memory, which is required to relocate cached food. This finding is again consistent with the hypothesis that feedback from niche construction, namely, the activities of hiding and retrieving food items, probably instigated selection for better spatial memories and orientation in the food-storing species, but not in the non-food-storing species. In general we would expect to find other examples of feedback from behaviorally modified natural selection pressures to selected phenotypic traits.

7.2.2.2 Patterns of Response to Selection May Depend on the Capacity for Niche Construction

Many counteractive niche-constructing behaviors regulate the environment in such a way as to buffer against particular natural selection pressures. As a consequence, potent niche constructors should be more resistant to genetic evolution in autonomously changing environments than less able niche constructors. For instance, if we assume that primate niche construction is more flexible than that of most other mammals, or that vertebrates generally have a more extensive niche-constructing capability than invertebrates, then a number of hypotheses follow. First, consider Vrba's (1992) hypothesis of "turnover pulses." We would expect advanced niche constructors such as primates to show a weaker evolutionary response to fluctuating climates than other mammals. Similarly, we would expect vertebrates to exhibit less response to fluctuating climates than invertebrates. Second, Bergmann's and Allen's rules (Gaston et al. 1998) suggest, respectively, that populations in warmer climates will have smaller bodies and bigger extremities than those in cooler climates. Again we would expect sophisticated niche constructors to show less correspondence to these rules than other animals, since they should be better able to regulate temperature through niche construction. Third, it should be possible to reverse the inference and to use the fossil record to infer something about the niche-constructing capabilities of animals. For example, the greater the organism's phenotypic (as opposed to extended phenotypic) response to environmental change, the more restricted its capacity for niche construction is likely to have been. Further predictions alone these lines are given in chapter 9.

7.2.2.3 Genetic and Phenotypic Signatures

In chapter 6 we described how yam cultivation by the Kwa appears to have left a genetic signature in this population, namely, a high frequency of the hemoglobin *S* allele and the concomitant sickle-cell disease. In theory, it is possible that genetic signatures for other niche-constructing activities, evident in paleological or archaeological records, could be identified and used as evidence for the presence or absence of a niche-constructing trait. If so, advances in molecular techniques could eventually aid this line of enquiry. This method may be particularly useful in the study of human evolution (see chapter 9), where it may be possible to identify genetic signatures relating to cultural events and then use these to trace the history of specific cultural activities, such as the use of fire, or the spread of agriculture, across human groups. In addition, genetic signatures may be sought in other species (e.g., fire-prone or agricultural plants) that might provide evidence of prey exploitation.

7.2.3 Methods for Detecting Niche Construction in the Wild

One of the principal differences between standard and extended evolutionary theory concerns the process of adaptation. According to standard evolutionary theory organisms acquire adaptations only by responding to autonomous natural selection pressures in their environments. To quote George Williams (1992): "Adaptation is always asymmetrical, organisms adapt to their environment, never vice versa." But according to extended evolutionary theory, organisms may acquire adaptations as above, or they may change their selective environments by positive niche construction to make their environments suit themselves. With niche construction there are therefore two evolutionary routes to the complementarity of organism and environment, and evidence for this second route would constitute support for our extended evolutionary theory.

When we observe a match between the characteristics of an organism and factors in its environment, how can we tell a posteriori whether this match came about through the classical selection of organisms or a process of construction of environments? Suppose the correspondence arose through natural selection. We would expect to see a direct relationship between the structural or other putative adaptive characteristics of the organism and the factors in the environment that constituted the hypothesized selection pressure. We should not expect to see any associated organism-driven changes in the relevant factors in the organisms' environment, at least since the time that the organism acquired its structural features. Nor should we expect to see evidence for evolutionary feedback in the form of structural or functional adaptations in the organism to these organism-constructed factors in the environment.

What might be expected if the match had resulted from a niche-constructing activity which engineered a suitable environment for the organism? First, there should be no direct relationship between a structural change in the organism and the relevant autonomous selection pressures. Instead there should be signs that an evolutionarily older structure or process has been coopted for a new function (Gould and Vrba 1982) or that a general adaptation, such as a capacity for learning, has been employed in a novel manner. Second, we should expect to see evidence for an associated organism-driven change in the environment. Third, we might expect to see evidence for further adaptations in the organism to these organism-constructed factors in the environment. Fourth, we might find evidence of evolution proceeding in the absence of autonomous changes in an environment. This would occur in cases of inceptive niche construction that are initiated by changes in organisms, instead of by autonomous changes in their environments. Fifth, we might sometimes expect to find no evolutionary response to autonomous changes in the species' selection pressures, since such changes could have been nullified by counteractive niche construction.

Some examples of niche-constructing organisms will illustrate this reasoning. Consider the case of the earthworm "kidney" (Turner 2000), which we introduced in chapter 1. Turner describes how earthworms are structurally very poorly adapted to cope with physiological problems such as water and salt balance on land: "When we apply physiological criteria to deciding what the proper habitat of an earthworm might be, it is hard to escape the conclusion that they are not really terrestrial. Rather, earthworms seem to belong in a freshwater habitat" (p. 103).

Earthworms have water-balance organs called nephridia which have much in common with the nephridia of freshwater oligochaetes, but few of the structural adaptations one would expect to see in an animal living on land. For instance, earthworms produce high volumes of urine, a characteristic of freshwater animals. This is the first clue that niche construction has been at work: bad structural adaptation, and a poor correlation between the characteristics of the organism and the factors in its environment. It raises the question: How do earthworms survive in a terrestrial environment? The answer is, of course, that earthworms utilize various kinds of niche construction. They select the edaphic horizon in soils, to which they are physiologically best suited, and expand this horizon through activities such as tunneling, exuding mucus, eliminating calcite, and dragging leaf litter below ground. This is clue number 2: we can observe one or more niche-constructing activities capable of compensating for the poor structural adaptation. Clue number 3 is the observation that earthworms are responsible for major changes in the soil environment, enhancing plant yields, reducing surface litter, aggregating particles, increasing the amount of topsoil as well as levels of organic carbon, nitrogen, and polysaccharides, and enhancing porosity, aeration, and drainage (Hayes 1983; Lee 1985). It is the soil rather than the worm that has changed to meet the demands of the worm's freshwater physiology.

A second example is given by plastron gills in aquatic spiders and beetles (Turner 2000). Clue number 1 is that these animals breathe air yet live under water with no structural adaptation equivalent to a gill. Clue number 2 is that they do this through niche construction, devising an underwater aqualung by carrying bubbles of air underwater. Clue number 3 is environmental change, the observation that their metabolism allows the bubbles to act as gills by taking in oxygen from the surrounding water, although in this case the effect on the environment may be too small to measure. Clue number 4 is evolutionary feedback; for instance, some of these beetles have evolved a flattening of the limbs closest to the body, and these act as hydrofoils that accelerate water past the bubble and inflate it.

7.2.3.1 Correlation with Environmental Factors

These "clues" are examples of general methods for detecting the action of niche construction in evolution that are, in principle, applicable to all organisms. For instance, perhaps the most common inference of natural selection in the wild derives from a correlation across a geographical region, between the trait of interest in an organism and the suspected environmental factor thought to be the source of selection favoring the trait (Endler 1986). Failure to detect such a relationship is generally interpreted as consistent with the null hypothesis that this environmental factor exerts no selection on the trait (Endler 1986). However, there is another evolutionary explanation, which is that the organism can afford poor structural adaptation of the trait to the environment because it compensates through niche construction. A straightforward procedure for investigating this possibility would be as follows.

Step 1: Search for a correlation between some organismal structure and environmental factors.

Step 2: If no relationship is found, investigate whether the organism exhibits niche construction that might compensate for poor adaptation of the structural trait.

Step 3: Investigate whether there is evidence for organism-driven modification of the selective environment.

Step 4: If so, search for evidence for evolutionary feedback in the form of structural or functional adaptations to the constructed environment.

We predict that correlations between structural traits and environmental factors that might otherwise be expected to be sources of selection will usually be weak or absent when organisms niche-construct in a manner that counteracts or circumvents this selection.

7.2.3.2 Comparisons between Closely Related Sympatric Species

The procedure detailed in section 7.2.3.1 can, with variants of step 1, be utilized to detect the action of niche construction in virtually all of the circumstances in which evolutionary biologists suspect that natural selection is operating. For instance, closely related sympatric species are predicted to exhibit character displacement, which can be revealed through comparison of sympatric and allopatric populations (Endler 1986). A failure to detect character displacement can be the first sign of a role for niche construction, if sympatric species act in ways that reduce competition. Thus, what would have been interpreted as a negative finding can instead be the impetus to investigate an alternative evolutionary explanation.

Step 1: Search for character displacement.
If none is found, proceed with steps 2–4 as above.

We predict that character displacement will not be found in situations where sympatric populations of organisms construct niches in a manner that reduces competition.

7.2.3.3 Comparisons between Unrelated Species Living in Similar Environments

A third source of evidence for natural selection is provided by convergent evolution of unrelated species in similar conditions (Endler 1986). Once again, a failure to detect convergent evolution can suggest a role for alternative forms of niche construction, if apparently similar autonomous selection pressures seem to have had different effects in the two populations.

Step 1: Search for convergent evolution.
If none is found, proceed with steps 2–4 as above.

We predict that convergent evolution will not be found when different populations exposed to similar selection pressures respond with different forms of niche construction in a manner that counteracts or circumvents their common selection pressures.

7.2.3.4 Long-Term Studies of Trait Frequency Distributions

A fourth source of evidence for the action of natural selection is stability of trait frequencies over time. If traits exhibit long-term stability this is often interpreted as evidence for the presence of stabilizing selection. However, there is another explanation in the form of counteractive niche construction, which buffers fluctuations in the selective environment.

Step 1: Search for long-term stability and for corresponding evidence for stabilizing selection. If there is no evidence for stabilizing selection, proceed with steps 2–4 as above.

We predict that there will be situations in which long-term trait stability results not from stabilizing selection that purges extreme phenotypes but from counteractive niche construction that modifies extreme environments.

7.2.3.5 Perturbations of Natural Populations

A fifth method for detecting natural selection is to monitor the response of trait distributions to natural or artificial perturbations. It is assumed that under natural selection, trait frequency distributions should not be at equilibrium immediately after an environmental perturbation. The null hypothesis here is that the population will not change after the perturbation, except through chance effects. Natural selection is assumed to have occurred if the population subsequently diverges from its perturbed trait frequency distribution more than expected by chance. Once again, however, a failure to detect such a divergence might constitute evidence for niche construction rather than for the absence of natural selection. For example, counteractive niche construction could negate any novel selection pressures.

Step 1: Search for divergence from new trait frequency following perturbation. If divergence is not found, proceed with steps 2–4 as above.

We predict that counteractive niche construction will act to prevent the divergence of trait frequency distributions following perturbations.

7.2.3.6 Optimality Models

A sixth method for detecting natural selection is to predict trait frequency distributions on the basis of known characteristics of the trait, assuming that natural selection will produce phenotypes that are in some sense optimal with respect to the environment. Major departure of the trait from its putative optimum might constitute evidence that there is no selection. Examples such as the earthworms and aquatic spiders illustrate that global optimization of structural traits is far from inevitable, and a failure of data to confirm the model might constitute the first evidence for niche construction.

Step 1: Test for predicted optimal structural or behavioral adaptations. If predictions are not confirmed, proceed with steps 2–4 as above.

We predict that niche construction will render structural adaptations unnecessary, sometimes resulting in poor correspondence between predicted functioning of what appear to be structural adaptations and their actual performance.

7.2.4 Testing the Predictions of Our Theoretical Models

The theoretical analyses described in chapters 3 and 6 illustrate some general features of the role of niche construction in evolution rather than make specific quantitative predictions about relevant population-genetic variables in particular populations. While it may be possible in the future to construct theoretical models that allow qualitative predictions about key variables, such as the amount of genetic variation in particular circumstances, the models that we have developed so far are unlikely to be useful in this regard. However, a few qualitative predictions might be made on the basis of these analyses, particularly in circumstances where there is an ecological inheritance. For instance, we would predict that niche-constructing populations ought to exhibit time lags in the response to selection, with the characteristic momentum and inertia effects, and that the size of these time lags ought to increase with the number of generations of niche construction that have affected the amount of the resource. Moreover, our models predict that the time lags will be greater when there is a primacy effect than when there is equal weighting or a recency effect. A comparison of models 1 and 2 suggests that the time lags are likely to be more consequential when the amount of the resource is better described as a weighted average than when it is an ecologically dynamic function. A comparison of models 1 and 3 suggests that greater time lags are likely to result from gene-based niche construction than from cultural niche construction.

7.2.5 Acquired Characters Can Play a Role in the Evolutionary Process

The semantic information that individuals acquire during their lives cannot be genetically inherited, and (unless culturally transmitted) is erased when they die. Nonetheless, processes such as learning can still be of considerable importance to subsequent generations because learned knowledge can guide niche construction. Thus we predict that acquired characteristics can play a role in evolution through their influence on the selective environment. For example, the Galápagos woodpecker finch creates a woodpecker-like niche by learning to use a tool to peck for insects (Tebbich et al. 2001), and we have argued that this acquired behavior probably created selection in favor of a bill able to manipulate tools rather than the sharp, pointed bill and long tongue characteristic of woodpeckers. This hypothesis is open to test using the methods described in section 7.2.3. For instance, as predicted in 7.2.3.1, these birds exhibit no correspondence between the shapes of their bill, tongue, and tail and the relevant factors in the environment, such as the presence of grubs under the tough bark of trees. Then step 2 suggests investigating whether the birds exhibit niche construction that compensates for the poor adaptation of their structural traits, and in this case whether the niche construction is learned. While in this situation it is apparent from step 2 that there is an organism-driven modification of the selective environment (step 3), in other cases whether the organism's learning has resulted in a change in the environment would have to be determined. Moreover, even in the instance of a learned niche-constructing behavior, researchers could search for evidence of evolutionary feedback in the form of structural or functional adaptations to the constructed environment (step 4). For example, we might speculate that these finches would still require the robust skull morphology characteristic of woodpeckers. In principle, all of the methods for detecting an evolutionary role for gene-based niche construction can be applied to acquired niche-constructing traits.

7.2.6 Selection Experiments: Population Response through Natural Selection or Niche Construction

While natural selection typically takes many generations to work, niche construction is individual based, and can therefore immediately be put to work by individual organisms. As a consequence niche construction is likely to permit much more rapid responses to changing natural selection pressures than conventional traits. Another quote from Turner is apt:

> Earthworms, when they came onto land, seem to have opted out of the retooling option and pursued a strategy of using ATP energy to work against soil weathering. Along the way, they essentially coopted the soil as an accessory organ of water balance. The advantage of adopting this strategy is clear: *it is accomplished much more rapidly than a retooling of internal physiology* (2000, p. 119, our italics).

Niche construction is likely to be a more rapid route to complementarity between an organism's features and the factors in its local environment than natural selection. Selection experiments may inadvertently constrain the niche-constructing responses open to the population, because they are usually designed to facilitate a genetic response to selection. In one selection experiment this has not occurred. Jones and Probert (1980) found that *Drosophila* with deleterious eye mutants could survive in a patchy laboratory environment through habitat selection, but would be selected against when no habitat choice was available. In our terms, these flies have carried out relocational niche construction, which has reduced the importance of classical natural selection. In general, we predict that if laboratory or other captive populations are exposed to artificial selection regimes in which they can respond through either niche construction or natural selection, then the population will respond predominantly through niche construction and, only when the former option is eliminated, through selection. Moreover, we anticipate that those members of the population that are least fit relative to the imposed selective regime will be the individuals that exhibit the strongest evidence for niche construction. Finally, if separate lines are permitted to respond through natural selection on the one hand, or niche construction on the other, we expect these lines to diverge genetically from the original population, but the former might exhibit structural adaptations while the latter might be selected for enhanced capacity to construct niches.

7.3 POTENTIALLY FRUITFUL APPLICATIONS OF NICHE CONSTRUCTION

7.3.1 Cooperation, Mutualism, and Altruism

Standard evolutionary theory provides two principal explanations for cooperation among organisms in a population: kin selection (Hamilton 1964) and reciprocity (Trivers 1985; Axelrod 1984), as exemplified by solutions such as tit-for-tat to prisoner's-dilemma-type games (Axelrod 1984). Our extended evolutionary framework, however, indicates that the suite of feasible evolutionary explanations for cooperation may be larger (Laland et al. 2000b).

An organism may act in ways that benefit another organism, if the second organism niche-constructs in ways that increment the first organism's relative fitness (Laland et al. 2000b). More precisely, our general *niche-construction statement on cooperation* suggests that an organism O_1 should act in ways that benefit another organism O_2 provided that the total niche-constructing outputs of O_2, or any of O_2's descendants, modify resources in the environment of O_1, or any of O_1's descendants, in such a way that fitness benefits to O_1 exceed the cost of O_1's action, and increase O_1's fitness relative to other members of O_1's population that do not niche-construct in the same manner (see Laland et al. 2000b, p. 171).

For example, if the organisms are relatives, cooperation should be more likely among kin that niche-construct in mutually beneficial ways, and less likely among kin that niche-construct in mutually detrimental ways, than would be predicted if these indirect effects were neglected. This reasoning can be encapsulated in a deliberately simplified and approximate expression, which illustrates how co-operation can result from a combination of altruistic and mutualistic relationships with kin, namely,

$$C_1 < B_2 r + B_{1(\mathrm{nc}2)} - C_{1(\mathrm{nc}2)}.$$

Here organism O_1 will cooperate with kin (O_2) provided the costs (C_1) of cooperation are less than the indirect fitness benefits specified by Hamilton's rule (the benefit to O_2 multiplied by the degree of relatedness, or $B_2 r$), plus the direct fitness benefits to O_1 that stem from the beneficial niche construction by O_2 and its descendants (or $B_{1(\mathrm{nc}2)}$), minus the direct fitness costs to O_1 that stem from the detrimental niche construction of O_2 and its descendants (or $C_{1(\mathrm{nc}2)}$). Cheverud (1984) makes a similar point using a quantitative genetic model. The same logic also applies to nonkin, where reciprocal altruism (which would be better described as mutualism) is a special case in which the organisms are unrelated individuals of the same species that directly modify each other's environment. Here, r is typically zero; and reciprocal altruism will invade a selfish population if

$$C_1 < B_{1(\mathrm{nc}2)} - C_{1(\mathrm{nc}2)} \text{ and } C_2 < B_{2(\mathrm{nc}1)} - C_{2(\mathrm{nc}1)}.$$

Intra- and interspecific mutualisms also fit this framework. In many cases, the niche-constructed by-products of several organisms are exploited by a population. For example, shoals of fish, flocks of birds, and herds of animals reduce the risk of predation relative to solitary animals merely because their combined presence changes the selective environment of each individual (Hamilton 1971). Coordinated fish driving by cormorants, and seal hunting by killer whales, provide examples of individuals coordinating their niche construction so that it results in more effective food acquisition for everyone (Connor 1995). Mutualistic interactions that result from the exploitation of incidental by-products are more evolutionarily robust than classically altruistic interactions because each individual acts selfishly, and, as a result, mutualistic interactions are probably less susceptible to selfish cheaters than are altruistic interactions.

It is even possible that niche construction could engender a form of group selection. For instance, some incidental by-products of individual niche construction will have consequences for the population, and differences in the niche construction expressed by members of different populations may generate variation in group productivity. In cases where these incidental by-products are unavoidable, selfish individuals that reap the benefits without doing the niche construction may not be able to invade. Here a kind of group selection may occur, but if it does, it will be subject to the same caveats as other proposed examples of group selection (Williams 1966), and we would describe its outcome as mutualism, not altruism.

An illustration of how the niche-construction statement on cooperation could be formalized is given in appendix 5, where we develop a model of the evolution of altruistic behavior in Hymenoptera.[1] The model is based on Uyenoyama and Feldman's (1980) population-genetic sex-linked model of altruism from sisters to siblings but relaxes the assumption of an equal sex ratio. We then use this formulation to derive an expression for the expected sex ratio, on the assumption that the ratio of investment in the two sexes is equal. We show that when altruistic niche construction is directed by workers exclusively to their sisters, the sex ratio is expected to exhibit a higher frequency of males than the 1/4 fraction or 1:3 ratio predicted by Trivers and Hare and is a function of the costs and benefits of altruism. An expression for this ratio is given by equation A5.5a in appendix 5. This may help to explain why the average sex-investment ratio χ across 40 species of monogynous ants is less female biased than the 1:3 ratio that Trivers and Hare predicted, namely, around $\chi = 0.37$ (Bourke 1997). While there are likely to be many other factors influencing sex ratios of social hymenoptera (Keller 1995; Sundstrom et al. 1996; Passera et al. 2001), the difference may in part be due to Trivers and Hare's omission of the costs and benefits of altruism. In contrast, when worker niche construction benefits both sexes, the sex ratio can be considerably more female biased than the 1:3 ratio, as niche construction for the general good of the colony selects for a more female-biased sex ratio. An expression for this ratio is given by equation A5.5b in appendix 5. Numerical analyses reveal that the expected proportion of males ranges from 0.2 to 0.4 for realistic

[1] Cheverud (1984) has constructed a quantitative genetic model, which includes maternal performance and parental care that could, if nest building were regarded as a form of parental care, be interpreted as making a similar theoretical argument to the model in appendix 5.

values of the costs and benefits of altruism and the degree to which workers bias investment toward sisters. The expressions for the expected sex ratio given in appendix 5 may also help to explain some of the variability in sex ratios observed among social insects. If we regard niche-constructing activities such as nest building and maintenance as altruistic, the different costs and benefits of niche construction in different species will result in different sex ratios in each case.

We can reverse the logic of the general niche-construction statement on cooperation to produce an equivalent *niche-construction statement on conflict*. Organisms should be prepared to act in a hostile manner toward other organisms that niche-construct in a manner detrimental to them (Laland et al. 2000b). More precisely, an organism O_1 should act in a hostile manner, to the disadvantage of another organism O_2, provided the total niche-constructing outputs of O_2, or of any of O_2's descendants, modify resources in the environment of O_1, or any of O_1's descendants, to the detriment of O_1, if the resulting reduction in the fitness costs to O_1 of O_2's outputs exceeds the cost of O_1's agonistic behavior, and increases O_1's fitness relative to other members of O_1's population that do not act in the same hostile manner (see Laland et al. 2000b, p. 144).

This reasoning might explain the evolution of aggressive behavior, including a form of reciprocal hostility, in which individuals and their descendants trade antagonistic acts. Organism *A* should actively harm organism *B* by investing in niche construction that destroys *B*'s selective environments, provided the fitness benefits that accrue to *A* in doing so are greater than the fitness costs.

7.3.2 Speciation and Peak Shift

Recent investigations into Sewall Wright's shifting-balance theory have demonstrated that drift-mediated peak shifts on a static adaptive landscape are possible (Barton and Rouhani 1987; Whitlock 1995), but that environmental fluctuations in the landscape are more likely to result in such peak shifts (Whitlock 1997). In fact, they "must account for orders of magnitude more phenotypic peak shifts than drift-mediated processes" (Whitlock 1997, p. 1045). When niche-constructing organisms change components of their environments, they may introduce different adaptive landscapes in the same way as do autonomous environmental fluctuations. Thus populations of organisms may sometimes eliminate peaks in adaptive landscapes simply by changing their own environments, or by substituting one adaptive landscape for another, through their phenotypic niche-constructing activities. Niche construction therefore offers a way of traveling across valleys in adaptive landscapes. A similar point has been made with respect to indirect genetic effects (Wolf et al. 1998). Moreover, given Whitlock's comments about the greater potency of environmental change than drift in this respect, the probability that organisms induce their own peak shifts through niche construction should be relatively high.

Niche construction may also play a role in sympatric speciation, particularly through habitat selection (Via 2001). Seger (1985, 1992) has shown that whenever assortative mating produces a better fit of a phenotypic trait to a resource distribution than random mating, then natural selection will be expected to favor such assortment, and this may create the opportunity for sympatric speciation if appropriate isolation barriers evolve. For example, if the pertinent phenotype of a prey population (e.g., seed size) has a uniform distribution, while the relevant character of a predator population (e.g., bill size) has a normal distribution, then natural selection may favor assortative mating for extreme phenotypes in the predator (e.g., either large or small bills). For instance, predators at the two tails of their phenotypic distribution may have a relative fitness advantage due to reduced competition for prey compared to conspecifics with intermediate phenotypes. Under such circumstances, providing predation does not significantly change the shape of the prey distribution, predators would increase their fitness further by mating assortatively, to produce offspring like themselves rather than offspring with the less fit intermediate phenotype. If selection were to fix this assortative mating behavior and isolating barriers came into operation, the predator population could eventually split into two new species.

One feature of niche construction is that it changes the shape of resource distributions. A modified resource distribution may favor assortative mating in a population that exploits that resource (Schluter 2001; Via 2001). For example, prey items often pose detection problems because of crypsis. There is evidence that some predators, particularly birds, develop a search image and focus their predation on the most common prey phenotype (Giraldeau 1997). Seger (1985, 1992) reasons that this may create a selection pressure that favors assortative mating among the rarer forms of prey. If this distorts an originally normal prey distribution to render it bimodal, then again by Seger's reasoning, assortative mating may be favored in the predator. Consistent with the above arguments, in a recent review Via (2001) notes that there are now considerable data suggesting that strong disruptive natural selection on habitat use, on characters associated with resource competition, and on habitat choice can facilitate sympatric speciation.

If through inceptive niche construction a subsection of a population begins to exploit a new resource, this may create the circumstances envisaged by Seger in which assortative mating among exploiters would be advantageous. For the woodpecker finch, for example, a plausible sequence of events is that ancestral finches started to exploit a new resource, namely, grubs in bark of trees, by their innovative tool use. Subsequent generations were selected for better tool-using skills, but if they mated with non-tool-users their offspring might not inherit genes associated with tool use. On the other hand, assortative matings with other finches having the same or better niche-constructing skills might have an increased number of surviving offspring relative to random matings. Thus selection could have favored assortative mating, together with other isolating mechanisms. The general point, which is worthy of further investigation, is that organisms may drive the process of speciation by creating resource distributions that favor assortative mating. This is because resource distributions are dynamic and vary as a function of their exploitation by niche-constructing organisms.

7.3.3 Evolutionary Rates and Punctuated Equilibria

Counteractive niche construction that mitigates a selection pressure may under some circumstances allow populations to maintain greater levels of genetic variation at those loci that would have been affected by selection had the population not expressed that particular niche construction, because it shields such variation from selection. For example, in mammals, genetic variation in the ability to deal with heat, through body size, or shape of ears or tail, may be exposed to less intense selection in populations that escape extreme temperatures by burrowing than in those that do not. However, if the counteractive niche construction breaks down, say when a new predator forces the mammals into the open, the presence of significant levels of variation in genes affecting heat exchange may facilitate rapid genetic evolution. In other words, for specific traits, counteractive niche construction may sometimes facilitate periods of evolutionary stasis, punctuated by rapid genetic change. Moreover, following such change, since niche construction, particularly by dominant species, modifies the selective environments of other species, subsequent niche construction could trigger a cascade of evolutionary events that realign ecosystems (Jones et al. 1997). Not all macroevolutionary patterns are likely to exhibit punctuated equilibria, and it is also possible that niche construction may sometimes reduce genetic variability (Jaenike and Holt 1991). Nevertheless, niche construction may provide a microevolutionary process capable of eventually producing punctuated macroevolutionary trends in particular traits (Eldredge and Gould 1972).

7.4 CONCLUSIONS

Niche construction generates a new class of predictions about the evolutionary process. Section 7.2 suggests tests for niche construction and, in particular, shows that it has the potential to form the foundation for a new empirical program. Section 7.3 illustrates that there are several other areas of evolutionary theory that niche construction has the potential to illuminate. In chapters 8 and 9 niche construction is shown to generate a number of predictions relevant to ecology and to the human sciences, and these may be tested with a suite of empirical methods similar to those described in this chapter.

CHAPTER 8

Testing Niche Construction 2

Empirical Methods, Theory, and Predictions for Ecology

8.1 INTRODUCTION

In chapter 5 we showed how our extended theory of evolution that includes niche construction can apply to ecosystem-level ecology. It does so primarily by allowing abiota, as well as biota, to be more readily incorporated into evolutionary models. This has the effect of lifting a restriction which, until now, has largely confined the use of evolution by ecologists to the population-community approach, while simultaneously preventing its use by process-functional ecologists (Holt 1995).

In this chapter we describe how our hypotheses concerning the implications of niche construction for ecosystem-level ecology can be tested with laboratory and field experiments and theoretical studies. We outline methods to detect significant ecological ramifications of niche construction and suggest some testable predictions. These include exploring the effects of canceling or enhancing niche construction, as well as more complicated ideas such as tracking the impact of a population's niche construction around an ecosystem and identifying EMGAs. We go on to discuss how the niche-construction perspective may shed light on various topics in ecology, including the stability of ecosystems, and whether ecosystems can be treated as cybernetic systems. We make no attempt to integrate these topics but instead treat them only as illustrative examples.

8.2. TOWARD A REFORMULATION OF EVOLUTIONARY ECOLOGY

At present, the theoretical framework for evolutionary ecology is based on a fusing of standard evolutionary theory with ecological models (Maynard Smith 1974; Roughgarden 1979; May 1981; Fox et al. 2001). In this section we will show how, in principle, evolutionary models that incorporate niche construction could be fused with similar ecological models, thereby providing evolutionary ecologists with a new conceptual tool. The development of such a body of theory is an ambitious program (see Getz 1999), and here we do not construct or analyze such models but merely point to how such theory could be formulated. Hence, the mathematical expressions given below are purely for purposes of exposition. There are two obvious places to start. First, the ecological models developed for ecosystem engineering could be evolution-*ized*, by tracking how different genotypes affect, or are affected by, niche construction. Second, our own population-genetics models, particularly model 2 in chapter 3, could be ecolog-*ized*, by introducing relevant ecological variables. We shall deal with each of these possibilities in turn and then consider other approaches.

8.2.1 Incorporating Evolutionary Genetics into an Ecosystem-Engineering Model

We begin with a straightforward extension of an ecological model proposed by Gurney and Lawton (1996). Gurney and Lawton's model describes the dynamics of a population that must modify its own habitat to survive. Their main assumptions and basic equations are as follows. At time t there are H units of habitat capable of supporting a niche-constructing (that is, habitat-modifying) population, composed of N individuals. Units of habitat are chosen so that a population of N individuals requires N units of habitat to maintain a constant size, so H is the instantaneous carrying capacity of the population. The dynamics of the population can be described by the logistic equation

$$\frac{dN}{dt} = rN\left[1 - \frac{N}{H}\right]. \tag{8.1}$$

They assume that habitat shortage results in increased mortality, and habitat excess produces increased fecundity. Habitat can be found in one of three states: (i) virgin habitat V, which is unusable by the niche constructors, but can be modified by them so it becomes usable; (ii) usable niche-constructed habitat H, but which decays into a degraded state; and (iii) degraded habitat D, which is unusable but will eventually recover to the virgin state.

The total habitat stock T remains constant and is given by $T = V + H + D$. Gurney and Lawton assume that a unit of usable habitat has a constant probability per unit time (δ) of decaying into the degraded state, that a unit of degraded habitat has a constant probability per unit time (ρ) of recovering to the virgin state, and that a single niche-constructing organism can produce new usable habitat at the rate $p(V, N)$ per unit time.

Hence, the dynamics of the system can be described using equation 8.1, together with

$$\frac{dH}{dt} = p(V, N)N - \delta H \tag{8.2}$$

and

$$\frac{dV}{dt} = \rho(T - V - H) - p(V, N)N. \tag{8.3}$$

Gurney and Lawton's system can be extended to become an evolutionary model through the application of methods that are well established within theoretical ecology (e.g., Roughgarden 1979; Fox et al. 2001). For illustration, we present one such extension, which considers how the rate at which niche constructors convert virgin to usable habitat is affected by genetic variation. As before, we assume that the capacity for niche construction is affected by alleles E and e at locus **E**. If we let W_1, W_2, and W_3 be the fitnesses, f_1, f_2, and f_3 the frequencies, and H_1, H_2, and H_3 (where $H = H_1 + H_2 + H_3$) the carrying capacities of genotypes *EE*, *Ee*, and *ee*, then we can write

$$W_i = 1 + r - \frac{r}{H_i}N \tag{8.4}$$

for $i = 1$–3, with mean fitness given by

$$\overline{W} = 1 + r\left(1 - N\sum_{i=1}^{3}\frac{f_i}{H_i}\right). \tag{8.5}$$

It follows that expression 8.1 can now be rewritten as

$$\frac{dN}{dt} = rN\left(1 - N\sum_{i=1}^{3}\frac{f_i}{H_i}\right). \tag{8.6}$$

For simplicity, let us assume that niche constructors convert virgin to usable habitat at a rate ($\alpha_i V$) for genotypes $i = 1,2,3$. Then dH_i/dt is now given by

$$\frac{dH_i}{dt} = \alpha_i V f_i N - \delta H_i \tag{8.7}$$

for $i = 1$–3. Also, 8.3 can be modified to take account of the effects of all three genotypes:

$$\frac{dV}{dt} = \rho(T - V - H) - VN\sum_{i=1}^{3}\alpha_i f_i. \tag{8.8}$$

Using standard recursions for genotype frequencies, computed using 8.4 and 8.5, together with equations 8.6–8.8, the evolution of the tendency to convert virgin to usable habitat in an ecological system can be explored.

An alternative extension would allow the genotypes to differ in their capacity to respond to niche-constructed usable habitat by altering their growth rates. Such a model would investigate how genetic variation at a recipient locus, now equivalent to our **A** locus, is affected by changes in the niche-constructed environment. Similar evolutionary models could explore other ecological aspects of niche construction.

Such models would be useful for any situation in which ecological and evolutionary responses to niche construction are likely to occur together. One example relates to the density-dependent regulation of population sizes (Ricklefs and Miller 2000). Ecologists usually assume that (i) the capacity of organisms to utilize environmental resources is a property of the organisms themselves, (ii) the capacity to supply organisms with resources is a property of environments, and (iii) the capacity of environments to supply organisms with resources determines the equilibrium population density that the environment can support, that is, the carrying capacity of the environment for that population. Usually ecologists also assume that (iv) the environment's capacity to supply organisms with resources does not depend on the capacity of those organisms to modify their environments (although exceptions are discussed in chapter 3). In the system described by equations 8.1, 8.2, and 8.3 assumption (iv) is no longer true. Here the amount of the key environmental resource, H, depends not only on the rate at which usable habitat decays into a degraded state and the rate at which degraded habitat recovers to a virgin state, but also on the per capita rate at which virgin habitat is converted into new usable habitat by niche-constructing activities. Hence, H now depends partly on feedback from the population, and therefore potentially on various population parameters including r, the intrinsic rate of increase of the population.

This feedback complicates the concept of density dependence. For example, an approach toward an equilibrium density of H individuals might be delayed by the effects of niche construction on the value of H. In some cases the population might never reach H and its size might only be regulated by density-independent factors. The opposite could also happen. A population could accelerate the rate at which it reaches H because of its negative niche construction. If, in addition, we were to assume that natural selection favors niche constructors that convert virgin to usable habitat at a faster rate, the density-dependent dynamics may become dauntingly complicated. On the plus side, however, the inclusion of niche construction might explain why the apparently simple idea that the size of a population is regulated by the capacity of its environment to supply it with resources has, until recently (Hanski 1999), proved so difficult to confirm (e.g., the arguments initiated by Lack [1954] and Andrewartha and Birch [1954]).

8.2.2 Incorporating Ecology into Evolutionary Models of Niche Construction

A corresponding extension is possible for our genetic models. Recall model 2 of section 3.3.3, which in the case of positive niche construction describes the amount R of a resource **R** in the environment at time t as

$$R_t = \lambda_1 R_{t-1} + \lambda_2 x_t + \lambda_3. \tag{8.9}$$

Here λ_1, λ_2, and λ_3 are coefficients that determine the degree of independent depletion, the effect of positive niche construction, and the degree of independent renewal, respectively. It is apparent that 8.9 already incorporates many of the environmental variables found in Gurney and Lawton's model. Indeed there are many parallels; for instance, if we were to rewrite 8.9 as

$$\frac{dR_A}{dt} = \lambda_2 xN + \lambda_3 - \lambda_1 R_A \tag{8.10}$$

and set $R_A = H$ and $\lambda_3 = 0$, the resemblance to 8.7 is striking. Equation 8.10 provides an ecological version of our formulation that tracks the absolute amount R_A of the resource in the environment, rather than the proportion that is available to the population. Combined with the set of gametic recursions given by A1.3 in appendix 1, and with an expression that gives the change in population size as a function of its size in the previous generation and its mean fitness (i.e., $N_{t+1} = \overline{W}N_t$), we again have the kind of system that can simultaneously explore the evolutionary and ecological ramifications of niche construction. We reiterate the need for such models at the ecosystem level. Both ecological and evolutionary responses to the feedback from prior niche construction would be expected to interact simultaneously with all three of the ecosystem currencies discussed in chapter 5. An example of this type of analysis is provided by Kerr et al. (1999), who explored the evolution and ecological impact of flammability in resprouting plants.

The theoretical models in chapter 3 and appendixes 1–3 may also offer further potential points of entry for modeling the ecological ramifications of niche construction. For instance, *primacy* and *recency* effects, which were introduced relative to R in model 1, give a temporal dimension to our analyses and relate closely to ecological variables such as the length of time a population has been present at a site (Jones et al. 1994). Similarly, the coefficient λ_1 in model 2, which is a measure of the spontaneous rate of depletion of the resource, could represent the durability of constructions, such as artifacts. The effect of the resource on the contributions of genotypes at locus **A** to two-locus fitnesses could easily be adjusted to become density dependent. However, as the impact of niche construction is often, although not always, localized to the vicinity of the constructor, some other aspects of niche construction may be better investigated using agent-based models that explicitly incorporate spatial structure into the analysis (Durrett and Levin 1994; Kerr et al. 2002). In some circumstances niche-constructing mutants should be able to invade a population by modifying their local environment, to their sole benefit, or to the detriment of their immediate neighbors, giving them a selective advantage that they would not have if their impact on the environment were experienced by all individuals in the population.

8.2.3 More Models for Evolutionary Ecology

In chapter 5 we showed that niche construction adds the biotic $\rightarrow$ abiotic and abiotic $\rightarrow$ abiotic links to the biotic $\rightarrow$ biotic and abiotic $\rightarrow$ biotic links considered in standard evolutionary theory.

Figure 8.1 illustrates five alternative scenarios based on three out of the four elementary interactive links in ecosystems. The noncontroversial link (i) biotic $\rightarrow$ biotic interaction is not considered here. Each of these scenarios illustrates one or more possible environmentally mediated genotypic associations, or EMGAs. In each case the

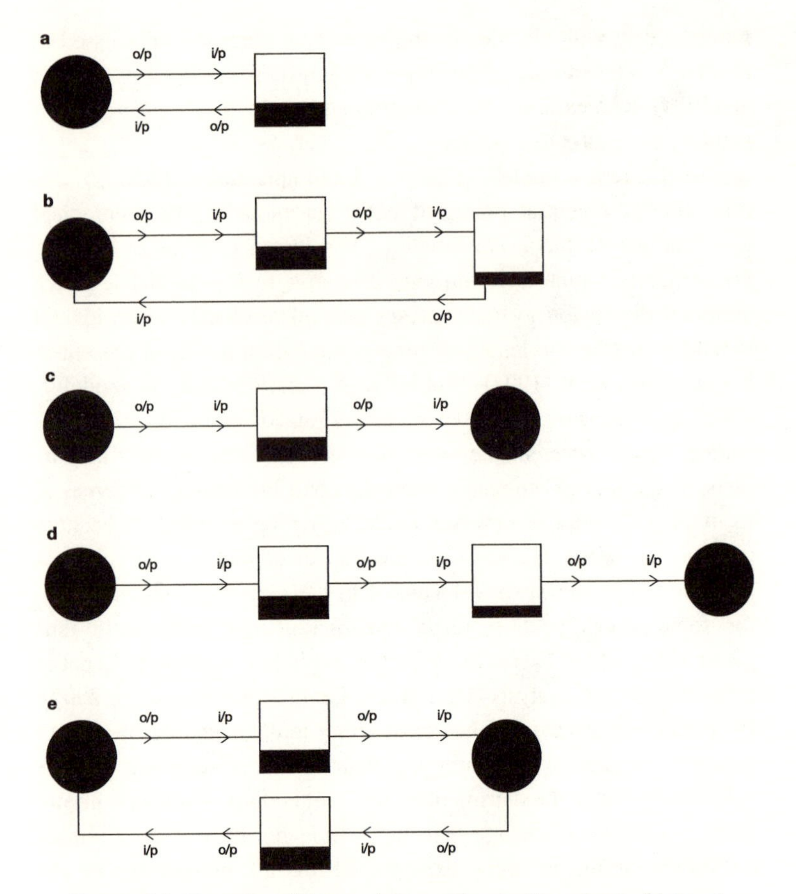

FIGURE 8.1. Examples of loops, chains, and webs formed out of interacting biota (circles) and abiota (squares) in ecosystems. Inputs and outputs are indicated by (i/p) and (o/p), respectively. Black coloration indicates the expression of genes in the ecosystem component. The shaded parts of the abiota indicate the extent to which the abiota have previously been driven into altered physical states by the niche construction of "upstream" biota. (a) A population of organisms expresses genotypes that modify an abiotic ecosystem component by niche construction. A modified natural selection pressure subsequently feeds back from this abiotic component to affect selection at different genes in the same population. (b) The same as (a) except that the feedback now occurs via two interacting abiota. (c) The same as (a) except that the modified natural selection pressure now feeds forward to a second population. (d) The same as (c) except that there are now two intermediate and interacting abiota in the chain. (e) The indirect coevolution of two niche-constructing populations via two intermediate abiota. In this example, the abiota do not interact directly with each other.

EMGA comprises an association between an **E**-locus and an **A**-locus genotype, mediated by an environmental component **R**. In cases (a) and (b) both the **E** locus and the **A** locus are in the same population. In cases (c), (d), and (e) the **E** locus and the **A** locus are in two different populations. The scenarios all include abiota, that is, they all involve link (ii) abiotic → biotic and link (iii) biotic → abiotic interactions, and some involve link (iv) abiotic → abiotic interactions as well. Neither links (iii) nor (iv) are included satisfactorily in models from standard evolutionary theory. In principle, however, both can be modeled by extended evolutionary theory. These five scenarios have therefore been chosen because they each represent part of an ecosystem that standard evolutionary ecology either does not include or includes only in a manner that may sometimes introduce distortions. In each case, we will suggest ways in which an extended evolutionary ecology could model the subsystem more effectively.

Figure 8.1a illustrates a case in which a single population expresses niche construction that modifies an abiotic resource in the population's environment, which feeds back to affect selection acting on the same population. At the population-genetic level this corresponds to our basic **E**-**R**-**A** models presented in chapter 3, where the niche construction is affected by genes at the **E** locus while the recipient locus corresponds to **A**. At the ecological level, this scenario is also well described by Gurney and Lawton's (1996) model, as well as many other ecological models (DeAngelis 1992). As we suggested in the previous section, this is the kind of scenario that could be explored using evolutionary ecology models, for example, by evolution-*izing* Gurney and Lawton's model, or similar systems, in the manner described by equations 8.6–8.8, or by ecolog-*izing* our own niche-construction models, for instance, in the manner described by equation 8.10.

This approach offers some potential advantages. It tracks the dynamics of abiota, which may be very different from the dynamics of any evolving population, allowing the abiota to affect both the ecological and evolutionary consequences of the niche-constructing activities of organisms. As we have seen, it also potentially facilitates an exploration of how evolving organisms may influence the carrying capacity of their own environments, or their population growth rates.

Figure 8.1b is the same as 8.1a, except that there are now two abiotic components, interacting via a link (iv) ecosystem interaction, mediating between the **E** and **A** loci. In evolutionary models this extra link should not matter much since the only evolutionarily significant effect of this population's niche construction is a change in the first abiotic component, which eventually has a knock-on effect on the second abiotic component. This in turn modifies a natural selection pressure stemming from the second abiotic component, which acts on the **A** locus in the original population. Thus scenario (b) could be modeled in the same way as (a), and the same points we made about (a) should apply again. If we label the absolute amount of the two abiotic resources as R_{A1} and R_{A2}, then from 8.10 we can write

$$\frac{dR_{A1}}{dt} - f[x, N, R_{A1}, \lambda_1], \tag{8.11a}$$

$$\frac{dR_{A2}}{dt} = g[R_{A1}, R_{A2}, \lambda_2], \tag{8.11b}$$

where x is the frequency of allele E that is expressed in niche construction, N is the number of individuals in the niche-constructing population, and λ_1 and λ_2 represent the rates of independent change in R_{A1} and R_{A2}, respectively.

Figure 8.1c is the same as 8.1a, except that, although all individuals in both these populations have genes **E** and **A**, the **A** locus is expressed in a different population from the **E** locus. Here, the **E** locus in the first population influences the evolution of a second population by expressing niche construction that modifies an intermediate abiotic resource **R**. A modified natural selection pressure from **R** then acts on the **A** locus in the second population. This time there is no feedback from either the second population or the intermediate abiotic resource to the first population. There is therefore no coevolution between these two populations. Instead figure 8.1c could represent donor control in an evolutionary context, for example, an evolutionary response by other fauna to the niche-constructing activities of a benthic "bulldozer" species in a lake or in marine sediments. In this case the dynamics of the abiotic resource might be described by a function like 8.11a, where the amount of **R** is given by R_{A1}, while the influence of the first population on the evolutionary ecology of the second might be specified by, for example, making the carrying capacity a function of the amount of the abiotic resource:

$$\frac{dH_2}{dt} = k[R_{Ai}, \tau], \tag{8.12}$$

where H_2 is the carrying capacity of the second population, and τ represents other processes independent of either population that affect H_2, and in this instance $i = 1$. Thus, if R_{A1} is affected by the expression of the **E** locus, so is H_2.

Figure 8.1d describes a scenario that is the same as that depicted in 8.1c, except there are now two interacting abiota. Here the dynamics of the two abiota and the second population are described by 8.11 and 8.12, where $i = 2$.

Figure 8.1e depicts the indirect coevolution of two populations either via a single intermediate abiotic resource, but at two different points in a cycle, or via two different abiotic resources. For example, two populations might be competing for the same abiotic resource (e.g., water, minerals, light), or they might have an indirect mutualistic relationship, for instance the indirect mutualisms investigated by Daufresne and Loreau (2001) between primary producers and decomposers that are mediated by material cycling. Alternatively, figure 8e might represent a predator-prey coevolutionary relationship that includes an intermediate abiotic refuge available for the prey, such as holes in the ground. We anticipate that, among host-parasite coevolutionary relationships, coevolution via intermediate abiota will be rare because, generally speaking, in this kind of coevolution, one population is a resource for the second population without other intermediates being involved. However, for many coevolutionary relationships, intermediate abiota should be quite important. For example, competition usually involves competing for "something," whether it be an intermediate biotic resource or an intermediate abiotic resource in a shared environment.

For instance, consider the following well-known two-population Lotka-Volterra competition model:

$$\frac{dN_1}{dt} = r_1 N_1 \left[1 - \frac{(N_1 + \alpha_{12} N_2)}{K_1}\right], \tag{8.13a}$$

$$\frac{dN_2}{dt} = r_2 N_2 \left[1 - \frac{(N_2 + \alpha_{21} N_1)}{K_2} \right]. \tag{8.13b}$$

Here K_1 and K_2 are the carrying capacities and r_1 and r_2 are the intrinsic growth rates of populations 1 and 2, respectively, α_{12} is a competition coefficient that measures the extent to which species 2 presses upon the resources used by species 1, and α_{21} is the corresponding coefficient for the effect of species 1 on species 2. However, if the two species compete over an intermediate resource, the amount of which is R_A, a more accurate depiction would make the competition functions dependent on the mutually exploited resource and treat R_A as a dynamic variable. For instance,

$$\frac{dN_1}{dt} = r_1 N_1 \left[1 - \frac{(N_1 + \varepsilon_{12} R_A)}{K_1} \right], \tag{8.14a}$$

$$\frac{dN_2}{dt} = r_2 N_2 \left[1 - \frac{(N_2 + \varepsilon_{21} R_A)}{K_2} \right], \tag{8.14b}$$

and

$$\frac{dR_A}{dt} = g[R_A, N_1, N_2, \lambda], \tag{8.14c}$$

where λ is the net rate of independent renewal or depletion of **R**. Here, conceptually similar models are beginning to emerge within evolutionary ecology (e.g., DeAngelis 1992), but they are far from mainstream.

Indirect mutualistic relationships could also be modeled in a similar way. Here, one potential advantage of our approach may be worth highlighting. Roughgarden points out that the theoretical issues raised by mutualism are typically difficult to investigate because "simple models for mutualism don't stand alone and quickly turn into models that must include other types of interactions too, all of which complicates the matter enormously" (Roughgarden 1998, p. 312). One problem is that when species help each other they generate an autocatalytic process that, in theory, might cause the size of each population to explode in finite time. This obviously cannot happen in reality, so in pursuit of greater realism "models of mutualism wind up building in a lot of density dependence to keep the autocatalysis under control" (Roughgarden 1998, p. 312).

Our niche construction approach may satisfy the "realism" criteria more naturally. For example, assume that the biotic component on the left of figure 8e is a plant population, and the one on the right is an ectomycorrhizal fungus population. Let us also assume that their mutualism is indirect because it operates via two intermediate abiotic nutrients. Thus mycorrhizae might niche-construct with the effect of enhancing the plants' ability to extract, say, phosphorus from the soil. The plants' niche construction might supply carbon to the mycorrhizae (Ricklefs and Miller 2000). Now there is no danger of autocatalysis because the capacity of the mychorrhizae to enhance the supply of phosphorus to the plants does not depend on their niche-constructing activities only, but also on other independent variables that determine the availability of phosphorus in the soil. However, if phosphorus is limited, so is plant growth, and so is the supply of carbon to the mycorrhizae. Hence, the population sizes of both these populations should be limited by the availability of intermediate abiota in a shared environment. We should therefore end up with a stable mutualistic relationship represented by equations 8.14a–8.14c, rather than with a "runaway" system. Obviously, this approach cannot deal with all cases of mutualisms (Bronstein 2001), but it may be able to account for some that have proved difficult to resolve.

8.3 METHODS FOR DETECTING THE EVOLUTIONARY AND ECOLOGICAL IMPACT OF NICHE CONSTRUCTION

8.3.1 Experimental Methods That Investigate the Consequences of Canceling or Enhancing a Population's Capacity for Niche Construction

Jones et al. (1997) proposed a general method for investigating experimentally the ecological effects of ecosystem engineering by comparing virgin, engineered, and degraded habitats. This method parallels the evolutionary method that we introduced in section 7.2.1 of the previous chapter. For instance, researchers might contrast the environments of experimental populations of niche-constructing organisms (engineered habitats) with the habitats of otherwise identical populations the impact of whose niche construction is experimentally negated (or degraded). Control conditions without subject organisms (virgin habitats) can be used to establish the nature and magnitude of the niche-constructed changes in the experimental environment and to provide the information needed to stop the effects of niche construction in the experimental condition. A further experimental condition could involve enhancing the effects of niche construction on the environment to provide a "superengineered" habitat.

8.3.2 Further Methods

The approach described in section 8.3.1 can be used to operationalize extended evolutionary theory and translate it into an empirical program. The easiest way to illustrate this point is to focus on a specific example. Figure 8.2 shows the flow of two nutrients, nitrogen and phosphorus, in two comparable but contrasting forest ecosystems, a 47-year-old Scots pine (*Pinus sylvestris*) plantation in England, and a 50-year-old mixed tropical forest in Ghana. The data are from Greenland and Kowal (1960) and Ovington (1962), and the figure is a textbook example of nutrient cycling in terrestrial ecosystems (Ricklefs and Miller 2000). This system constitutes a model that is sufficient to illustrate our points.

Figure 8.2 reduces both these forest systems to three compartments: living plant biomass (the trees), organic detritus (e.g., litter), and mineral soil. It then compares the flow of nitrogen and phosphorus through these three compartments in the two cases. The compartment sizes are represented by spheres of different sizes, with the size of each sphere proportional to the amount of the nutrient in that compartment. The arrows between the compartments, plus their associated numerical values, indicate the rates at which material in each compartment is transferred to the next compartment.

The most obvious differences between the two forests are as follows. (i) The tropical forest trees sequester more nitrogen and phos-

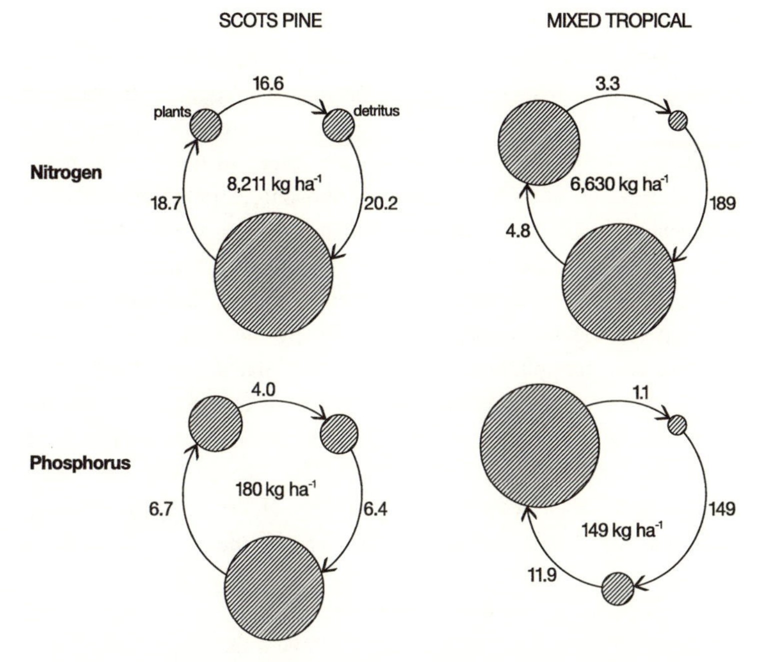

FIGURE 8.2. The flow of two nutrients, nitrogen and phosphorus, in two contrasting forests, a 47-year-old Scots pine (*Pinus sylvestris*) plantation in England, and a 50-year-old mixed tropical forest in Ghana. Here, both forest systems are reduced to three compartments only, the living plant biomass, organic detritus, and mineral soil. The compartments are illustrated by spheres of different size, each proportional to the amount of nutrient in it. The arrows between the compartments, plus their associated numerical values, indicate the rate at which the material in each compartment is transferred to the next compartment. This model suggests that tropical plants sequester more nitrogen and phosphorus than do temperate plants. It also shows that the rates of transfer of nitrogen and phosphorus among the three compartments are of the same order of magnitude in England but, in Ghana, the regeneration of both nitrogen and phosphorus in the detrital compartment is more rapid than it is in England. The niche-constructing activities of the constituent organisms may be partly responsible for the differences between these systems. Further details are given in the text. This figure is based on figures 13–15 in Ricklefs and Miller (2000, p. 264). Data are from Greenland and Kowal (1960) and Ovington (1963). (From *Ecology*, by Robert E. Ricklefs and and Gary L. Miller © 1973, 1979 by Chiron Press, Inc., 1990, 2000 by W. H. Freeman and Company. Used with permission.)

phorus, by retaining more of these nutrients, than do the temperate forest trees. This is particularly noticeable in the case of phosphorus, most of which is retained in the plants. (ii) Whereas the rates of nitrogen and phosphorus transfer are approximately the same among all three compartments in the temperate forest, in the tropical forest the regeneration of both nitrogen and phosphorus in the detrital compartment proceeds far more rapidly than it does in the temperate forest. Conceivably, the structural and functional differences between these two forest systems can to some extent be accounted for by external forcing functions, for example, by the differences in temperature and precipitation in England and Ghana (e.g., Leith 1973, 1975). However, the structure and function of ecosystems do not depend only on external forcing functions, but also on their internal properties (Ricklefs and Miller 2000). This leaves open the possibility that the niche-constructing activities of the constituent organisms may be partly responsible for the differences between these systems.

To consider this question further, the nitrogen cycles shown at the top of figure 8.2 can be represented in terms of their principal ecosystem links and compartments (fig. 8.3). To do so, however, we must identify more of the biota that are likely to be present in these cycles.

To the three compartments in figure 8.2 have been added detritivores, which convert organic nitrogen in plant litter into ammonia in the soil, and soil bacteria that convert ammonia to nitrite or nitrite to nitrate deposits in the soil. It is these soil nitrates that are extracted by the plants roots, probably with the help of mycorrhizae (omitted here). All the organisms illustrated in figure 8.3 niche-construct in one or more ways that affect the nitrogen cycle: the plants by generating detritus, the detritivores and the soil bacteria by collectively and progressively mineralizing nutrients. We can now ask some empirical questions about the extent to which the biota in figure 8.3 may be exerting control over this nutrient cycle by their niche-constructing activities.

One question that has often been asked in this connection is: *Does it make a difference to the structure and function of an ecosystem which species are present in particular compartments or at particular positions in its cycles?* (Pimm 1999; Law 1999; Diaz and Cabido 2001; Kinzig et al. 2001). Vitousek (1986) has suggested that this

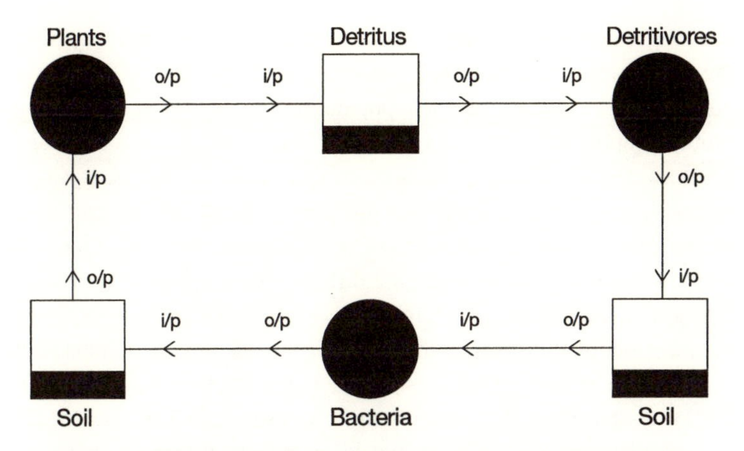

FIGURE 8.3. The same nitrogen cycles as those shown in figure 8.2 illustrated in terms of elementary ecosystem components. In this figure detritivores and bacteria have been added to the plants shown in figure 8.2. Circles indicate biota. Squares indicate abiota. Black coloration indicates the expression of genes in the ecosystem component. The shaded parts of the abiota indicate the extent to which the abiota have previously been driven into altered physical states by the niche construction of "upstream" biota.

question can be used as an explicit test of the importance of species properties in controlling ecosystem function. If the answer to this question is "no," then the ecosystem is controlled primarily by its energy and matter flows. That would be the case, for instance, if external forcing agents were responsible for the differences between the two forest systems in figure 8.2. Conversely, if the answer is "yes," then there must be at least one niche-constructing activity expressed by a particular species that is important to the control of a vital aspect of the ecosystem. This would be the case if the differences between the English and Ghanaian forests could be attributed to alternative forms of niche construction on the part of the two populations of plants, or the detritivores, or the bacteria. If a species were sufficiently important to the structure or function of the ecosystem as a whole, it might be regarded as a dominant species.

From the perspective of extended evolutionary theory, semantic information flows round entire ecosystems as a consequence of the niche construction of organisms. This raises the possibility of indirect coevolutionary events in an ecosystem chain or cycle stemming from the succession of EMGAs that are established by each of the niche-constructing activities of each population. As a consequence, any biotic component in any ecosystem could be responsible for the differences found in any particular compartment of the ecosystem, rather than for just its own dynamics or those of the compartment immediately "downstream" of itself. For instance, the Scots pine and mixed tropical forests in figure 8.2 differ in the proportion of organic nitrogen in the plants. This might reflect genetic differences between the plant populations, say in their capacity to take up soil nitrates, but it may also reflect differences between the bacteria (Balser et al. 2001) or detritivore population's niche construction that has knock-on consequences (fig. 8.3). It may even reflect differences between the niche-constructing activities of the two plant populations, such as the amount of litter produced, which subsequently feed forward around the entire cycle.

It follows that in any chain of interactive links involving both biota and abiota, any biotic component in the chain must be able to tolerate, or benefit from, the niche-constructing activities of all the biota upstream of itself regardless of the numbers of intermediate abiota. Similarly, any biotic component may "challenge" the niche of any biota downstream of itself, again regardless of the numbers of intermediate abiota. It follows that in any ecosystem cycle, such as those shown in figures 8.2 and 8.3, the closure of any chain means that there are likely to be some widespread coevolutionary consequences arising from the niche-constructing activities of each biotic component in the cycle. Some of these consequences will be insignificant, but some are likely to be ecologically and evolutionarily important.

Thus, extended evolutionary theory may offer a fresh perspective on the question of whether biota can exert control over ecosystems. Now what matters is whether the removal, or replacement, or addition of a particular niche-constructing species from a particular position in an ecosystem has the effect of removing or changing one or more EMGAs that are responsible for exerting some control over the ecosystem (see chapter 5). Again, this suggests that specific changes in ecosystems such as exclusions, invasions, or successions, or the structure and function of ecosystems as a whole, cannot be understood in terms only of energy and matter flows (Reiners 1986). It is also necessary to include the influence exerted by EMGAs and by the semantic-information flows they support. On this basis we offer the following general predictions, some of which are extremely intuitive, but others are less obvious.

1. If a species carries genes that allow it to tolerate or benefit from the niche-constructing activities of upstream biota, then it should be able to invade an ecosystem chain or cycle at that point (unless it is outcompeted by the present incumbent).

2. If the species also carries genes that cause it to niche-construct in ways that generate an "ecosystem service" (or knock-on consequences) that is similar to the service currently provided by another species at the same point in the ecosystem and that affects the downstream compartments (including abiota), then its invasion should be "benign" in the sense that it should not change the ecosystem's structure or function. In this case the invading species could potentially substitute for the one currently in place, and it would replace the incumbent if it is a superior competitor, for example, if it is better able to tolerate or benefit from the niche-constructing activities of the upstream biota at that point.

3. If a species is removed but another species is already available to fill the same position in the ecosystem, the second species should be able to compensate for the removed species, provided that it carries genes that are capable of expressing similar niche-constructing activities and of supporting a similar EMGA (assuming that it is also genetically able to tolerate the niche construction of upstream components).

4. If a species invades and replaces an incumbent, and niche-constructs in a different manner, generating different EMGAs, then the invading species is likely to be a "malign" invader, in the sense that it is likely to contribute to perturbation or destruction of the ecosystem to some degree. All downstream biota may be affected by the invasion, even if separated by abiotic links in the chain.

5. If no new species invades and the incumbent species remains in place but starts to niche-construct in a new way, then its novel niche construction will constitute an "invasion" by itself, and points 2 and 4 above should reapply. This might happen for a variety of reasons: for instance, there could be a new mutation in the incumbent species,

or if the incumbent is animal, it might learn, and socially transmit, a new niche-constructing behavior that gives it access to a new environmental resource. The use of tools by Galápagos woodpecker finches to extract grubs from trees is a possible example. Novel human cultural niche-constructing behavior provides others (see chapters 6, 9, and 10).

6. If an incumbent species remains in place but starts to niche-construct in a new way, then its novel niche construction could change the set of candidate species that are capable of invading an ecosystem. For example, an incumbent's novel niche construction could enable a species that was hitherto unable to invade to do so or, on the contrary, it could prevent a species that formerly was a potential invader from invading.

7. If no species is available that can substitute for a species currently in place then the species in place will be a "dominant" or "keystone" species, meaning that if it is removed, a whole ecosystem cycle, or even a whole ecosystem, will be perturbed.

One way of testing these predictions, whether in the field in or the laboratory (e.g., Naeem and Li 1997; Weatherby et al. 1998; Kerr et al. 2002), would be to identify the genes (or corresponding phenotypes) involved in each EMGA at each point in a cycle. This would entail (i) identifying a nutrient of interest; (ii) identifying the gene(s) (or corresponding phenotypes) in the next upstream biotic component of the ecosystem that affect(s) the nutrient through its niche construction; (iii) identifying the gene(s) (or corresponding phenotypes) in the next downstream biotic ecosystem component that are subject to modified natural selection pressures directly or indirectly because of the upstream biota's niche construction; (iv) and repeating, link by link. There are, however, two complications that must be borne in mind. One is that there is no simple transitivity. The gene(s) (or corresponding aspects of the phenotype) selected by modified natural selection pressures at step (iii) are most unlikely to be the same gene(s) (or corresponding aspects of the phenotype) that subsequently modify natural selection pressures for the next downstream biota at the next step (ii) and affect the same nutrient chosen at step (i). The second is that it may not be practical to work at the genetic level, in which case researchers may have to work with changes in phenotypes. However, the same logic should still apply.

In addition, researchers could experimentally manipulate the niche construction of a particular species and monitor the downstream consequences. For instance, they could remove or deplete the niche-constructing population (e.g., clearing trees) or dilute or cancel out its effects on the environment (e.g., removing litter). An alternative procedure for comparative studies would be to exchange the niche-constructed products of two populations, so that the "wrong" niche construction is introduced at one point in the cycle. Another possibility would be to add more of a specific niche-constructed product, or enhance the process of niche construction in some way (e.g., adding extra litter; Schwilk 2002) so more product is generated. In some cases it may even be possible to introduce a new niche-constructing activity into an ecosystem artificially, without introducing new organisms, for instance, by making a new resource available to one population or, more ambitiously, by training a new niche-constructing behavior in a group of animals. With respect to the latter idea we are encouraged by Sol and Lefebvre (2000), who found that avian species with relatively large brains and a high frequency of foraging innovations in their area of origin tend to be more successful invaders in the islands of New Zealand than species with smaller forebrains and lower innovation frequencies. So it might be possible to teach some large-brained avian innovators some new "tricks." All of these manipulations are theoretically possible, although in practice they may be very difficult.

Potentially, the impact of natural perturbations on ecosystem functioning might also be traced backward around the links of constituent biota and abiota. For instance, to use the example in figure 8.3 again, suppose a change is detected in one component of the cycle, say a sudden increase in the amount of bacteria that transform soil nitrite into soil nitrate. The obvious first step would be to investigate whether there has been a change in soil nitrite levels. If not, perhaps the change reflects an evolutionary episode on the part of the bacteria. In this case we might investigate whether it is a consequence of a corresponding evolutionary change or an increase in the number of bacteria that transform soil ammonia into soil nitrite. If a change is detected in this population too, we can trace it back further, for instance, by asking whether there has been any increase in soil ammonia, any alteration in the niche-constructing activities of the detritivores, or any change to the plants, each of which could have indirect downstream effects via the abiotic soil. The same logic applies to experimentally induced perturbations in ecosystem functioning, where one niche-constructing component could be modified, and its downstream effects traced forward around the ecosystem, even across abiotic bridges. If our reasoning is correct, there will be a detectable ecological or evolutionary response somewhere in the ecosystem.

The dynamics of these interactions could have interesting features too. For example, consider the effects of time lags. A nutrient with a long residence time in an abiotic compartment, but which enters that compartment owing to the niche-constructing activities of specific organisms, might act as a new or modified source of natural selection only much later, perhaps after many generations, when it has accumulated sufficiently to become accessible to other organisms. The same logic applies to ecological successions where, again, upstream biotic compartments could trigger evolutionary as well as ecological events further downstream, perhaps indirectly via EMGAs.

Finally, which of the components in an ecosystem exerts the greatest control on its structure cannot be taken for granted. Jones et al. (1994, 1997) point out how difficult it can be to identify the most potent niche constructors in an ecosystem on the basis of conventional ecological approaches only, where potency depends on the number and type of resource flows that are modulated by different niche-constructing organisms and on the number of other species that are affected by the flows that they alter. For example, Jones et al. describe the niche-constructing activities of diatoms that dominate areas of sandy shorelines in the Bay of Fundy. The diatoms produce carbohydrate exudates, which bind the sand and stabilize its movement. Amphipods are the dominant grazers of diatoms in these areas, and where the amphipods are abundant the stabilization of the sand is reduced. However, sandpipers are the dominant predators on the amphipods, and where sandpipers are present they reduce amphipod grazing and thereby promote the restabilization of the habitat by the diatoms (Daborn et al. 1993). In this system, which Jones et al. call a "coupled engineering and trophic cascade," it looks superficially as though the sandpipers must be the key species in the system. However, they are not because they are not the key niche constructors. The key niche constructors are the diatoms.

8.4 INDIRECT EFFECTS IN ECOSYSTEMS: SOME NEW APPROACHES TO OLD QUESTIONS

The flows of semantic information through ecosystems that are entailed by extended evolutionary theory, and the potential capacity of niche-constructing populations to influence the evolution of many (or all) the other populations in their ecosystem via EMGAs, also introduces more general questions. Do ecosystems have unforeseen properties that are currently unrecognized because EMGAs have not been incorporated into theoretical and empirical analyses? Questions such as "does coevolution promote greater stability in ecosystems?" and "are ecosystems self-regulating?" have hitherto been investigated from the perspective of standard evolutionary theory. These are questions that have been asked throughout the history of ecology but have still not received satisfactory answers (O'Neill et al. 1986; McGlade 1999a).

From our perspective such questions could not have been satisfactorily answered because the theoretical approaches applied have been too restrictive. As we saw in chapter 5, it is not possible to provide answers under the rubric of process-functional ecology in terms of energy and matter flows only. Nor is it possible to provide answers only under the rubric of population-community ecology, since standard evolutionary theory does not include evolutionarily driven changes in abiota, nor does it model the indirect coevolution of populations via abiota. The most we can hope for from conventional approaches are probably either negative answers derived from insufficiently analyzed scenarios or positive answers that are likely to appear either mystical or intractable because, in the absence of niche construction, they cannot explain how energy and matter flows in ecosystems can be physically affected by an evolutionarily generated flow of semantic information.

Extended evolutionary theory potentially surmounts these restrictions and could therefore encourage the reconsideration of some of these questions in a new light. We have no intention of trying to answer any of them now and indicate only how they might be approached in the future, both theoretically and empirically, in ways that might eventually lead to more satisfactory answers.

8.4.1 Stability in Ecosystems?

The first question is deceptively simple. Does coevolution promote greater stability in ecosystems? The traditional view is that it ought to, because coevolution should lead to mutually accommodated and coadapted species. Roughgarden (1979) discussed this complexity-stability issue with respect to community structure and pointed out that, up to that time, the issue had usually only been considered in terms of pure ecological models. However, the pure ecology was itself equivocal and still is today. For example, Elton (1958) originally argued that complexity implies stability. May (1974) then showed that, while increased diversity may stabilize communities, it tends to destabilize populations. Later, King and Pimm (1983) found that ecosystem resistance increases as the number of competitive interactions between species increases, and in apparent support McNaughton (1985) and Tilman and Downing (1994) found some evidence from field studies that species-rich systems are indeed more resistant to change than species-poor systems (Pimm 1999). However, Pfisterer and Schmid (2002) then reported that, when an array of experimental grassland plots was perturbed by a simulated drought, it was species-poor systems rather than species-rich systems that were more resistant to this perturbation. A possible reason why there are still no clear answers to biodiversity-ecosystem function questions in ecology may be because some previous approaches have been too restricted (Kinzig et al. 2001). For example, Pacala and Tilman (2001) point out that it is vital to run experiments for many years because of transients: "A transition from one behavior early in a biodiversity-ecosystem functioning experiment to another later on is expected. . . . Exponential growth dominates early, while competition dominates later on" (Pacala and Tilman 2001, p. 165).

Intuitively, the addition of coevolutionary relationships to these ecological scenarios might be expected to improve stability, at least in communities, but Roughgarden (1979) found this was generally not the case. At that time, however, there was little attempt to incorporate instances of indirect coevolution, via intermediate abiota, whose physical states could have been changed by niche-constructing organisms. More recently the potential importance of indirect coevolutionary relationships in ecosystems has become more widely appreciated, and this has led to a better understanding of some of the issues involved. For example, Diaz and Cabido (2001) emphasized that species richness is not an adequate surrogate for functional diversity and that the latter is more important than the former. Similarly, Pfisterer and Schmid (2002) suggested that their own results may have been caused by their simulated drought disrupting "niche complementarity" (species whose niches complement each other functionally are held to be more efficient in the use of resources) and that species-rich systems may sometimes be more prone to this sort of disruption than species-poor systems (Schmid et al. 2001; Naeem 2002). Both these ideas are consistent with our emphasis on the role of EMGAs in ecosystems. Other work has drawn attention to the fact that many indirect coevolutionary relationships may be operating simultaneously in an ecosystem. For example Daufresne and Loreau (2001) have shown that primary producers and decomposers in ecosystems can be linked by both mutualistic and competitive indirect interactions simultaneously, and in ways that raise further questions about the persistence of ecosystems. Mazancourt et al. (2001) reported a similar finding with respect to plant-herbivore relationships, and raised similar questions.

A number of authors (Tilman and Lehman 2001; Naeem 2001; Pacala and Tilman 2001; Balser et al. 2001) are now calling for the introduction of new concepts, approaches, and mechanisms to this debate. Niche construction and EMGAs are obvious candidates. Coevolutionary models that include niche construction may help answer some previously intractable questions. For instance, what may now be required are investigations that explicitly take account of the role of EMGAs in ecosystems, since indirect associations between genes, in different populations, involving either biotic or abiotic intermediates, could provide some important and hitherto neglected forms of feedback that may affect coevolutionary dynamics and stability of ecosystems.

8.4.2 Are Ecosystems Cybernetic?

Similar uncertainty surrounds this question, which may be interpreted as: Do ecosystems contain an information network (Engelberg and Boyarsky 1979)? If they do, then the structure and functioning of ecosystems are at least partly governed by feedback loops and by associated flows of information between their constituent organisms, which function in ways that contribute to the regulation and control of flows of energy and matter through ecosystems. This proposal has been made several times by different ecologists (McNaugton and Cougenhour 1981; Patten and Odum 1981; Margalef 1995; Straskraba 1995). It has also been opposed by others, notably by Engelberg and Boyarsky (1979), and critically evaluated by DeAngelis (1995).

One preliminary source of confusion is that the "information" referred to by both proponents and opponents of cybernetic ecosystems has typically not been the evolution-generated semantic information that we have been discussing here, but merely structural information or bits (e.g., Margalef 1963, 1995; Odum 1988). However, structural information is unlikely to have been the primary source of control in ecosystems (see Stuart 1985, and chapter 4). Any control is far more likely to depend on the niche-constructing properties of organisms, and therefore on the expression of the semantic information encoded in genes. A lack of consensus as to what is meant by information has probably been one obstacle to developing or resolving the arguments. This obstacle could be removed by assuming that the information most relevant to the control of ecosystems is the semantic information that is acquired and expressed by populations of niche-constructing organisms as a consequence of their prior evolution. Under this assumption we cannot expect cybernetic ecosystems to emerge from the laws of physics and chemistry, or from thermodynamics and stoichiometry only. Instead they must partly depend on the evolution of organisms.

A second major obstacle is the presence of nonevolving abiota in ecosystems. One of Engelberg and Boyarsky's most telling criticisms is that ecosystems include abiota as well as biota: "It is probably fair to say that the dominant interaction between the different populations of an ecosystem is the exchange of brute (informationally non-specific) matter and energy" (Engelberg and Boyarsky 1979, p. 322). As long as evolutionary ecology is based on standard evolutionary theory rather than extended evolutionary theory, the presence of "brute matter and energy" in ecosystems is likely to remain an insurmountable obstacle to the cybernetic-ecosystem hypothesis. As we saw in chapter 5, according to standard evolutionary theory the presence of abiota in ecosystems must interrupt any putative flow of semantic information through ecosystems. If evolutionary ecology is revised on the basis of extended evolutionary theory, however, this second obstacle also disappears. This is because extended evolutionary theory grants organisms the capacity to express some of the semantic information they carry in their genes (or brains) in their environments (including in environmental abiota) via their niche-constructing activities. This allows organisms to actively shift abiota into new physical states that could not be achieved under the laws of physics and chemistry alone. Potentially this is sufficient to connect the expression of semantic information by niche-constructing organisms to the acquisition of revised semantic information by populations of organisms in response to modified natural selection pressures. The result would be a continuous flow of semantic information through ecosystems, through both biota and abiota, making the cybernetic-ecosystem hypothesis far more plausible. Thus, ecosystems may be regarded as cybernetic in the sense that the interplay of material cycles and energy flows is partly under informational control, and this information generates regulatory feedback in the ecosystem, with no discrete controller required. This perspective generated by our extended evolutionary theory does not, however, imply that cybernetic ecosystems will necessarily be self-organizing, mutually compatible, or stable.

8.4.3 Are Ecosystems Complex Adaptive Systems?

One of the problems with the above discussion is that, although empirical researchers may approach ecosystem cycles link by link, or EMGA by EMGA, in reality it may be difficult to string them all together into a complex system (see above). Essentially the same problem confronts theoretical ecologists, who are readily able to derive equations specifying the dynamics of the ecosystem components, but for whom the system dynamics become intractable as the number of components increases. This raises the possibility of simulation, and here we will do no more than point to the potential of exploiting one approach, John Holland's (1995), to the understanding of complex adaptive systems (which Holland refers to in shorthand as "cas") by indicating how extended evolutionary theory can supply the syntax necessary to apply Holland's approach to ecosystems. This may aid investigators who want to travel this route to explore whether ecosystems demonstrate any self-designing or self-regulating properties as a consequence of the evolution of their biotic components and as a consequence of the effect of biota on abiota through niche construction.

Holland (1995) treats ecosystems as an example of a complex adaptive system, with other examples including the central nervous system, business firms, and New York City. Complex adaptive systems are systems composed of interacting agents, where each agent is described in terms of rules. Here, there are two kinds of agents to consider: biotic and abiotic ecosystem components. The rules that define biota are the laws of physics and chemistry, plus the information-processing rules that ultimately stem from the Darwinian algorithm. The rules that define abiota are, in part, the laws of physics and chemistry. However, unlike standard evolutionary theory, extended evolutionary theory allows abiota to be forced into new physical states, beyond any states achievable under dead-planet conditions, by the semantically informed, goal-directed activities of niche-constructing biota. Abiota can thus have design imposed on them by informed biota.

The rules at the heart of the semantic information-processing activities of biotic agents are assigned fitnesses or "credits" according to their outcomes for the agents. If a rule works well, this increases the probability that it will be used again. Here, "works well" means increasing the fitness of organisms, and hence being favored by natural selection. Similarly, if a rule works badly it is less likely to be used again. New rules can also be generated de novo by random mutations, while old rules can eventually be completely eliminated by selection.

Agents are connected to each other by flows. In ecosystems, between-agent flows involve the three universal ecosystem currencies:

energy, matter, and semantic information. It is at this point that Holland's (1995) genetic-algorithm approach can be made compatible with extended evolutionary theory by allowing the assignment of "fitness credits" to genes in biota to be influenced not only by autonomous selection pressures in environments, but also by selection pressures that have previously been modified by niche construction. This creates a flow of semantic information around entire ecosystems, which may turn a complex physical system not merely into a complex *adaptive* system (cas), but into a complex constructive-adaptive system (ccas).

The rules that define agents in a complex constructive-adaptive system all use the same syntax. We will illustrate this syntax here exclusively in terms of Holland's most basic rule, an IF [ ] THEN rule. The IF [ ] part of this rule is sensitive to inputs from an agent's environment via the agent's detectors, while the THEN part is responsible for governing the agent's choice of actions or outputs to its environment. A link will occur in a complex constructive-adaptive system whenever the input to an IF [ ] part of one agent's rule is sensitive to the prior actions specified by the THEN part of the rule of some other agent. In this way chains or cycles of between-agent links establish message-carrying communication channels between agents, and hence flows of semantic information throughout a system.

The point of contact between extended evolutionary theory and complex systems theory is that the syntax that governs our EMGAs is the same syntax that governs the links between agents in complex constructive-adaptive systems. EMGAs connect the informational content of genes in one population that is expressed in their niche construction (the THEN part of Holland's rule) to the informational content of genes, potentially in another population, that are subject to selection pressures modified by that niche construction (the IF part of Holland's rule). If extended evolutionary theory allows ecosystems to be effectively simulated as complex constructive-adaptive systems, it may prove possible to start exploring the extent to which ecosystems really are self-organizing and self-regulating.

8.4.4 Ecosystems: Superorganisms or Superconstructions?

The theory developed here might also illuminate one other old dispute that has "haunted" ecosystem-level ecology for nearly a century.

Are ecosystems superorganisms? If ecosystems are even partly self-designing or self-regulating, is that sufficient to turn them into superorganisms? To put the question crudely: Are ecosystems alive?

This is a strange question given that ecosystems incorporate non-living abiota. However, it has often been asked by some ecologists who advocate treating ecosystems holistically. The history of this superorganism tradition can be traced back to Clements (1916), Smuts (1926), Elton (1927), and Phillips (1931, 1934, 1935). Counterarguments can also be traced back to the same period, notably those from Gleason (1917, 1926). An influential recent example is Lovelock's (1979, 1988) Gaia hypothesis, in some versions of which the whole globe is considered to be a candidate superorganism. We will therefore use Gaia as our principal example here.

In our approach to ecosystems they are neither alive nor not alive. Rather, they have the intermediate status of "constructions" because they are collectively constructed by the activities of all the informed, functional, niche-constructing biota they contain, as well as by the reactions of abiota. For example, beaver dams are not alive, but neither are they put together by the laws of physics and chemistry alone. Rather, they are artifacts that owe their relatively high degree of organization to the semantic information that is expressed by niche-constructing beavers. Beaver dams could not exist without the semantic information that is encoded in the DNA of beavers and expressed in their dam building and maintenance behavior.

We suggest that ecosystems are constructions in a similar sense. Ecosystems, too, are not alive, but neither could they exist on dead planets because they can be constructed only by the collective activities of living organisms. Like beaver dams, ecosystems depend for their existence, at least in part, on the expression of whatever semantic information is acquired, via the Darwinian algorithm, by all the populations they contain, and that is expressed by these same organisms through multiple niche-constructing activities. Ecosystems are threaded with EMGAs that carry evolutionarily generated flows of semantic information into and out of both biotic and abiotic ecosystem components and that, when expressed by niche-constructing organisms, physically influence and change energy and matter flows in ecosystems.

This, we suggest, is probably what lies behind the illusion that ecosystems are alive. The expression of semantic information by niche-constructing organisms may grant ecosystems much of their impressive structural and functional integration, but it cannot grant them life. So it is a mistake to regard ecosystems as superorganisms. Instead, ecosystems are superconstructions, "super" because a multitude of constituent organisms contribute to their construction, and "constructions" because ecosystems are products not just of energy and matter flows, but also of complex collections of informed niche construction.

But this claim that ecosystems are superconstructions requires qualifications. One is that our beaver example could mislead. Beaver dams are obvious artifacts; so much so, that they can easily give the impression of being intentionally designed by intelligent animals for anticipated functions. However, most niche construction is entirely noncognitive in character, so intentionality cannot be ascribed to it, even when we humans can ascribe function to it.

A second qualification is that, since not all niche construction imposes design, regulation, or functional integration on the local environments of organisms, there is no reason to expect that all ecosystems will be well designed, regulated, and functionally integrated. At this stage it is conceivable that ecosystems are the rather diffuse and messy constructions of a multitude of uncoordinated activities. We anticipate that, where niche construction does exert control over ecosystems via engineering webs, it may allow the kind of functional integration that renders ecosystems orderly, and which at the largest scale is exemplified by the Gaia hypothesis. However, under different circumstances, the feedback generated by niche construction may introduce new sources of dynamic complexity or instability into ecosystems, which could generate chaotic interactions. Nonetheless, there is one claim of which we are certain: to the extent that ecosystems do exhibit orderliness or functional integration, it is partly because the collective niche-constructing activities of their constituent organisms are coordinated by a multitude of indirect forms of evolutionary feedback. This feedback is due to the EMGAs that permeate the ecosystem, allowing genes in the component populations to exert influences on each other as well as to impose order on abiotic ecosystem components.

All versions of the Gaia hypothesis that insist that "she" is alive are therefore misleading, mystical, and counterproductive. However, at the heart of the Gaia hypothesis there are two other simpler ideas that are potentially valuable. The first is that organisms alter their environments in ways that may benefit themselves. The second is that the many ways in which organisms do alter their environments may ultimately contribute to some very large-scale self-regulating feedback cycles that have had the effect of providing the services necessary for life over a very long time.

CHAPTER 9

Testing Niche Construction 3

Empirical Methods and Predictions for the Human Sciences

9.1 INTRODUCTION

Human cultural processes are exceptionally potent compared to protocultural processes in other animals, perhaps because of the more cumulative nature of human culture relative to the traditions of other animal species (see below). In chapter 6 we presented two gene-culture coevolutionary models that showed how cultural niche construction could modify natural selection. These models therefore raise the possibility that human evolution may be unique in this respect. Human cultural niche construction is also potent relative to the noncultural means by which humans construct niches. For these reasons, even though not all human niche construction is cultural (fig. 6.2), we focus here on cultural niche construction.

First we describe different types of feedback from cultural niche construction and suggest how each might generate a different detectable trace or "signature." Next, we use this notion of signature to illustrate how some of the methods described in chapter 7 can be employed to detect significant evolutionary ramifications of human cultural niche construction. Then we discuss possibilities for further conceptual advances in the social sciences when the extended evolutionary framework we developed in chapter 6 is substituted for the standard evolutionary perspective. These topics include hominid evolution, cooperation and conflict, demographics, and other aspects of the human sciences.

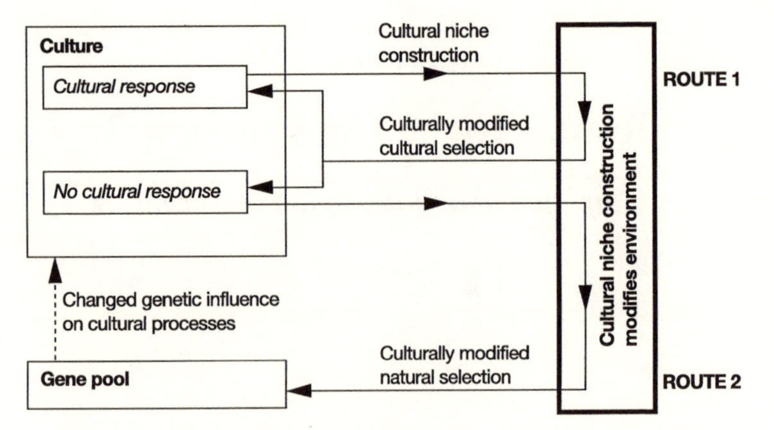

FIGURE 9.1. Culturally transmitted niche construction may result in modification of the environment and produce two forms of feedback. Route 1 represents cases in which cultural niche construction modifies selection pressures to which the population responds with further cultural change. Individuals adopt different cultural practices that are functional under the modified conditions, and biological evolution is minimal relative to cultural evolution. However, if the culture cannot adapt to the culturally modified environment rapidly enough, genetic change may become relatively important. Route 2 represents this case. Culturally modified natural selection may feed back to affect the genetic composition of the population. These genetic changes may or may not influence the population's capacity for cultural niche construction. If they do, cultural niche construction may drive repeated cycles of biological evolution. Here, to simplify this figure, processes operating at the ontogenetic level have been omitted.

9.2 SIGNATURES OF CULTURAL NICHE CONSTRUCTION

Figure 9.1 (derived from fig. 6.3, in chapter 6) illustrates the two principal routes by which a human population could respond to its own cultural niche construction. Route 1 comprises an adaptive cultural response to a change in an environment that was itself induced by cultural niche construction. It is illustrated by the feedback loop at the top of figure 9.1. Here an environment that has previously been changed by cultural niche construction modifies other cultural selection and transmission processes in a way that causes an adaptive cultural change in a human population. For example, suppose cultural niche construction changes a human environment by polluting it. Subsequently, the polluted environment may stimulate the invention and spread of a new technology to cope with this pollution. This is described by the route-1 loop comprising a culturally induced environmental change modifying cultural selection and transmission processes, which now favor the spread of a new technology, as well as the spread of some associated human behavioral changes, at the expense of alternatives (Cavalli-Sforza and Feldman 1981). Route-1 feedback thereby causes an adaptive cultural response to a culturally induced change. Provided that this response is sufficiently effective to counteract the change in the environment, route 1 would be confined to the cultural level alone. It should have no effect on human genetics. Route 1 is thus the stuff of history. It is familiar to archaeologists, historians, and most human and social scientists, whose concern is with human cultural affairs, and not with human genetics.

In route 2, the lower of the two feedback loops shown in figure 9.1, both human cultural and human genetic processes are involved. Route 2 applies whenever human cultural processes fail to express a sufficiently effective response (via route 1) to an environmental change that has previously been induced by cultural niche construction. In such cases, and to the extent that cultural processes cease to buffer culturally induced environmental changes, the latter are likely to give rise to culturally modified natural selection pressures. There may then be changes in allelic frequencies in human populations. For example, suppose there is no technology available to deal with a new challenge created in an environment by cultural niche construction, or suppose that the available technology is not exploited, possibly because it is too costly or because people are unaware of the impact that their own cultural activities are having on their environments. If such a situation persists for a long enough time, then genotypes that are better suited to the culturally modified environment could increase in frequency.

Occasionally a route-2 response to cultural niche construction may also have consequences for cultural processes themselves. The dashed arrow, at the bottom left in figure 9.1, represents the possibility that genetic predispositions, or aptitudes expressed throughout development, might affect what humans learn and how they behave, and thereby influence human cultural processes. When cultural niche construction modifies natural selection, those alleles that are favored by the modified natural selection are likely to have little or no direct influence on the expression of human culture itself. However, at some periods of human evolution, cultural niche construction could have modified natural selection in a manner that led to genetic changes that did affect the expression of human culture.

In principle, these different ways in which human populations can respond to cultural niche construction should have qualitatively different consequences, and they should therefore leave different signatures of past cultural niche-constructing events. For example, from the route-1 feedback loop we should expect to find cultural signatures in the form of cultural changes that can only be explained in the light of prior cultural niche construction. A simple prehistorical illustrative example is fishing. Fishing probably required the prior or simultaneous cultural invention and manufacture of fishing implements such as nets and fishhooks, and later (about 12 thousand years ago) of backward-pointing barbs that were likely used as harpoons (Klein 1999). For archaeologists these artifacts are cultural signatures of the cultural activity of fishing. Another example is the arrival of humans in Australia by water between 60 and 40 thousand years ago, which appears to signal a prior human ability to construct boats (Klein 1999).

In contrast, we should expect to find genetic signatures from the route-2 feedback loop, either in the form of population genetic changes or phenotypic changes that clearly depend on genetic changes. These genetic signatures should refer to changes in allelic frequencies in human populations, or perhaps in other species, for instance, in domestic animals or crops, that are explicable only in terms of the modification of natural selection pressures by prior human cultural niche construction.

Two examples of culturally induced genetic signatures are shown in figure 9.2. They are both caused by the modification of natural selection by agricultural niche construction. The first (fig. 9.2a), taken from Durham (1991), refers to the Kwa-speaking yam (or cassava) cultivators in West Africa whom we introduced in chapter 6. This figure illustrates a genetic change that occurred as a consequence of these cultivators (black circles) making clearings in tropical rain forests to enable them to grow their crops. The clearings

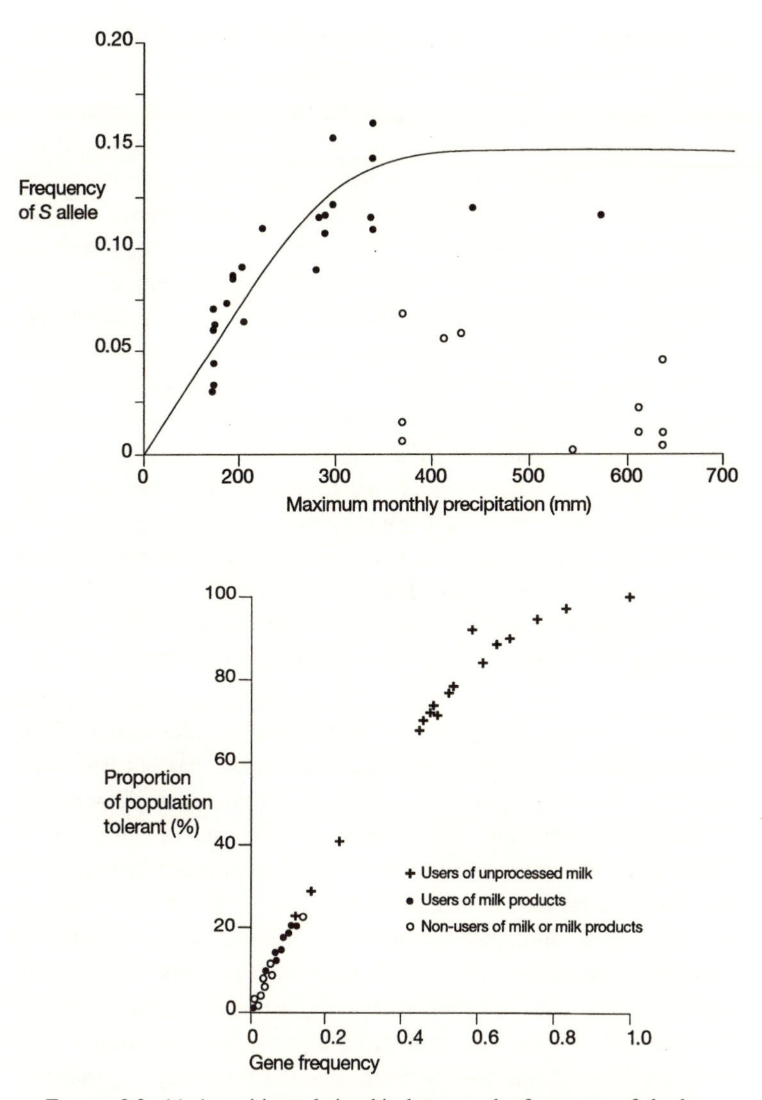

FIGURE 9.2. (a) A positive relationship between the frequency of the hemoglobin *S* allele and the maximum monthly precipitation can be seen for yam cultivators (filled circles) but not other closely related peoples. (From Durham [1991], fig. 3.9.) (b) Populations of unprocessed-milk users typically exhibit higher levels of tolerance and higher frequencies of lactose-absorption genes than users of milk products, which in turn exhibit higher frequencies than populations of nonusers. (From Ulijaszek and Strickland [1993], fig. 2.8.)

increased the amount of standing water, which provided better breeding grounds for malaria-carrying mosquitoes, which increased the prevalence of malaria, which apparently modified natural selection on these Kwa speakers, in favor of an increase in the frequency of the hemoglobin *S* allele, because, in heterozygotes the *S* allele confers some protection against malaria. Figure 9.2a also shows that the greater the amount of monthly precipitation, the stronger was the modified natural selection in favor of the *S* allele, but only in the yam cultivators. The open circles in figure 9.2a are from comparison populations of rice-cultivating Kwa speakers, whose environments are similar but whose agricultural practices are different and who apparently adopted agriculture much more recently than the yam cultivators. They do not show the same effect.

The second example (fig. 9.2b), from Ulijaszek and Strickland (1993), illustrates the genetic consequences of the human construction of a pastoral lifestyle, that then apparently selected for genes for lactose absorption in human adults (Flatz 1987; Scrimshaw and Murray 1988; Feldman and Cavalli-Sforza 1989; Durham 1991; Holden and Mace 1997; Hollox et al. 2001; Enattah et al. 2002). In some analyses the genetic basis of lactose tolerance is treated as a single gene for the enzyme lactase. In other treatments the nature of the genes conferring the ability to metabolize lactose has not been well specified. For our purpose it is enough to consider a single gene and its interaction with the cultural trait of milk use, as illustrated in figure 9.2b. This figure shows the relationship between lactose tolerance and the frequency of a gene that confers the ability to absorb lactose in three culturally different human populations. They are (i) users of unprocessed milk (black crosses), (ii) users of processed milk products such as butter, whey, yoghurt, and cheese (black circles), and (iii) nonusers of milk or milk products (open circles). Figure 9.2b shows that lactose tolerance and the frequency of genes that confer lactose tolerance by increasing lactase activity are highest in populations that have a long history of dairying and consumption of fresh milk and lowest in populations that do not use milk or milk products (Durham 1991; Ulijaszek and Strickland 1993; Hollox et al. 2001). It also suggests that the cultural processing of milk products, by breaking down the lactose, reduced the intensity of natural selection for genes that increase lactase activity, illustrating how routes 1 and 2 in figure 9.1 can interact.

The hypothesis that adult tolerance to lactose is advantageous in dairying cultures has had a checkered history. For example, the hypothesis was formerly opposed by an alternative suggestion that adult lactose tolerance is primarily an adaptation to reduced exposure to the sun at high latitudes (Durham 1991). A phylogenetic analysis by Holden and Mace (1997), based on Pagel (1992), however, found no support for this latitudinal hypothesis but did find strong evidence that adult lactose tolerance must have arisen independently up to three times in cultures that keep animals for milk, but did not arise in nondairying cultures. Holden and Mace's analysis also indicated that the dairying, which first appeared approximately 8,000–6,000 years ago, came first, and that lactose tolerance was favored subsequently. Hence it would seem that it was indeed the cultural practice of herding that selected for this particular genetic change. Genes for adult lactose absorption appear to constitute a genuine genetic signature of a past niche-constructing practice, namely, pastoralism and dairy farming.

In a third scenario according to which feedback from prior cultural niche construction could affect human evolution (the dashed arrow in fig. 9.1), a genetic response to cultural niche construction directly influences a population's capacity to express culture. Here we would expect to see an association between genetic and cultural signatures. A candidate example is the expensive tissue hypothesis of Aiello and Wheeler (1995) (see below). These authors noted reductions in hominid gut size that corresponded to increases in relative brain size. They suggested that the cultural practices of hunting and scavenging probably led to an increase in meat consumption, making the large guts associated with a herbivorous diet unnecessary. If so, that could have allowed more energy to be devoted to brain tissue. Enlarged neocortices may have subsequently allowed hominids to express more sophisticated cultural niche construction, leading to the cultural practice of cooking, thereby permitting a further reduction in gut size and a corresponding further increase in brain size.

There may have been many episodes of cultural niche construction that selected for genetic changes in our ancestors and that subsequently enhanced their capacity for culture. For example, Wrangham et al. (1999) proposed that the origin of cooking may have been much earlier than Aiello and Wheeler (1995) suggest. Cooking may have originated with *Homo erectus*, possibly as long as 1.6 million years ago (see also Rowlett 1999). Wrangham et al. (1999) then considered other possible phenotypic signatures of cooking including reduced tooth size and, more speculatively, reduced sexual dimorphism, in addition to reduced gut size. These ideas were opposed by Brace (1999), who suggested a more recent origin of cooking. From our point of view these controversies are valuable because they encourage the empirical search for signatures of past human cultural niche construction. With this in mind we now describe how our hypotheses concerning the evolutionary role of niche construction might be tested with experiments and theoretical analyses.

9.3 METHODS FOR DETECTING THE EVOLUTIONARY IMPACT OF HUMAN CULTURAL NICHE CONSTRUCTION

In principle, the methods for detecting the evolutionary impact of niche construction described in section 7.2 apply as much to humans as they do to every other species. In practice, some are ruled out by ethical considerations. For instance, we cannot manipulate human environments for experimental purposes without regard to the consequences for the humans who live in them, and this reduces the range of experimental manipulations permissible in humans. We can, however, take advantage of "natural experiments" which, in the human case, are often better documented than they are for other species. For example, in section 7.2.1, we suggested how the ecological and evolutionary consequences of experimentally canceling or enhancing a population's capacity for niche construction may be detected or inferred. Such canceling and enhancing events do sometimes happen naturally, and when they do they can be exploited to test hypotheses concerning niche construction. We offer some examples of these kinds of investigation below and discuss other methods for detecting signatures left behind by past human cultural niche construction.

9.3.1 Comparative Analyses That Explore Trait Evolution in Hominids

In section 7.2.2 we described how, for any clade of organisms, it should be possible to differentiate phenotypic traits (or recipient characters) that may have become adaptive in environments that had been subject to niche construction, and how to use comparative methods to accomplish this (e.g., Harvey and Pagel 1991). We showed how such methods could be used to determine whether changes in a selected recipient character correlate with a particular niche-constructing activity, whether the niche-constructing activity is typically ancestral to the recipient character, and whether the recipient character in question is derived. There should be a significant relationship between the pertinent environmental state and the recipient character only when the niche-constructing activity is also present. The same logic applies at the cultural level, and the same methods can be applied to hominids, or to contemporary human populations, where they may shed light on the relationship between different kinds of cultural niche construction and their consequences. In this section we provide examples of genetic and cultural signatures of past inceptive cultural niche construction.

A first example is the cultivation of fava beans (*Vicia fava*) for approximately 8,000 years, by different ethnic groups in circum-Mediterranean regions (Jackson 1996). Fava beans contain the toxic glycosides vicine and convicine, as well as some other antimetabolites which, in susceptible individuals, produce the metabolic disease known as favism. Humans who carry variants of a gene that express reduced enzyme activity of the red blood cell enzyme glucose-6-phosphate dehydrogenase (G6PD) are particularly susceptible to favism and ought to have been selected against by natural selection. In spite of this, the frequency of G6PD deficiency is quite high in Mediterranean people. The solution to the riddle emerged when comparative studies revealed the role of malaria, caused by *Plasmodium falciparum.* Malaria was found to be inhibited in G6PD-deficient red blood cells, which apparently granted the human carriers of this deficiency a selective advantage in malaria-prone areas that partly offset the selective disadvantage due to their susceptibility to favism (Golenser et al. 1983; Jackson 1996). So this is another example where cultural niche construction, here the cultivation of fava beans, interacted with natural selection, and apparently led to a changed genetic response in a human population by route 2 (fig. 9.1).

A different example is the controlled use of fire. The origins of the control of fire by hominids are uncertain and possibly date back to 1.5 million years ago (Jackson 1996; Klein 1999). The oldest secure evidence comes from a *Homo erectus* site in China dated between 500 and 240 thousand years ago (Klein 1999). Subsequently, the control of fire by hominids has had many consequences, for both human genetic and cultural processes. One series of consequences appears to have stemmed from cooking. Heat detoxifies many plant allelochemicals, which means that cooking can make inedible foods edible (Jackson 1996; Wrangham et al. 1999; Ambrose 2001). The changed diets that cooking permitted probably promoted a succession of subsequent anatomical changes in hominids. For example, Wood and Brooks (1999) point out that *Homo ergaster* is the first hominid to dispense with a large digestive tract and that its relatively small jaw and teeth are consistent with a diet requiring lower bite forces and less chewing. They suggest that such changes are probably related to the processing of food outside the mouth, through cooking. Another less obvious consequence suggested by Klein (1999) is that the pelvic narrowing observed in *H. ergaster* must have reduced gut volume, but this could have occurred only if food quality improved simultaneously, which again suggests cooking.

As noted above, opportunities for detecting genetic signatures of human cultural niche construction may also occur in other species with which humans interact. Plant and animal species that show divergences from their wild relatives after their domestication by humans are obvious examples. Even here, however, there can be surprises. For example, molecular data suggest that the domestication of dogs probably occurred several times, perhaps much earlier than the archaeological evidence currently suggests and almost certainly earlier than the onset of agriculture. It may have occurred as long ago as 135 thousand years (Vilà et al. 1997; Pagel 1999). Other less obvious cases may be found when the selection imposed by hominid cultural niche construction has been inadvertent. For example, when did hominids first wear clothes? As cloth, fur, and skins rot quickly, there are no fossil remains to answer this question, but a lowly louse may help. There is one human louse, the body louse, which lays its eggs exclusively in human clothes. In principle, it should therefore be possible to find molecular evidence to establish when this body louse initially diverged from another closely related human louse, the head louse, thereby indicating a date by which hominids must have been wearing clothes (Ehrlich 2000). Other useful genetic or phenotypic signatures of human cultural niche construction may possibly be found in other species too, among them the fire-prone plants in Australia (Kershaw 1986) or animal prey species exploited by humans (Stiner et al. 2000).

A salient event that illustrates some cultural signatures of prior cultural niche construction (route 1) was the origin of agriculture based on the domestication of plants and animals. Diamond (1998) argues that the original domestication of plants was largely opportunistic and depended primarily on the available stock of potentially useful plants in different parts of the world. The domestication of animals, however, was more complicated because, in at least two continents, the Americas and Australia, the absence of large mammals suitable for domestication was probably itself partly a consequence of the prior hunting activities of preagricultural humans. Human hunters, using tools and weapons, were apparently a major cause of animal extinctions (Klein 1999), including the extinction of some animals that might have been domesticated by later agriculturists (Diamond 1998). When agriculture did start, its consequences for both human genetics and human cultural practices were dramatic. The spread of agriculture eventually had enormous effects on human demography, human settlement patterns, human life expectancies, and human social structures and methods of government. The advent of agriculture may also have precipitated new sex roles, social systems, and mating systems (Hrdy 1999). It also introduced humans to new diseases, particularly the so-called "crowd" diseases such as measles, tuberculosis, flu, and smallpox, many of which originate from domestic animals (Diamond 1998). Later, the persistence of these diseases in Old World communities led to the evolution of some degree of resistance to them. Infectious diseases then greatly aided the Spanish conquests of the Incas in Peru in the sixteenth century, because the Spanish carriers of these diseases were generally already resistant to them, while the Incas were not (Diamond 1998).

9.3.2 Cultural Niche Construction in lieu of Morphological Adaptations

In addition to finding genetic signatures due to inceptive cultural niche construction, as above, we might also find the opposite, an absence of expected genetic or phenotypic signatures in response to natural selection because of counteractive cultural niche construction. In section 7.2.2.2 we described how the patterns of response in organisms to natural selection can be changed by both counteractive and inceptive niche construction. We also argued that under counteractive niche construction, more sophisticated niche constructors should exhibit weaker structural responses to independent changes in their environments than less able niche constructors. If we now assume that hominid cultural niche construction is more potent than that of other mammals, and that in more technically advanced hominid cultures there is a greater capacity for counteractive niche construction than in less technically advanced ones, then a number of further hypotheses emerge.

First, we would expect hominids to show less morphological response to fluctuating climates than other mammals. We would also expect more technologically advanced hominids to exhibit less response to fluctuating climates than those that are less technologically advanced. Second, recall that Bergmann's and Allen's rules suggest, respectively, that populations in warmer climates should have smaller bodies and bigger extremities than those in cooler climates (Gaston et al. 1998). Again, we would expect hominids to show less correspondence to these rules than other mammals. We would also expect more technically advanced humans to exhibit less correspondence to these rules than those who are less technically advanced. Klein's (1999) comparison of Neanderthal versus modern Inuit (or Eskimo) adaptations to cold climates appears to provide some support for this view. Third, by the same logic, we would expect an inverse relationship between robust morphology and the capacity for expressing counteractive niche construction. This idea is consistent with Foley's (1988) observation that there is a trend away from robustness and sexual dimorphism in modern humans. Fourth, it should be possible to reverse the inference, and to use the fossil record to infer something about the niche-constructing capabilities of animals, including hominids. We suggest that the greater the hominids' phenotypic (as opposed to extended phenotypic) response to environmental change, the more restricted must have been their capacity for counteractive niche construction. Fifth, we might expect more technically advanced hominids to exhibit a broader habitat range than those that are (or were) less technically advanced, as their capacity for counteractive niche construction would enable them to survive better in novel habitats. Here, the hypothesis of Sol and Lefebvre (2000), that more innovative species with larger brains are more successful invaders of novel habitats than less innovative species with smaller brains, is probably relevant. Sol and Lefebvre tested their hypothesis in birds only, but the same idea may apply to hominids too.

These ideas have in common that an independent environmental change is assumed to happen first, and that without counteractive niche construction we would expect to see a genetic or phenotypic response to changing natural selection pressures. But under counteractive niche construction we should not. Instead, we should sometimes find evidence that counteractive cultural niche construction reduced or overwhelmed the effects of natural selection. This introduces a "dog that didn't bark" kind of signature of cultural niche construction.

In section 7.2.3.1 we described how a common method for inferring natural selection was to establish a correlation, for example, across a geographical region, between the trait of interest in an organism and the suspected environmental factor thought to be the source of natural selection on the trait. We went on to describe how niche construction might explain the failure to find such a relationship, namely, that organisms can afford an apparently weaker structural match of the trait to the environment if they are able to compensate through niche construction. Again, if all hominid species had been potent niche constructors, such correlations should be difficult to find. In principle, therefore, it should be possible to carry out steps 2–4 of the procedure outlined in section 7.2.3.1, even with extinct populations. It may well be possible to detect whether a hominid population exhibits niche construction that may have compensated for poor adaptation of some structural trait, whether there is evidence for organism-driven modification of the selective environment, and whether there is evidence for evolutionary feedback in the form of structural or functional adaptations to the constructed environment. In practice, this step-by-step procedure is likely to be very difficult because of the impoverished nature of the hominid archaeological record but it may be possible in some cases.

The above reasoning also leads us to expect different patterns of structural evolution in hominids from those that are characteristic of mammals in general. Consider, for example, the extent to which two temporally and geographically overlapping hominid populations, say Neanderthals and modern humans in the Levant around 70–50 thousand years ago, were in contact or competition. A standard evolutionary perspective would lead to the prediction that closely related interacting sympatric species ought to exhibit character displacement to reduce competition (Endler 1986). Were researchers to compare the morphology of the two hominid populations and find no evidence for character displacement, they might conclude that the populations were not sympatric at any point in time. But this could be wrong. An alternative explanation is that sympatric species may niche-construct in ways that reduce the competition between them, allowing them to remain sympatric (e.g., Milton 1991). In section 7.2.3.2 we suggested a procedure for testing this hypothesis.

Many of these ideas have been proposed before (e.g., Lewin 1998; Klein 1999) although some of them are difficult to reconcile with evolutionary explanations because they invoke niche construction. In this respect a theory of evolution that explicitly incorporates niche construction could offer a better rationale for developing some new and more detailed predictions that might be empirically testable.

9.3.3 Using Theoretical Models

A third approach is to test predictions generated by gene-culture coevolutionary models that explicitly incorporate niche construction. In this respect, the models presented in chapter 6 are of limited use, because, like the models in chapter 3, they are very general. Their purpose is to provide a broad theoretical framework rather than immediately testable predictions. However, they may form a framework in which to discuss specific human niche-constructing activities in terms of testable hypotheses. To an extent this has already started to happen, although the terminology used by different authors is sometimes different, and the niche-construction component is sometimes only implicit.

One example concerns the control of irrigation by farmers in Bali (Lansing et al. 1998). Balinese farmers manage their fields and irrigation systems in organizations called "subaks." Subaks are local associations of farmers that set cropping patterns and pest control policies, for up to three crops a year, of either rice or vegetables. The average size of a subak is 92 farmers and 42 hectares. As the amount of water available for each subak is influenced by the irrigation of upstream neighbors, the subaks seek to avoid water shortages by staggering their planting cycles relative to those of their neighbors. However, pest control is aided by subaks planting the same crops at the same time as their neighbors. Thus, choosing the best cropping pattern involves complicated trade-offs between water availability, pest damage, and the policies of neighboring subaks. Lansing et al. (1998) have suggested an evolutionary model of this situation that allows these farmers both to modify their environments by irrigation and cropping decisions and to receive feedback from the modified environments in the form of changed cultural and natural selection pressures. They call this process "system-dependent selection" rather than "niche construction," but it is essentially the same idea.

9.4. APPLICATIONS OF NICHE CONSTRUCTION

9.4.1 Hominid Evolution

Archaeologists and anthropologists currently seek to reconstruct the evolutionary history of modern humans from fossil and molecular data in the context of standard evolutionary theory (fig. 1.2a). By ignoring niche construction, this encourages the idea that human evolution must have been directed solely by independent natural selection pressures, that is, by selection pressures that have not been modified by niche construction. These selection pressures may sometimes have arisen from hominid social interactions (e.g., Byrne and Whiten 1988; Durham 1991; Dunbar 1993; Foley 1995), but they do not usually include other sources of selection in the external environment that might have been modified by ancestral hominid niche constructors. When human adaptation is assumed to be the result of both natural selection and niche construction (figure 6.2c), the suite of hypotheses about the causes, rates, and processes of evolutionary change is considerably enlarged.

9.4.1.1 Processes of Human Evolution

Consider the possible ways in which a new evolutionary episode might be initiated in hominid evolution. For illustrative purposes only, consider a suite of explanations for the divergence of the lineages leading to the Pongidae and Hominidae families in the late Miocene, and assume two ancestral allopatric populations. The subsequent divergence of these two populations could have been triggered by any of the following events. First, each population could have been exposed to different external environments with different selection pressures, leading to allopatric speciation in the manner proposed in standard evolutionary theory. Alternatively, both populations may have been exposed to the same novel selective pressures, say a changed habitat, but only one population, say the ancestors of the Pongidae, was able to respond with counteractive niche construction, say by retreating to a still unchanged habitat, while the Hominidae ancestors remained where they were and became adapted to the new environment. Third, both populations may have responded to the same novel selection pressures with counteractive niche construction but in different ways, each subsequently generating a different array of novel modified selection pressures which fed back on themselves. Fourth, the divergence might have been caused by inceptive niche construction on the part of one population, initiated by a change in members of the population, possibly because of a mutation or a new learned behavior, which subsequently caused, rather than resulted from, a change in the environment. For instance, this might have entailed one population's discovering a new habitat or devising a new form of niche construction (West-Eberhard 1987; Bateson 1988; Plotkin 1988). Cultural or protocultural processes, say on the part of the Hominidae, may have produced selection for some biological innovation (Feldman and Cavalli-Sforza 1976; Boyd and Richerson 1985; Wilson 1985). Fifth, the ancestors of both lineages may have initiated different kinds of inceptive niche construction, again with no key environmental event triggering their divergence.

This enlarged suite of possibilities for hominid evolution suggests that some new traits might overcome their own fitness disadvantage through niche construction. One possible example that we have already touched on is the evolution of the large human brain. The mass-specific metabolic rate of the human brain is about nine times that of the average metabolic rate of the human body as a whole, but there is no elevated basal metabolic rate in humans that would compensate for it. Aiello and Wheeler (1995) found that this was possible because the size and metabolic cost of the human gut, in particular of the gastrointestinal tract, are considerably reduced relative to the guts of other similar-sized primates. They hypothesized that our ancestors could afford a reduction in gut size because they used their brains to improve their diets in proportion to their loss of gut, probably in two different episodes: the first coinciding with the appearance of the genus *Homo*, approximately two million years ago, and supported by increased meat eating, the second coinciding with the appearance of archaic *Homo sapiens* during the latter half of the Middle Pleistocene and supported by the cultural invention of cooking (Wrangham et al. 1999). This is an example of how a trait, the human brain, might have evolved by overcoming its fitness disadvantage through inceptive niche construction. In fact, this hypothesis may not be restricted to humans. Reader and Laland (2002) found that a number of measures of behavioral plasticity (the incidence of innovation, social learning, tool use) are strongly correlated with species' relative and absolute "executive" brain volumes in primates. There appears to have been selection for an increase in those regions of the primate brain that facilitate inceptive niche construction.

9.4.1.2 Rates of Evolution

Niche construction may also have influenced the rate of hominid evolution. Much attention has focused on how cultural transmission affects evolutionary rates. Allan Wilson and colleagues have argued that changes in the hominid niche, resulting from complex social behavior and cultural (or protocultural) transmission, may have generated a "behavioral drive," which accelerated morphological evolution by fixing a greater proportion of genetic mutations (Wilson 1985). Wilson noted that there is a positive relationship between relative brain size and rate of anatomical evolution among vertebrates, which he argued was consistent with his behavioral-drive hypothesis. However, theoretical analyses suggest that cultural processes are able both to accelerate and decelerate evolution (Feldman and Cavalli-Sforza 1976). These apparently contradictory findings make better sense in the light of our new perspective, because as models 3 and 4 in chapter 6 demonstrate, cultural niche construction can either counteract or accelerate evolutionary change.

It is widely recognized that culture can also shield low-fitness genetic variants from selection (Boyd and Richerson 1985; Feldman and Laland 1996), as illustrated by route 1 in figure 9.1. For instance, improved levels of health care and sanitation are examples of culturally mediated counteractive niche construction that mitigate selection against individuals with some gene-related disorders or susceptibilities, who may survive and reproduce in the modified environment. In fact, ontogenetic processes may also damp out natural selection, for example, if individuals develop antibodies that counter disease or learn to avoid parasites or predators (Bateson 1988; Plotkin 1988). In addition, the recent culturally enhanced mobility of humans facilitates greater mixing of genes between populations, reducing differences, and slowing down the divergence of populations. Moreover, a new culturally induced environmental change may be responded to exclusively by a new cultural adaptation (route 1 in figure 9.1).

However, if cultural innovations modify natural selection pressures, then genetic change may follow (route 2 in fig. 9.1). If, as seems likely, the rate of change of cultural niche construction is rapid relative to independent changes in the environment, biological evolutionary rates may be accelerated. A number of gene-culture coevolutionary models have found that, because cultural transmission may homogenize a population's behavior, and because culturally transmitted traits can spread through populations more rapidly than genetic variants, cultural processes can generate atypically strong selection

(Feldman and Laland 1996). Thus route 1 typically slows while route 2 may accelerate rates of genetic evolution among humans and their "cultural" ancestors.

9.4.1.3 The Evolutionary Roots of Culture

A first step toward an understanding of the roots of human cultural processes is to address the evolution of social learning. Over the past 15 years a variety of mathematical analyses have explored the adaptive advantages of social learning, relative to learning asocially, or of expressing an unlearned pattern of behavior that has evolved genetically (Boyd and Richerson 1985; Rogers 1995; Cavalli-Sforza and Feldman 1983; Aoki and Feldman 1987; Bergman and Feldman 1995; Laland, Richerson, and Boyd 1996; Feldman et al. 1996). Despite a plurality of methods, this body of theory has reached a consensus as to when social learning should be favored. When environments change very slowly, adaptations should be more likely to occur at the level of population genetics because there are only modest demands for knowledge updating, which can be met by genetic systems responding to gradually changing selection pressures. In contrast, if environmental change is fast, or when there are sudden environmental shifts, tracking by individual learning should be favored, as should a process of horizontal (within-generation) transmission of information. In such environments, the genetic system will change too slowly to cope, while social learning from the parental generation is likely to be too error-prone, as individuals would receive outdated information. Under intermediate rates of environmental change, however, social learning from parents should be favored. Here, "intermediate" entails changes that are not so fast that parents and offspring experience different environments, but not so slow that appropriate genetically transmitted behavior could evolve instead.

The term "social learning," as currently applied to animals, describes a heterogeneous array of processes, with a variety of functions, found in a broad array of vertebrate and invertebrate species. A more narrow use of the term would restrict it to those processes that might reasonably be regarded as homologous to processes operating in human social learning and that mediate a general capacity to acquire information from others, regardless of the nature of the information, its function, or the sensory modality employed. Within the narrow category of social learning, humans probably transmit more information from one generation to the next than all other species. Protocultural species typically depend primarily on horizontal transmission based on social enhancement, rather than on imitation or teaching (Galef 1988; Laland et al. 1993). A comparative perspective thus implies that the earliest forms of social transmission were probably horizontal. In contrast, humans appear to acquire large amounts of information from their parents and from the parental generation. For instance, a study of Stanford University students revealed that the religious and political attitudes were strongly consistent between parents and offspring (Cavalli-Sforza et al. 1982). The same has been reported to apply in nonindustrial societies. Among Aka pygmies, an African group of hunter-gatherers, there was evidence for parent-to-child transmission of many customs (Hewlett and Cavalli-Sforza 1986), while among horticulturists in the Democratic Republic of Congo the young acquire knowledge about foods primarily from their parents (Aunger 2000c). These findings suggest that the lineage leading to *Homo sapiens* evolved for increasing reliance on vertical and oblique cultural transmission.

The theoretical analyses described above imply that a shift from transient horizontal traditions toward increased transgenerational cultural transmission reflects a greater constancy in the environment over time. Such a shift is difficult to reconcile with the traditional evolutionary perspective, as there is no evidence to suggest that environments have become more constant over the last few million years, but rather the opposite. Moreover, even if they had, other protocultural species would be expected to show more transgenerational transmission, too. However, the increasing reliance of hominids on vertical and oblique transmission is consistent with our perspective, as a significant component of the hominid selective environment is likely to have been niche-constructed, in which case transmission of more cultural information to their offspring would be advantageous.

Here we offer only speculative examples. For instance, insofar as hominids hunted rather than scavenged, by tracking or anticipating the movements of migrating or dispersing prey, populations of hominids may have increased the chances that a specific food source was available to them, that the same tools used for hunting would always be needed, and that the skin, bones, and other materials from these animals would always be at hand to use in the manufacture of further tools. Such activities would create the kind of stable social environment in which related technologies, such as food preparation or skin processing, would be advantageous and could be repeatedly socially transmitted from one generation to the next. It is also possible that, once started, transgenerational cultural transmission may have become an autocatalytic process with greater culturally mediated environmental regulation leading to increasing homogeneity of environment as experienced by parent and offspring, favoring further transmission of information. For example, if new material cultural traits respond to, or build on, earlier material cultural traits, niche construction could set the scene for an accumulating material culture. This could lead to offspring learning higher-order packages of cultural information from their parents and other adults, as appears to be the case in preindustrial societies (Hewlett and Cavalli-Sforza 1986; Guglielmino et al. 1995).

The stress above is on the passage of information from parents to offspring, or vertical transmission. However, the mode of transmission is not central to our hypotheses. Our arguments, which require only transgenerational transmission, actually relate to the temporal distance between the transmitting and receiving individuals. For instance, the time between parents and offspring learning a particular skill is usually a generation, while peers or siblings learn skills at more similar times. It is this temporal factor that is important. If the environment, including the social environment, is stable from one generation to the next, it should pay one generation to learn the same things as the preceding generation, irrespective of whether the transmission is from parents, teachers, or peers.

9.4.1.4 Material Culture

A central characteristic of the niche-construction perspective is its emphasis on how the modifications that organisms make to their environment can feed back to affect subsequent evolution. In chapter 6 we extended our evolutionary framework to incorporate knowledge-gaining ontogenetic and cultural processes, arguing that niche construction at each level could affect information gain at the same and other levels. However, according to the triple-inheritance version of gene-culture coevolution that we introduced in chapter 6, humans inherit not only genetic information and cultural knowledge from their parents and ancestors, but also modified environments. It follows that if human niche construction is dominated by cultural processes, then human material culture must itself be a part of our human ecological inheritance and be an important component of each generation's legacy from its ancestors. Although culture is often viewed as ideational and cognitive in nature, an important aspect of cultural expression, material culture, has had, and continues to have, a major influence on human environments.

This line of reasoning has recently been developed further in a series of articles by Aunger (2000a,b, 2002). Aunger argues that the increasing complexity of Western culture cannot be only a response to what individuals are learning directly from the people around them. Much of what we know is stored outside our heads, in books, on computer disks, in databases, and in a multitude of other artifacts. Our physically constructed environment is a storehouse of cultural information, and also a mediator of communication. Each child inherits a rich ontogenetic niche, filled with sophisticated artifacts, and much of its learning is a result of interactions with these artifacts. For Aunger, the standard cognitivist conception of culture ignores an important part of culture—the evolution of artifacts:

> We must include in our definition of culture not only what we learn socially, but also the implicit knowledge we acquire or simply make use of, that derives from artefacts, perhaps placed there by earlier generations. . . . These bits of "crystallized intelligence" influence cultural activity because they serve as tools for thought and behaviour. . . . We will almost certainly underestimate the power of culture and mistake our own place in the world, if we do not take into account the reality of our highly reconstructed environments—the cultural niche—into which we have put much of our "intelligence" (Aunger 2002, p. 310).

We agree with Aunger that artifacts are important in human cultural learning, but in terms of our theory they are components of a general process whereby organisms choose and construct features of their local environments and in the process leave a material legacy to their descendants and those around them. This legacy affects not only how subsequent generations evolve, but also how they develop and what they learn. As we said in the preceding section, the more an organism constructs the environment of its offspring, the greater should be the advantage of transmitting cultural information across generations. Thus the characteristics of the material culture are likely to influence the extent and content of the transgenerational cultural transmission.

9.4.2 Human Altruism, Cooperation, and Conflict

In section 7.3.1 we described how standard evolutionary theory provides two principal explanations for cooperation in organisms, kin selection and reciprocity. These ideas have been used to explain the evolution of cooperation in nature but neither is sufficient to explain all human cooperation (Boyd and Richerson 1985; Laland et al. 2000a,b). Kin selection is restricted to kin, while the evolution of cooperation based on reciprocity is probably limited to small groups with repeated pairwise interactions. This is because in large interacting social groups the costs of reciprocation rise faster than the benefits (Boyd and Richerson 1988). It is also hard to account for certain forms of human altruism, such as military heroism, in terms of reciprocity or kin selection (Boyd and Richerson 1985). Our evolutionary framework, however, indicates that the suite of processes that may be regarded as feasible evolutionary explanations for human cooperation is considerably larger than just kin selection and reciprocity. The general statement in chapter 7 describing the circumstances under which we would expect cooperation to arise applies equally to humans.

We suggested that much cooperation in nature was best described as selfish mutualistic interactions that result from the exploitation of incidental by-products, with organisms frequently investing in others so as to enhance the benefits that eventually return to themselves, rather than as altruistic interactions. Much of human cooperation might also be a result of the exploitation and investment in the by-products of others, in other words, in terms of mutualisms resulting from human niche construction. For example, barter and exchange are mutualistic interactions in which individuals or organizations trade products for more desirable alternatives. Moreover, human individuals and institutions invest, metaphorically, or even literally, in other individuals or institutions in order to enhance their own returns.

Another explanation for cooperation that is gaining considerable attention is the revised group-selection hypothesis proposed by Wilson and Sober (1994) (see also Sober and Wilson 1998). Wilson and Sober pointed out, as have others (Uyenoyama and Feldman 1980; Uyenoyama et al. 1981), that both kin selection and reciprocal altruism can be formulated as group selection. However, Wilson and Sober argue that social groups, or indeed any other higher-level entities, can be vehicles of selection. They propose a nested hierarchy of vehicles for genes (individual, group, metapopulation) in which each level also includes a population of lower-level units. Selection then acts at the lowest level for which there are fitness differences. Thus, if there are no fitness differences between individuals, there should be a "frame shift" (p. 592) to selection at the group level.

The original or naive group-selection hypothesis failed primarily because the processes that maintain group differences and select between groups must typically be weak compared with the processes that break down group differences and select within groups (Williams 1966). If group-level adaptations are based on the cooperation of altruists, then any individual who refuses to cooperate can reap the benefits without paying the costs, and selfish strategies should be favored by natural selection. It is still not clear whether the group selection of genetic variation can surmount this obstacle to account for large-scale human cooperation.

Boyd and Richerson (1985) introduced the idea of cultural group selection. They emphasized a "when in Rome do as the Romans do" conformity, where individuals adopt the behavior of the majority of their group. The significance of this "conformist transmission" is that it minimizes behavioral differences within groups while maintaining differences between groups. Thus, group selection of cultural variation "for the good of the group" is possible because, if an altruistic cultural trait becomes frequent in a group, the transmission process could subsequently discriminate against selfish individuals. Under group selection on cultural variation, it is not genes that are selected for, but rather groups of individuals expressing a particular culturally transmitted idea. As cultural selection between groups may favor beliefs that benefit the group at the expense of the individual, Boyd and Richerson provided a new explanation for human cooperation.

Several properties of cultural inheritance, as opposed to genetic inheritance, make Boyd and Richerson's idea attractive. One is that cultural inheritance, unlike genetic inheritance, may depend on more than two "cultural parents." It is therefore possible for individuals to be sensitive to the most frequent cultural traits in their society and to conform to them. Second, group selection of cultural variants can be faster than group selection of genetic variants because cultural death does not imply the physical death of all the people in a culture. A threatened or defeated people may adopt the cultural traits of a conquering people, either voluntarily or under duress. Thus, unlike the group selection of genes, here migration will not weaken the process. Third, symbolic group marker systems, such as totem animals, human languages, and flags, make it considerably easier for human groups to maintain their identities, and to resist imported cultural traits, than it is for local gene pools, or demes, to maintain their identity by resisting gene flow (Boyd and Richerson 1985). Bowles (2000) stresses how conventions can generate and maintain cultural differences between groups and that cultural processes other than conformist transmission can also act in this manner. Fourth, cultural transmission of information about cheaters (e.g., gossip) reduces the efficacy of noncooperative strategies (Dugatkin 1992). To the extent that social processes such as gossip, pressure to conform, and conventions operate to maintain group identities and prevent cheating, they act to restrict cultural rather than genetic variation.

Conformist transmission may potentially be exploited by powerful individuals, groups, or institutions, which dominate the dissemination of information through societies to promote their own interests. In preagricultural egalitarian societies this was probably not very important since in such societies inequalities of power and wealth are typically both temporary and minimal (Cohen 1998). However, in post-agricultural societies that display rankings, and in complex civilized states that display class stratifications, significant economic inequalities occur, and power networks develop (Cohen 1998). In these societies powerful and coercive cultural parents may stand to gain considerably from persuading other less powerful humans to conform, perhaps by recruiting extra assistance in modifying environments in ways that benefit them rather than the helpers. These processes can be amplified by tool use, for instance, by the technology of modern media, by weapons, by art, or by deceit. Religious, commercial, and political propaganda, for example, may all be used to persuade, trick, or coerce conformity from individuals against their personal interests in favor of the interests of a dominant class of cultural transmitters.

The niche-construction statement on conflict in section 7.3.1 should also extend to the human cultural level, with the qualification that at this level other processes may be operating as well. Group selectionists commonly focus on the positive repercussions of group selection (that is, within-group altruism) and neglect the negative repercussions (that is, between-group selfishness, hostility, and conflict). Group selection does not directly favor altruistic individuals so much as selfish groups (Laland et al. 2000a). The group-level traits most effective in promoting group replication may also engender outgroup hostility, intergroup aggression and conflict, fear of strangers, slanderous propaganda concerning outsiders, and so on. The same processes that underlie the best of human motives may also favor the worst attributes of human societies

9.4.3 Demographics and Life Histories

In section 7.3.1 we suggested that an individual may be more likely to invest in a cooperative exchange with a second individual that niche constructs in a manner that increments the relative fitness of the former than with individuals that niche-construct in ways that reduce the former's fitness. If we regard parenting as a cooperative act, parents should invest in more children if those children are subsequently likely to provide them with incidental benefits through their niche construction, or in fewer children to the extent that children do not provide such benefits or provide incidental costs. This kind of reasoning may help to explain phenomena such as the demographic transition and differential investment in sons and daughters.

The radical shift in human reproductive behavior in the late nineteenth century, known as the demographic transition, constitutes a major challenge to evolutionary approaches to human behavior (Borgerhoff Mulder 1998). There are two principal features that require explanation: the overall decline in the number of offspring that parents produced despite increasing resources, and the observation that rich families reduced their fertility earlier and to a greater degree than poor families. While there are many hypotheses to explain these features, none is entirely satisfactory (Borgerhoff Mulder 1998). The niche-construction perspective potentially offers new explanations that might contribute to this debate. For example, parents in pre-transition societies reap a number of incidental benefits from their offspring's niche construction; their children contribute work on the family farm, help in the family home, produce income through child labor, and, later, offer protection from hostile outsiders, and care for parents in old age. In such circumstances, parents might benefit by having as many children as possible, and consequently wealthier parents might be expected to have, on the average, more offspring than poorer parents. Post-transition societies, however, differ in that they typically include governmental and societal institutions and regulations, such as hospitals and health care, schools, a police force, a judicial system, regulations against child labor, state benefits and pensions, and so forth. One consequence of these social developments is that the incidental benefits that accrue to parents from children become devalued as the same benefits are provided by society. Conversely, the costs of having children either remain the same or increase. This transition would thus be expected to tip the balance in favor of parents' investing in fewer children. Perhaps because wealthier parents are better able to exploit the incidental benefits provided by society, have less need of the security provided by a large number of children, or have greater costs associated with child rearing, such parents might be expected to reduce their fertility earlier and more than poorer parents.

Similar reasoning would lead us to predict that parents should invest more in sons or daughters to the extent that offspring of each sex subsequently contribute incidental benefits by their niche construction, some of which may be real but others merely perceived.

9.4.4 Phenogenotypes and Multiple-Process Adaptation

9.4.4.1 Units of Transmission and Selection in Human Evolution

Traditionally, genetic and cultural evolutionary processes have been studied separately even if occasionally by members of the same university departments. The separate methodologies for the two processes were developed essentially in isolation from one another, although aspects of one have occasionally informed the other. For example, human geneticists have often used archaeology or linguistics to confirm their ideas about population history. It is probably fair to say that scholars of cultural variation have either ignored or actively opposed attempts to incorporate the science of human genetics into their scholarship.

An exception was in the area of behavior genetics, which emerged in the late 1960s after a post–World War II hiatus during which the reaction to the evils of Nazi racism and its eugenic basis profoundly affected studies of human behavior. However, behavior genetics focused on the statistical partitioning of standing human variation in qualitative traits without a mechanistic basis for either the genetic or environmental components it purported to evaluate. This led Cavalli-Sforza and Feldman (1973a) to attempt a reformulation of the transmission of human behaviors from parents to children that used both the machinery of classical genetics and the notion of vertical transmission familiar to epidemiologists. This allowed the study of complex human traits to have a dynamic or evolutionary framework, not just a statistical representation. In this framework the natural unit, the individual, is viewed as having a genotype and a phenotype. The phenotype might vary continuously or discretely with different rules of transmission for each. The combination of genotype and phenotype is called the phenogenotype, and our claim is that this is a natural unit for human evolutionary studies.

If the genotype has no influence on the way in which the variants of the cultural trait are transmitted or acquired or on their relative fertility and mortality, then we can use a purely cultural-transmission scheme to describe how the trait passes among individuals of the same or different generations (Cavalli-Sforza and Feldman 1981). Likewise if, for example, the trait is a blood group, Mendelian genetics suffices to describe its transmission and any cultural component can be omitted. These may be regarded as two extremes of a transmission process that in general would allow the rules of transmission and acquisition of a cultural trait to be influenced by genes that would, of course, be subject to Mendelian transmission.

Superimposed on this transmission system are the processes of natural selection, which may determine differential rates of fertility or mortality among the different phenogenotypes. As pointed out by Cavalli-Sforza and Feldman (1981), for a cultural trait, it might be more appropriate to invoke the notion of cultural selection, which suggests different rates of acquisition of, or preferences for, the different cultural variants. Again, it is conceivable that these rates or preferences may have a genetic influence. When the transmission among or selection on phenogenotypes cannot be separated into purely genetic and purely cultural components, the resulting dynamics are called gene-culture coevolution. For instance, Feldman and Cavalli-Sforza (1989) studied the coevolution of a gene for lactose tolerance and the cultural trait of milk usage. Here, for example, one of the phenogenotypes was a milk-using individual homozygous for the lactose-tolerance allele.

If the cultural transmission occurs independently of any genetic variation, then the gene-culture coevolutionary dynamic is the superposition of a purely cultural system, whose dynamics might resemble, for example, those of an epidemic (horizontal transmission) or of an infection passed from parent to child (vertical transmission) on the separate biological system. Conversely, with true dependence between genes and the cultural trait under study, the transmission among phenogenotypes and natural and cultural selection becomes much more complex. The use of phenogenotypes in models 3 and 4 in chapter 6 highlights some of the interesting dynamical consequences of this complexity.

Sociobiologists think about evolution from the perspective of the gene: those characteristics that have been favored by selection are the expression of the selfish genes (Dawkins 1989) that were best able to increase their representation in the next generation. For advocates of gene-culture coevolutionary theory, the logic is the same, but the replicator is different: instead of the selfish gene, there is the phenogenotype. Those human characteristics that have been favored in the combined regime of natural selection and cultural transmission are the expression of the phenogenotypes that were best able to increase their representation in the next generation. As an intuitive shorthand, a phenogenotype can be thought of as a human with a package of genes and experiences. In this sense, the phenogenotype approach reestablishes the organism (or rather classes of organism) as the central unit of human evolution, not as vehicle but as replicator. In fact, what is really replicated is a biocultural complex, with a composite array of semantic information (acquired through multiple processes, and stored at different levels) and inherited resources. But in the search for simple conceptual and formal models that can shed light on the dynamics of such systems, we regard the phenogenotype approach as the best method currently on the market.

The use of phenogenotypes does not imply that the adoption of cultural traits is necessarily constrained by genes. A phenogenotype is a *class* of individuals, with a combination of a genotype and a variant of a cultural trait, and represents the basic evolutionary unit when both biological and cultural transmission and/or selection processes are operating in association. Phenogenotypes are the natural units in dual- (fig. 6.1b) or triple- (fig. 6.1c) inheritance systems, as the selection (whether natural or cultural selection) occurs *between* phenogenotypes. Treating cultural and biological processes as independent, while frequently a reasonable approximation (Boyd and Richerson 1985), on other occasions is problematic because it ignores any nonrandom associations between genotypes and cultural traits (Feldman and Cavalli-Sforza 1984; Feldman and Zhivotovsky 1992). If genes and culture are evolving together, we have to track phenogenotypes if we are to understand this coevolution.

Neither does the use of phenogenotypes imply that genotypes and cultural traits are locked together in a single individual. When an individual adopts a new cultural variant its phenogenotype changes, and as a consequence of cultural selective processes, phenogenotype frequencies may change within generations. Phenogenotypes are no more than a *currency* for tracking how genotype and cultural trait frequencies change over time in dual- or triple-inheritance models. That is, if genes and culture are evolving together, we have to track phenogenotypes if we are to understand this coevolution. This is true even when cultural traits are subject to evolutionary processes acting independently of genes. They are the simplest entities that can take account of any interactions between the two types of transmission, genetic and cultural, and the two types of selection, natural and cultural, including interactions explicitly mediated by cultural niche construction (figure 9.1).

9.4.4.2 Multiple-Process Adaptation

Controversy has surrounded the sociobiological postulate that human beings typically behave in ways that maximize their inclusive genetic fitness (Sahlins 1976; Montagu 1980; Segerstrale 2000). It is plausible that the processes that underlie culture are adaptations and that socially learned or culturally transmitted information must initially have increased average reproductive success. Mathematical models that have explored the evolution of social learning support these assumptions by revealing that the capacity for social learning cannot be favored unless it generally increases some measure of fitness. The same must be true of knowledge-gaining ontogenetic processes. So what characteristics of human cultural processes, if any, could allow humans to behave in maladaptive ways, or to transmit maladaptive information?

One of the most important findings to emerge from gene-culture coevolutionary theory is that there are a variety of mechanisms by which culture can lead to the transmission of information that results in a fitness loss relative to alternatives. Cavalli-Sforza and Feldman (1981) provided theoretical confirmation of the intuitive notion that cultural traits associated with a viability or fecundity deficit may still increase in frequency in a population if there is strong conversion of individuals to the same trait. Boyd and Richerson (1985) found that if individuals adopt the behavior of influential or successful members of their society, maladaptive cultural variants can spread, despite having a substantial viability disadvantage. Other gene-culture models reach the same conclusion (Feldman and Laland 1996).

Our perspective suggests that, in each generation, populations of organisms persistently construct or reconstruct significant components of their environments. This means that, as they evolve, organisms may, in effect, drag part of their own environments along with them, thereby transforming their own adaptive landscapes. If ontogenetic processes, culture, and counteractive niche construction in general have consistently damped out the need for a genetic response to changes in the population's environment, hominid populations may have become increasingly divorced from their ecological environments. At the same time, our primate and hominid ancestors may increasingly have responded to novel selection pressures initially generated by inceptive niche construction and subsequently dominated by cultural traditions. In particular, components of the social environment, for example, traits related to family, kinship, and social stratification, may have been increasingly transmitted from one generation to the next by cultural inheritance to the extent that contemporary human populations may have become largely divorced from local ecological pressures. Support for this argument comes from Guglielmino et al.'s (1995) study of variation in cultural traits among 277 contemporary African societies, in which most of the traits they examined correlated with cultural (linguistic) history, rather than with ecological variables. If this study is representative, then socially transmitted cultural traditions are a lot more important than most evolution-minded researchers studying human behavior would admit.

Current use by some evolutionary psychologists of the concept of the environment of evolutionary adaptedness (EEA) is unsatisfactory in another respect. It treats humans as passive victims of selection rather than as virtuoso niche constructors. Human beings cannot be exclusively adapted to a past world and not at all adapted to modern life; if so, we would be extinct. Our capacity to interact with worlds that we continuously modify by niche construction reflects the fact that humans are very adaptable creatures. The flexible nature of our learning and culture allows us to survive and flourish in a broad range of environments, including those that are heavily niche-constructed by modern humans.

Despite what appears to be a continuous process of adaptation, could we humans, nonetheless, drive ourselves to extinction? There are two reasons for supposing that this is a possibility. First, culture greatly enhances the human capacity for both positive and negative niche construction. For example, science-based technology is having an enormous impact on the human environment. It has made many new resources available via both agriculture and industry, it has influenced human population size and structure via hygiene, medicine, and birth control, it has drastically changed human warfare, it is drastically reducing biodiversity, and it may already have resulted in the degradation of large areas of our global environment. These are all potential sources of modified natural selection. Second, human cultural processes can work much faster than human genetic processes and may possibly generate new environments faster than the human genetic processes can respond to them (via route 2 in fig. 9.1). In these circumstances, human culture might drive either local or global self-induced extinctions.

In many respects, this restates an old evolutionary problem. Many relatively long-lived species have encountered rates and types of environmental change, whether self-induced or independent, that exceeded the capacity of their genes to handle, and as a result have gone extinct (Lawton and May 1995). Clearly, one way in which humans could prevent self-induced extinctions of this kind is by adapting to culturally induced environmental changes through rapid responses at some nongenetic level, especially through further cultural change (via route 1 in fig. 9.1).

Unfortunately there are well-known difficulties with this kind of solution for humans. First, a human population may not recognize the source of a novel, culturally induced natural selection pressure and may therefore make no attempt to counteract it at the cultural level. This was the case with the Fore of Papua New Guinea, who maintained a cannibalistic tradition despite the fact that it perpetuated the deadly disease kuru (Cavalli-Sforza and Feldman 1981; Durham 1991). Second, the required corrective technology may not always be available or may be too costly to introduce. As mentioned earlier, the Kwa could have responded to the increased natural selection due to malaria by cultural control of this disease, but they lacked the technology to do so. Third, the feedback from cultural niche construction may be indirect, which may make it difficult to recognize any longer-term negative consequences of the niche construction. Rogers (1995) points out that the adoption of wet rice cultivation in Madagascar seems to have had a range of diffuse indirect effects, some of them manifest only several generations later. Fourth, responding to cultural change with further cultural change always risks introducing a runaway situation, in which each new solution generates the next problem at an ever-accelerating rate. The phenomenon of antibiotic resistance is a recent example (Ewald 1994; Palumbi 2001a,b). We conclude that cultural niche construction is so powerful that it could potentially drive our species to extinction.

CHAPTER 10

Extended Evolutionary Theory

10.1 THE NICHE-CONSTRUCTION PERSPECTIVE

There are two ways of thinking about niche construction in evolution. One is very familiar; everyone with an interest in natural history knows that organisms possess adaptations that allow them to choose and construct aspects of their worlds. Niche construction is a widely recognized product of evolution. However, in this book we have invited the reader to think about niche construction differently by regarding it not as just a product of evolution, but as a co-contributor, with natural selection, to the evolutionary process itself. From this perspective, niche construction is more than just the expression of adaptations, it is a broader evolutionary process of which the usual notion of adaptation is just a part.

Why are we advocating this alternative? Standard evolutionary theory enjoys the advantage of being relatively simple, which allows shortcuts to the development of hypotheses and the drawing of conclusions. The added complexity of treating niche construction as a process must earn its keep by providing new insights and testable hypotheses. So how does extending evolutionary theory by incorporating niche construction enhance our understanding of the evolutionary process?

In this last chapter we will attempt to synthesize the arguments from all the earlier chapters in a final advocacy of the merits of extended evolutionary theory.

10.2 CONSTRUCTIVE AND EXTERNALIST EXPLANATIONS OF EVOLUTION

Godfrey-Smith (1996, p. 131) distinguishes among three basic types of explanation for the relationships between organisms and environments. He calls them "externalist," "internalist," and "constructive."[1]

"Externalist" refers to explanations of the internal properties of organic systems in terms of environmental properties. Internalist explanations describe one set of internal properties in a system in terms of another set of internal properties in the same system. These internal properties could either be in organisms or in environments. Constructive explanations interpret environmental properties in terms of the properties of organisms. Godfrey-Smith's distinctions, especially when used in conjunction with Lewontin's pairs of differential equations, 1.1 and 1.2 versus 1.3 and 1.4 (see chapter 1), provide us with a convenient way of summarizing the essential differences between how standard evolutionary theory and our extended evolutionary theory explain organism-environment relationships.

The standard view is that niche construction should not be regarded as a distinct process in evolution because the manner in which organisms modify their environments does not redirect the pressures of natural selection in any significant way but is merely a component of the expression by contemporary organisms of past natural selection. Hence, the ultimate cause of niche construction is the same as that of all the other properties of organisms, the natural selection of genotypes by environmental selection pressures. At the heart of standard evolutionary theory we therefore find Godfrey-Smith's (1996) externalist explanations. The properties of organisms are explained relative to the properties of environments, but the properties of environments are not explained relative to the properties of organisms. Instead, the latter are usually accounted for only in terms of internalist explanations in the environments themselves. This is the same

[1] Godfrey-Smith's use of these terms is idiosyncratic. Different disciplines and different people, ranging from Marxists to postmodernists, assign different meanings to these same terms. We therefore emphasize that in this chapter we are using these terms exclusively in Godfrey-Smith's sense.

point that Lewontin (1983) made when he summarized standard evolutionary theory by equations 1.1 and 1.2.

We agree with Lewontin that this standard conceptualization is a distortion for at least two reasons. The first and less important reason is that no aspect of the phenotype and no level of variation in populations can be regarded as fully determined by naturally selected genotypes. Genes may be regarded as determining proteins, but their influence on phenotypes is more diffuse. Besides, not all of the genes that influence niche construction are likely to have been subjected to prior natural selection anyway. The pattern of alleles at these loci could be due to other causes, for instance, mutation, drift, or selection of linked genes. In addition, the information that is expressed by niche-constructing organisms could be acquired instead of genetic in origin. For instance, individual animals may learn appropriate niche-constructing behavior either as a result of their own independent experience, or as a result of social interaction. Thus evolution could be mediated by changes in the niche-constructing activities of organisms rather than by autonomous changes in natural selection pressures, and therefore by properties of organisms rather than of environments. The effects of genes on a phenotype, whether the phenotype is the carrier of the genes or another individual, are mediated by developmental (including environmental) processes, and to leapfrog those processes is tantamount to denying that development exerts any meaningful influence on the phenotype.

The second and more fundamental reason why the standard theory is a distortion is that the selective environments of organisms are not independent of those organisms but are themselves partly products of their prior or possibly current niche-constructing activities. Therefore, the properties of environments cannot be independent of the properties of organisms, and the process of natural selection cannot be wholly unaffected by the process of niche construction. Hence, the simplifying assumption that the ultimate cause of niche construction lies exclusively in properties of environments must be inaccurate. The externalist explanations of evolution provided by standard evolutionary theory, according to which properties of organic systems are explained exclusively in terms of independent properties of environments, are insufficient and likely to limit our understanding of how evolution works.

A more complete view of the process of evolution requires a combination of both externalist explanations based on natural selection and constructive explanations based on niche construction. This is still a long way from being the fully constructivist approach that Godfrey-Smith (1996) calls the "asymmetric constructivist view" (p. 133). An asymmetric constructivist position would merely exchange an exclusively externalist explanation of evolution, which explains the properties of organisms in terms of the properties of environments, for an exclusively constructivist explanation. That would simply exchange one kind of mistake for another and is not what we are advocating. Instead, we are proposing a mix of externalist and constructivist explanations, according to which natural selection is partly dependent on the niche-constructing activities of organisms, and niche construction is largely dependent on prior natural selection pressures, including those that are, or have been, biotically modified. This is the solution that Godfrey-Smith's (1996) analysis encourages us to defend. He calls it interactionist. It is also the solution that Lewontin (1983) advocates with his second pair of differential equations (see eqs. 1.3 and 1.4). Thus, we seek to describe evolutionary change in terms both of the properties of environments that affect organisms and of the properties of organisms that affect environments.

This solution, however, still leaves open the possibility that even if the assumptions of extended evolutionary theory are more accurate than the assumptions of standard evolutionary theory, it may still not make sense to switch from standard evolutionary theory to extended evolutionary theory. There are two reasons, the first of which was recognized by Lewontin (1983) himself. To cite Godfrey-Smith (1996, p. 141) again:

> For Lewontin, the imposition of an externalist explanatory regime in biology was one of Darwin's central achievements. This enabled evolutionary theory to advance beyond (internalist) views in which changes undergone by species were understood as analogous to the unfolding of a developmental program. Darwin located the ordering mechanisms of evolutionary change in the environment instead.

The huge advances in biology due to Darwin and to standard evolutionary theory are there for everyone to see. So perhaps we should heed Hillaire Belloc's advice and not "let go the hand of nurse," at least not prematurely. The second reason is the one we have repeatedly acknowledged. Compared to extended evolutionary theory, standard evolutionary theory undoubtedly enjoys the advantage of theoretical simplicity, and that should not be given up lightly. So, even though standard evolutionary theory may be insufficient because of what it distorts or omits, this incompleteness may still not justify going beyond a very successful, simple, and proven theory of evolution. We need to review the balance sheet once more to make sure. So we will end with two questions: What do we stand to lose if we stick with standard evolutionary theory in spite of its limitations described in the first nine chapters? What do we stand to gain if we substitute extended evolutionary theory for the standard theory?

10.3 DEFICIENCIES OF STANDARD EVOLUTIONARY THEORY

10.3.1 The Problem with Adaptation

The first thing we lose by sticking with standard evolutionary theory is the more comprehensive understanding of adaptation that Lewontin described and recommended in his (1983) attack on adaptationism.

If high-school students were asked to classify an earthworm using standard classificatory methods, it would rapidly become apparent that the earthworm has no business living in soil (Turner 2000). As we saw in chapter 7, earthworms are structurally very poorly adapted to cope with physiological problems such as water and salt balance on land, and they would seem to be better suited to a freshwater habitat. They can survive in a terrestrial environment only by constructing a more suitable niche through activities such as tunneling, exuding mucus, eliminating calcite, and dragging leaf litter below ground.

Earthworms coopt the soils that they inhabit and the tunnels they build to serve as accessory kidneys that compensate for their poor structural adaptation. For instance, by producing well-aggregated soils, the worms weaken matrix potentials and make it easier for them to draw water into their bodies (Turner 2000). However, as we saw earlier, in the process, earthworms dramatically change their environments. They reduce surface litter, aggregate soil particles, increase levels of organic carbon, nitrogen, and polysaccharides, enhance plant yields, and improve porosity, aeration, and drainage. The results of earthworm activity highlight a problem with the concept of "adaptation." In this case it is the soil that does the changing, rather than the worm, to meet the demands of the worm's freshwater physiology. So what is adapting to what?

This kind of phenomenon explains why, for so many years, Lewontin has been arguing that there is something wrong with our concept of adaptation (Lewontin 1982, 1983, 2000). Standard evolutionary theory short-changes the active role of organisms in constructing their environments. From the standard evolutionary perspective, which is externalist, the niche-constructing traits of earthworms can only be described as adaptations by earthworms to their soil environments in response to natural selection. Earthworms have apparently taken an alternative evolutionary trajectory to many other terrestrial creatures in solving their water-balance problem by behavioral means. Yet standard evolutionary theory would describe the adaptive match between organism and environment as having come about through the action of autonomous natural selection pressures from the environment.

From the niche-construction perspective offered by extended evolutionary theory, this standard account of adaptation is misleading, and even wrong. For example, without ancestral niche construction by many organisms, including earthworms, topsoil would not exist, so how could earthworms adapt to it? The ancestors of contemporary earthworms must have contributed significantly to the construction of the soil environments to which their descendants are now adapted. In fact, there are two logically distinct routes to the complementary match between organisms and their environments. Either organisms can change to suit their environments, or environments can be changed by the organisms to suit them, and probably their descendants. In the earthworm case, there is no denying that in contemporary populations the match between earthworms and their soil environment is brought about at least in part by the second route, that is, through earthworm-induced changes in the soil. The problem is that in standard evolutionary theory the contribution of this second route to the complementary "adaptive" match between organisms and environments is usually ignored. For instance, how are we to explain that earthworms have "freshwater" kidneys (Turner 2000), but are adapted to soil environments?

10.3.2 Changing the Evolutionary Process

We would also lose many insights into how the evolutionary process works. For example, consider the difference it makes if natural selection stems from autonomous components of environments or from niche-constructed components of environments. The difference can be summed up in one word: feedback. If organisms evolve in response to selection pressures modified by their ancestors, there is feedback in the system. The examples in chapter 2 demonstrate that niche construction can significantly modify natural selection pressures. Many organismal traits would seem to have evolved as a consequence of prior niche construction, and each is an example of feedback in evolution. It is well established that systems with feedback behave quite differently from systems without feedback (Robertson 1991), and by neglecting this feedback, the standard evolutionary perspective must at least sometimes misrepresent how evolution works.

The models in chapter 3 illustrate the differences that this feedback can make to the evolutionary process. First, traits whose fitness depends on alterable sources of selection (recipient traits) coevolve with traits that alter sources of selection (niche-constructing traits). This results in very different evolutionary dynamics for both traits from the dynamics that would occur if each of these traits had evolved in isolation. For example, feedback from niche construction can cause evolutionary inertia or momentum, lead to the fixation of otherwise deleterious alleles, support stable polymorphisms where none are expected, eliminate what would otherwise be stable polymorphisms, and influence linkage disequilibrium. So these models clearly demonstrate that the dynamics of evolution are quite different when feedback is introduced.

A second difference is ecological inheritance. According to standard evolutionary theory, organisms inherit only genes from their parents. Our extended evolutionary perspective, however, stresses two legacies that organisms inherit from their ancestors: genes, and a modified environment with its associated modified selection pressures. As the examples in chapter 2 document, ecological inheritance is likely to be extremely widespread. Also, as the analyses in chapters 3 and 6 showed, because of the multigenerational properties of ecological inheritance, niche construction can generate many different evolutionary dynamics. Moreover, since niche-constructing organisms can modify the environments of other species as well as their own, the feedback to the niche constructors may be indirect and operate via any number of ecological components, including those that are abiotic, of ecosystems (chapter 5). This means that the feedback from niche construction may manifest itself in diverse and sometimes cryptic forms. In short, wherever there is ecological inheritance, the evolutionary process is likely to operate in a different manner from that described by standard evolutionary theory.

A third difference is that acquired characteristics can play a (non-Lamarckian) role in the evolutionary process as they influence the selective environment through niche construction. Returning to a now familiar example, the Galápagos woodpecker finch's use of spines is not inherited, but rather learned afresh by each individual (Tebbich et al. 2001), yet nonetheless almost certainly created a selection pressure favoring a bill able to manipulate tools rather than the sharp, pointed bill and long tongue characteristic of woodpeckers. If so, it is an example of an acquired character influencing evolution.

For most vertebrates, the role of acquired characteristics in evolution is likely to be fairly restricted, but there is every reason to believe that acquired characters may have been important in hominid evolution. The models that we described in chapter 6 revealed circumstances under which cultural transmission could overwhelm natural selection, accelerate the rate at which a favored allele spreads, produce novel evolutionary events, and perhaps have triggered hominid speciation. We argue that, because cultural processes typically operate faster than natural selection, cultural niche construction is likely to have more profound consequences than gene-based niche construction.

10.4 ADVANTAGES OF AN EXTENDED EVOLUTIONARY THEORY

What are the benefits that the extended evolutionary perspective can offer? These relate mainly to phenomena that either are neglected by standard evolutionary theory, or, as with the evolutionary role of abiota in ecosystems, fall largely outside its scope.

10.4.1 Benefits for Evolutionary Biology

The explicit introduction of niche construction into evolutionary theory generates new hypotheses and suggests new empirical work. For instance, in chapter 7 we sketched a set of testable empirical predictions that distinguish between standard and extended evolutionary theory and indicated the kind of data that would be more consistent with extended evolutionary theory than with standard evolutionary theory. Of course, the results of such experiments or observations could prove us wrong. We also outlined a series of methods that could detect significant evolutionary ramifications of niche construction.

One general point is worth making about possible future empirical programs. Although extended evolutionary theory is more complicated than standard evolutionary theory, there is no reason why any empirical program carried out to explore extended evolutionary theory should be more complicated or difficult than any empirical program that would fall under the rubric of standard evolutionary theory. In general, extended evolutionary theory should merely raise new questions, not new empirical difficulties.

10.4.2 Benefits for Ecosystem-Level Ecology

Extended evolutionary theory may also contribute to some related disciplines as well as to evolutionary biology, including disciplines that currently either cannot fully benefit from standard evolutionary theory because they lie beyond its scope or suffer from standard evolutionary theory's limitations. For instance, in chapter 5 we suggested that niche construction may have important implications for ecosystem-level ecology, because a niche-construction perspective could shed evolutionary light on several problems that have hitherto been considered within the domain of ecology only. One reason is that niche construction modulates and may partly control the flow of energy and matter through ecosystems. We argued that niche construction can be characterized as the expression of semantic information, accrued through natural selection. We have also characterized natural selection, including natural selection that has been modified by prior niche construction, as a process that results in the acquisition of semantic information by organisms. Thus, we have the prospect of semantic information flowing through entire ecosystems. Genes in the same or different populations may interact with each other via both biotic and abiotic components of the environment to form environmentally mediated genotypic associations (EMGAs). The recognition that niche construction may instigate flows of semantic information through ecosystems may shed light on several classic problems in ecology.

The work of ecologists who study ecosystem engineering (Jones et al. 1994, 1997), which is a synonym for niche construction, illustrates the utility of this perspective. Jones et al. pointed to several ecosystem phenomena that cannot be understood in terms of energy and matter flows only. They stressed the critical role played by the creation of physical structures and other modifications of their environments by niche-constructing organisms that partly control the distribution of resources for other species in ecosystems.

Conversely, by ignoring niche construction, standard evolutionary theory limits the scope of interactions that might be perceived to occur between biotic and abiotic components in ecosystems and ignores forms of feedback that probably play a role in coevolutionary scenarios and ecosystem dynamics. This is probably one reason why it has hitherto proved so difficult to integrate process-functional and population-community ecology, or the former with standard evolutionary theory (O'Neill et al. 1986; Holt 1995). With niche construction, and in terms of the flow of semantic information through ecosystems, the idea of evolutionary control webs emerges. Chapter 8 sketches how these insights can be employed in ecological research and points to ways of integrating ecological and evolutionary models of niche construction.

10.4.3 Benefits for the Human Sciences

In chapter 6 we explored the repercussions of an extended evolutionary theory for the human sciences. Here the theory acts as a hypothesis-generating framework around which human scientists can structure evolutionary approaches to their disciplines. While the processes involved in human evolution are very complex, this conceptual model reveals particular subprocesses, or suggests specific hypotheses, that are subject to empirical test and can be developed into formal models. The predictions in chapter 9 and the formal models in chapter 6 illustrate how this can be done.

We also discussed the two principal reasons why many human scientists have hitherto found it difficult to use evolutionary theory. One is that standard evolutionary theory appears to have too little to offer them. Human scientists are predominantly interested in human behavior and culture, rather than genes, and as a consequence they have little use for standard evolutionary theory. Our extended evolutionary framework may be more appealing to human scientists because it includes additional roles for phenotypes and artifacts in evolution. The other reason why human scientists have difficulty with evolutionary theory is that contemporary adaptationist accounts, as in evolutionary psychology, are frequently simplified to the point of distortion. Adding niche construction makes evolutionary theory more complicated, but for human scientists it may eliminate some of these egregious distortions. In this respect the potentially illuminating nature of niche construction is illustrated by the material in chapter 9, where we show how our framework could be applied in the human sciences and tested empirically.

10.4.4 Other Potential Benefits Not Yet Considered

Two other potential benefits that may arise from a shift from standard evolutionary theory to extended evolutionary theory are worth mentioning.

10.4.4.1 Developmental Biology

The first is the provision of extra connections between evolutionary and developmental biology. We highlighted one of these connections in chapter 4, namely, the interdependence of natural selection and niche construction, neither of which can work without the other. Niche construction is an individual-based selective process that arises from the development and life histories of individual organisms and allows them the opportunity to gain sufficient "chosen" or "selected" energy and material resources from their environments to survive and reproduce. Niche construction thereby contributes both to the building of the next generation of a population and to changes in the niche-constructing organisms' own selective environments, as well as in the environments of others. Conversely, natural selection is a selective process that can only work when there is variation in the populations. However, the principal consequence of natural selection is that it allows ontogenetic processes to be informed by semantic information encoded in naturally selected genes according to the Darwinian algorithm and genetic inheritance. Moreover, it is only because ontogenetic processes can be informed by natural selection that individual organisms can survive and reproduce. The result is an intimate interplay between phylogenetic and ontogenetic processes in biology. Neither process on its own can account for either the evolution of populations or the development of individual organisms. Together, they might account for both.

A second possible source of connections between developmental and evolutionary biology arises from our EMGA concept. In chapter 5 we discussed EMGAs primarily in connection with ecosystem-level ecology. It may seem a long way from ecosystems to developmental biology, but the EMGA concept might have some utility for developmental biology as well. In their book on phenotypic plasticity Schlichting and Pigliucci (1998, p. 333) made the following provocative statement (our italics):

> The DRN (developmental reaction norm) also *leads us to eliminate the dichotomy between internal and external environments*. In fact, we view the developmental process itself as a reaction norm. *Gene expression may be constitutive or determined by the cellular environments to which the genes are exposed*. The ensuing phenotypes will be dependent on these patterns of gene expression as well as on the interactions of the gene products mediated by the other features of the environment. We offered as an example the change in the environment of cells and gene products during the

transition to three-dimensionality—previously all cells were in contact with the external environment, now some are never exposed to those conditions. Any new phenotypic characteristics of this internal cell population will be observed as pleiotropic effects of the original genetic complement.

A message for us in this paragraph is that, despite appearances to the contrary, there may be a high degree of continuity between the EMGAs established by niche construction, and that are mediated by biota or abiota in the external environments of populations, and the interactions among genes that are caused, on the one hand, by their expressed products and, on the other hand, by the modified selection of other genes in response to those products in developing organisms.

Dawkins (1982) offered one highly suggestive example. He contrasted two kinds of houses built by two different animals. One is the stone, or stick, "houses" built by caddis fly larvae from material they glean from river beds. These houses are clearly external artifacts and are therefore extended phenotypes. The other kind is the shells of snails. Are snail shells phenotypes, or are they extended phenotypes? Clearly they are houses of a kind, but they are houses that are normally regarded as part of the bodies of snails, rather than separate from them, and therefore as phenotypes, rather than as extended phenotypes. Snail shells constitute a borderline case, particularly because, as Dawkins pointed out, a snail's shell is inorganic and dead. The shell's chemical substance is directly secreted by living cells in the snail, but it then remains attached to the snail's body after it has been constructed as part of the snail's body or internal environment. In fact the status of a snail's shell appears to be very similar to the status of the web in our spider example (chapters 1 and 2) except that the snail shell sits on top of the snail, whereas the spider sits on its web.

What happens when this boundary is fully crossed and the product of a gene is not expressed in the external environments of organisms at all, but remains inside them in the manner indicated by Schlichting and Pigliucci? Logically, we end up with a kind of internalized EMGA. There are differences, of course, between the EMGAs of this book, which are always mediated by the external environment, and those that interest developmental biologists. For example, organisms must interact with their external environments to gain the resources they need to develop. But there are parallels which could throw some additional light on, say, the evolution of developmental reaction norms (Schlichting and Pigliucci 1998). In fact, one of the first researchers to recognize the role of niche construction in evolution was a developmental biologist, namely, Waddington (1959).

10.4.4.2 Conservation Biology

Another possible benefit of extending standard evolutionary theory that we have not explored in this book relates to conservation and restoration biology, relative to human-induced perturbations in ecosystems (McGlade 1999b). Probably because of the cumulative nature of human cultural inheritance, human cultural niche construction is a uniquely powerful source of perturbation in ecosystems (Laland et al. 2001). We are currently witnessing anthropogenic changes and associated human-induced losses of biodiversity in ecosystems on unprecedented scales (Tilman 2000; Purvis and Hector 2000; Gaston 2000; McCann 2000; Chapin et al. 2000; Margules and Pressey 2000). However, the power of human cultural niche construction could be used by culturally well informed humans to arrest, reverse, or counteract at least some of the human-induced damage to diverse ecosystems that our species is currently causing. To be well informed, however, it is first necessary to reach a better understanding of the relationship between human cultural niche construction, as described in chapters 6 and 9, and the consequences of niche construction for ecosystem-level ecology, as described in chapters 5 and 8.

Although human cultural niche construction is uniquely powerful, it should, nevertheless, be manifest in ecosystems in the same ways as all other kinds of niche construction, not only by perturbing them, but also by contributing to their regulation, protection, and control. This could promote a more "benign" human management of ecosystems, in response to anthropogenic change. However, the necessary connections between human cultural niche construction and its consequences for ecosystems are unlikely to be made on the basis of standard evolutionary theory for two reasons. First, standard evolutionary theory does not incorporate the general point that local environments, and in a more diffuse sense ecosystems, are partially engineered by organisms. This means not only that niche-constructing organisms regulate environmental states, but also that the evolution of organisms must drive environmental changes, including abiotic changes. Extended evolutionary theory does recognize this point (chapter 4). Second, standard evolutionary theory can neither account for the contribution of human cultural niche construction to human gene-culture coevolution (chapter 6), nor can it be applied to ecosystem-level ecology (chapter 5). Extended evolutionary theory can do both.

The practical implications of extended evolutionary theory for conservation can be encapsulated in two points. First, there is now wide recognition that it is likely to prove insufficient to concentrate conservation efforts on "saving" endangered species, for instance, through captive breeding and reintroduction, if the causes of their decline are not addressed. Consequently, the goal of many conservation efforts has switched toward preserving ecosystems. Yet, given financial limitations and other practical constraints, it is not immediately obvious how conservation efforts should be targeted in order to achieve this stated goal. Ecosystems are difficult to protect without an understanding of the processes that control and regulate them, and affect their resilience. In this respect, a niche-construction perspective could prove helpful. Conservation efforts may be most effective if they ensure the survival of the key engineers in an ecosystem, and the methods described in chapter 8 and elsewhere (Jones et al. 1997) suggest tools that might help to identify these species. Second, it is plausible that the most effective means to preserve ecosystems is not to focus on saving particular species, even the most important engineers, but rather to concentrate on preserving their engineering effects, some of which may be abiotic.

10.5 CONCLUSION

That is the balance sheet. In our view there are enough advantages provided by the niche-construction perspective to justify our claim that evolutionary theory would benefit by extending the classical neo-Darwinian processes of natural selection and genetic inheritance to include processes of niche construction and ecological inheritance.

As Lewontin's (1983) equations indicate, we need better descriptions of the relationship between the properties of evolving organisms and their coevolving environments. To achieve that, however, we shall have to recognize that evolution depends not on one, but on two general selective processes: natural selection and niche construction.

APPENDIX 1

Model 1a: A Simple Model of Niche Construction for Diploids

Consider an isolated population of randomly mating, diploid individuals, and two diallelic loci **E** (with alleles E and e) and **A** (with alleles A and a). In generation t, the frequencies of the four gametes (EA, Ea, eA, and ea) are given by u_{1t}, u_{2t}, u_{3t}, and u_{4t}, respectively, so that the frequencies of alleles E and A are $x_t = u_{1t} + u_{2t}$ and $p_t = u_{1t} + u_{3t}$, respectively. In each generation, the amount R of the resource **R** created or destroyed as a consequence of niche construction is assumed to be a function of the frequency of allele E at time t. The simplest such function is linear, in which case, over n generations of niche construction, the cumulative effect at time t is

$$R_t = \sum_{i=t-n+1}^{t} \pi_i x_i, \qquad \text{(A1.1)}$$

where π_i represents the weight attributed to the impact of the ith generation's niche construction. We define three special cases of A1.1, characterized by the weighting given to π_i.

(i) *Equal weighting:* All n generations of niche construction equally influence the frequency of the resource in the environment at time t. Thus

$$\pi_t = \pi_{t-1} = \pi_{t-2} = \cdots = \pi_{t-n+1} = \frac{1}{n},$$

and the frequency of the resource at time t is given by

$$R_t = \frac{1}{n} \sum_{i=t-n+1}^{t} x_i. \tag{A1.2a}$$

(ii) *Recency effect*: Recent generations of niche construction have a greater effect than earlier generations on the frequency of the resource at time t. Thus

$$\pi_t > \pi_{t-1} > \pi_{t-2} > \cdots > \pi_{t-n+1}.$$

A simple example is expressible in terms of a parameter μ that represents the decrement in relative weighting of each generation with time:

$$\pi_t = \frac{1}{\varphi},\ \pi_{t-1} = \frac{\mu}{\varphi},\ \pi_{t-2} = \frac{\mu^2}{\varphi}, \ldots, \pi_{t-n+1} = \frac{\mu^{n-1}}{\varphi},$$

where $0 < \mu < 1$, and $\varphi = (1 - \mu^n)/(1 - \mu)$, and hence the frequency of the resource at time t is given by

$$R_t \frac{1-\mu}{1-\mu^n} \sum_{i=t-n+1}^{t} \mu^{t-i} x_i. \tag{A1.2b}$$

(iii) *Primacy effect*: Earlier generations of niche construction have a greater effect on the frequency of the resource at time t than more recent generations. Here

$$\pi_t < \pi_{t-1} < \pi_{t-2} < \cdots < \pi_{t-n+1}.$$

As in (ii), a simple example employs the parameter μ to represent an increment in the relative weighting of each generation with time:

$$\pi_t = \frac{\mu^{n-1}}{\varphi},\ \pi_{t-1} = \frac{\mu^{n-2}}{\varphi},\ \pi_{t-2} = \frac{\mu^{n-3}}{\varphi} \ldots, \pi_{t-n+1} = \frac{1}{\varphi},$$

where $0 < \mu < 1$, and $\varphi = \frac{1-\mu^n}{1-\mu}$. Here the frequency of the resource at time t is given by

$$R_t = \frac{1-\mu}{1-\mu^n}\sum_{i=1}^{n}\mu^{i-1}x_{t-n+i}. \qquad \text{(A1.2c)}$$

Note that in all three cases, if $n = 1$, then the frequency of the resource at time t is equal to the frequency of allele E in that generation.

Genotypic fitnesses are given in table 3.1 and are assumed to be functions of a fixed viability component and a frequency-dependent viability component. The fixed component is given by the α_i and η_i terms, which are the fitnesses of single-locus genotypes in the standard two-locus multiplicative viability model. The frequency-dependent component is a function of the frequency R of the resource, which is given in terms of the frequencies of allele E over n generations by equations A1.1 and A1.2 above. The terms involving ε express the deviations from constant viabilities due to niche construction, where the size of ε scales the amount of niche construction, with $-1 < \varepsilon < 1$. We analyze this model for the cases $f = 2$, 1, and 0.5, which represent simple examples of relationships between the contribution of locus **A** to genotypic fitnesses and **R**. It is assumed that the double heterozygotes have equal fitness. The fitness structure assumes that the resource level is experienced equally by all members of the population (i.e., spatial homogeneity).

The entries in table 3.1 give rise to the four standard gametic recursions

$$Wu_1' = [u_1(u_1w_{11} + u_2w_{21} + u_3w_{12} + u_4w_{22})] - rw_{22}D, \qquad \text{(A1.3a)}$$

$$Wu_2' = [u_2(u_1w_{21} + u_2w_{31} + u_3w_{22} + u_4w_{32})] + rw_{22}D, \qquad \text{(A1.3b)}$$

$$Wu_3' = [u_3(u_1w_{12} + u_2w_{22} + u_3w_{13} + u_4w_{23})] + rw_{22}D, \qquad \text{(A1.3c)}$$

$$Wu_4' = [u_4(u_1w_{22} + u_2w_{32} + u_3w_{23} + u_4w_{33})] - rw_{22}D, \qquad \text{(A1.3d)}$$

where r is the recombination rate, $D = u_1u_4 - u_2u_3$ is the linkage disequilibrium between loci **E** and **A**, the sum of the right-hand sides of A1.3a–A1.3d is W, and the primes indicate frequencies in the next generation.

NO EXTERNAL SELECTION

We take the case $R_t = x_t$ in which only the current generation of organisms affects the resource. With $\alpha_i = \eta_i = 1$ $(i = 1,2)$ we have $w_{11} = w_{12} = w_{13} = 1 + \varepsilon x_t^f$, $w_{21} = w_{22} = w_{23} = 1 + \varepsilon x_t^{f/2}(1 - x_t)^{f/2}$, $w_{31} = w_{32} = w_{33} = 1 + \varepsilon(1 - x_t)^f$. Recalling that $x_t = u_{1t} + u_{2t}$, $p_t = u_{1t} + u_{3t}$, we may rewrite the system (A1.3) with $q = 1 - p$ as

$$p' - p = \frac{\varepsilon pq[px^{f/2} + q(1-x)^{f/2}][x^{f/2} - (1-x)^{f/2}]}{W}, \tag{A1.4}$$

$$x' - x = \frac{\varepsilon D[px^{f/2} + q(1-x)^{f/2}][x^{f/2} - (1-x)^{f/2}]}{W}, \tag{A1.5}$$

$$D' = \frac{D\{1 + \varepsilon x^{f/2}[px^{f/2} + q(1-x)^{f/2}]\}\{1 + \varepsilon(1-x)^{f/2}[px^{f/2} + q(1-x)^{f/2}]\}}{W^2}$$
$$- \frac{rD[1 + \varepsilon x^{f/2}(1-x)^{f/2}]}{W}, \tag{A1.6}$$

where p', x', D' refer to generation $t + 1$ and p, x, D to generation t. The value of W is $1 + \varepsilon\,[px^{f/2} + q(1 - x)^{f/2}]^2$.

Notice first that when $x = 1/2$ then $x' = x$ and $p' = p$ so that $D' = D(1 - r)$, which is the classical recursion for linkage disequilibrium between two genes in the absence of selection. Consider the change in the frequency of A over one generation. From A1.4, if $\varepsilon > 0$ then $p' > p$ if $x > 1 - x$; i.e., $x > 1/2$. We now show that $x' > 1/2$ if $x > 1/2$ and $\varepsilon > 0$. From A1.4, we have

$$x' - \frac{1}{2} = x - \frac{1}{2} + \frac{\varepsilon D[px^{f/2} + (1-p)(1-x)^{f/2}][x^{f/2} - (1-x)^{f/2}]}{W},$$

and since $\varepsilon > 0, f > 0$, and $D > -1/4$ we have

$$x' - \frac{1}{2} > x - \frac{1}{2} - \frac{[x^{f/2} - (1-x)^{f/2}]}{4}. \tag{A1.7}$$

We must show that the right-hand side of A1.7 is positive when $x > 1/2$. In other words, we must show that

$$4x - 2 > x^{f/2} - (1 - x)^{f/2}, \qquad \text{(A1.8)}$$

when $x > 1/2$. Note that if $1 > x > 3/4$, then the left side of A1.8 is greater than 1 and the right-hand side is less than 1 for $f/2 > 0$. It remains to prove A1.8 for $1/2 < x < 3/4$. For ease of writing use $a = f/2$ and first consider $a \leq 2$. Rewrite A1.8 as

$$2x - x^a > 2(1 - x) - (1 - x)^a. \qquad \text{(A1.9)}$$

Since $x > 1 - x$ when $x > 1/2$, if we can show $2x - x^a$ is increasing in x then A1.9 is true. But the derivative of the left-hand side of A1.9 is $2 - ax^{a-1}$, which is positive if $a/2 < x^{1-a}$ for $x > 1/2$. But if $a < 1$,

$$\frac{a}{2} < \frac{1}{2} < \left(\frac{1}{2}\right)^{1-a} < x^{1-a},$$

since $x > 1/2$. Hence $2x - x^a$ is increasing in x and the left-hand side of A1.9 exceeds the right-hand side. It follows that A1.7 and A1.8 are true. If $a = 1$ then obviously A1.9 is true. If $2 \geq a > 1$ then $2 - ax^{a-1} \geq 2\,(1 - x^{a-1}) > 0$ since $x < 1$ and $1 > a - 1 > 0$. It remains to show that A1.8 is true for $a > 2$ and $1/2 < x < 3/4$. Obviously,

$$4x - 2 - x^a + (1 - x)^a \qquad \text{(A1.8}')$$

vanishes at $x = 1/2$ for all a. The derivative of (A1.8′) with respect to x is

$$4 - a[x^{a-1} + (1 - x)^{a-1}]. \qquad \text{(A1.10)}$$

For $4 > a > 2$, $x^{a-1} + (1 - x)^{a-1} < 1$ and A1.10 is positive, so that A1.8′ is positive and A1.8 is true. We are left to deal with $a > 4$. For this case, consider again $2x - x^a$ in A1.9 but with $a \geq 4$. To complete the proof, we require $2 > ax^{a-1}$ for $a > 4$. Since we require $x \leq 3/4$, consider the inequality $2 > a(3/4)^{a-1}$ for $a \geq 4$. It is obviously true at $a = 4$, and the right-hand side is decreasing as a function of a for $a \geq 4$. This completes the proof.

Hence, when $\varepsilon > 0$ the sequence $\{p_t\}$ must increase when

$x_0 > 1/2$ since x_t is then greater than 1/2 for all t. If $\varepsilon > 0$ and $x < 1/2$, then $\{p_t\}$ is a decreasing sequence: In the former case $p_t \to 1$ and the A allele fixes while in the latter $p_t \to 0$ and allele a fixes. It is also possible to show that $|D'| < |D|$ so that the association between the genes drops, in fact to zero, quite rapidly. Analogous arguments can be made for $\varepsilon < 0$. To summarize, for $\varepsilon > 0$ we have $p_t \to 1$ if $x_0 > 1/2$ and $p_t \to 0$ if $x_0 < 1/2$ with the results reversed if $\varepsilon < 0$. With $D = 0$ at equilibrium, the A and a fixation boundaries are neutral curves with respect to x. In fact, if the equilibrium value of x is $\hat{x}$, then when $\varepsilon > 0$ and $\hat{x} > 1/2$, $p = 1$ is approached at the asymptotic rate of $[1 + \varepsilon(1 - \hat{x})^{f/2}]/(1 + \varepsilon\hat{x}^{f/2})$ per generation. Symmetric considerations apply when $\varepsilon < 0$. Thus in this simple case we have a global picture of the stability which is summarized in figure 3.1.

SOME GENERAL BOUNDARY RESULTS WITH SELECTION

(a) Chromosome Fixations

Local linear analysis in the neighborhood of the four chromosomal fixations produces the following sets of eigenvalues for the case $f = 1$. In each case, if all eigenvalues are less than unity, that fixation is locally stable and the rate of approach to fixation is controlled by the largest of the three eigenvalues. For any fixation with at least one eigenvalue greater than unity, that fixation will be locally unstable.

$$\begin{aligned}
EA,\ \hat{u}_1 &= 1:\ \alpha_1/(\alpha_1\eta_1 + \varepsilon),\ (\eta_1 + \varepsilon)/(\alpha_1\eta_1 + \varepsilon),\\
&\qquad (1 - r)/(\alpha_1\eta_1 + \varepsilon),\\
Ea,\ \hat{u}_2 &= 1:\ 1/\alpha_1,\ 1/\eta_2,\ (1 - r)/\alpha_1\eta_2,\\
eA,\ \hat{u}_3 &= 1:\ 1/\eta_1,\ 1/\alpha_2,\ (1 - r)/\eta_1\alpha_2,\\
ea,\ \hat{u}_4 &= 1:\ \alpha_2/(\alpha_2\eta_2 + \varepsilon),\ (\eta_2 + \varepsilon)/(\alpha_2\eta_2 + \varepsilon),\\
&\qquad (1 - r)/(\alpha_2\eta_2 + \varepsilon).
\end{aligned} \tag{A1.11}$$

Note that only for fixations on EA and ea are the eigenvalues dependent on the niche-construction parameter ε. The denominators $\alpha_1\eta_1 + \varepsilon$ and $\alpha_2\eta_2 + \varepsilon$ are assumed to be positive. Thus increasing $\varepsilon > 0$ makes $\hat{u}_1 = 1$ and $\hat{u}_4 = 1$ more stable, in the sense that in each case the first and third eigenvalues decrease as ε increases. However, if $\alpha_1 > 1$ (for $\hat{u}_1 = 1$) or $\alpha_2 > 1$ (for $\hat{u}_4 = 1$), increasing $\varepsilon > 0$ actually increases the middle eigenvalue for each but, of course, cannot cause them to increase above 1 if their values at $\varepsilon = 0$ are less than 1.

(b) Allele Fixations

We may also examine properties of the allele-fixation "edges." There are four of these, corresponding to fixation of E or e with A and a segregating or fixation of A or a with E and e segregating. First, with E fixed, on the *EA*-*Ea* edge we have $\hat{u}_1 = 1$ (fixation on *EA*), $\hat{u}_2 = 1$ (fixation on *Ea*), and there is also the possibility of a polymorphic equilibrium

$$\hat{u}_1 = \frac{\alpha_1(1-\eta_2)}{\alpha_1(2-\eta_1-\eta_2)-\varepsilon}, \hat{u}_2 = \frac{\alpha_1(1-\eta_1)-\varepsilon}{\alpha_1(2-\eta_1-\eta_2)-\varepsilon}, \quad \text{(A1.12)}$$

which is locally stable in the *EA*-*Ea* edge if $\eta_2 < 1$ and $\varepsilon < \alpha_1(1 - \eta_1)$. Similarly on the e-fixation edge, besides $\hat{u}_3 = 1$ and $\hat{u}_4 = 1$ we have the possibility of the polymorphic equilibrium

$$\hat{u}_3 = \frac{\alpha_2(1-\eta_2)-\varepsilon}{\alpha_2(2-\eta_1-\eta_2)-\varepsilon}, \hat{u}_4 = \frac{\alpha_2(1-\eta_1)}{\alpha_2(2-\eta_1-\eta_2)-\varepsilon}, \quad \text{(A1.13)}$$

which is stable if $\eta_1 < 1$ and $\varepsilon < \alpha_2(1 - \eta_2)$.

On the A-fixation edge $\hat{u}_1 + \hat{u}_3 = 1$ and the a-fixation edge $\hat{u}_2 + \hat{u}_4 = 1$ we have, in addition to the chromosomal fixations, the possibility of polymorphic equilibria

$$\hat{u}_1 = \frac{1-\alpha_2}{2-\alpha_1-\alpha_2}, \hat{u}_3 = \frac{1-\alpha_1}{2-\alpha_1-\alpha_2} \quad \text{(A1.14)}$$

with A fixed and

$$\hat{u}_2 = \frac{1-\alpha_2}{2-\alpha_1-\alpha_2}, \hat{u}_4 = \frac{1-\alpha_1}{2-\alpha_1-\alpha_2} \quad \text{(A1.15)}$$

with a fixed. These will be stable if $\alpha_1,\alpha_2 < 1$, and are the same one-locus two-allele equilibria obtained classically in the absence of niche construction.

(c) Interior Stability of Fixation at the A/a Locus

Consider fixation of allele A with E and e segregating, namely, equilibrium A1.14, i.e., $\hat{u}_2 = \hat{u}_4 = 0$. In the neighborhood of this equilibrium, the local stability matrix for perturbations in u_2, u_4 is $\mathbf{S} = \|S_{ij}\|$ with

$$S_{11} = (\hat{W})^{-1}\{\alpha_1\hat{u}_1 + \hat{u}_3(1 - r) + \varepsilon(\hat{u}_1\hat{u}_3)^{f/2}[\hat{u}_1 + \hat{u}_3(1 - r)]\}, \tag{A1.16a}$$

$$S_{12} = (\hat{W})^{-1}r[1 + \varepsilon(\hat{u}_1\hat{u}_3)^{f/2}]\hat{u}_1, \tag{A1.16b}$$

$$S_{21} = (\hat{W})^{-1}r[1 + \varepsilon(\hat{u}_1\hat{u}_3)^{f/2}]\hat{u}_3, \tag{A1.16c}$$

$$S_{22} = (\hat{W})^{-1}\{\hat{u}_1(1 - r) + \alpha_2\hat{u}_3 + \varepsilon(\hat{u}_1\hat{u}_3)^{f/2}[\hat{u}_1(1 - r) + \hat{u}_3]\}, \tag{A1.16d}$$

where

$$\hat{W} = \eta_1\left\{\frac{1-\alpha_1\alpha_2}{2-\alpha_1-\alpha_2}\right\} + \varepsilon\left[\frac{1-\alpha_2}{2-\alpha_1-\alpha_2}\right]^f. \tag{A1.16e}$$

The eigenvalues of $\mathbf{S}$ can be seen to be

$$\lambda_1 = \frac{W^* + \varepsilon(\hat{u}_1\hat{u}_3)^{f/2}}{\eta_1 W^* + \varepsilon\hat{u}_1{}^f},\ \lambda_2 = \frac{W^* + \varepsilon(\hat{u}_1\hat{u}_3)^{f/2} - r[1 + \varepsilon(\hat{u}_1\hat{u}_3)^{f/2}]}{\eta_1 W^* + \varepsilon\hat{u}_1{}^f}, \tag{A1.17}$$

where $W^* = (1 - \alpha_1\alpha_2)/(2 - \alpha_1 - \alpha_2)$.

Clearly, if $\varepsilon = 0$, then λ_1 and λ_2 are less than unity if $\eta_1 > 1$ and greater than unity if $\eta_1 < 1$, and this is true for all $r < 1/2$. If $\varepsilon > 0$ and $\eta_1 > 1$ then the stability of $(\hat{u}_1,\hat{u}_3)$ may depend on whether $\hat{u}_1 > \hat{u}_3$, and r. If $\varepsilon > 0$, $\eta_1 > 1$, and $\hat{u}_1 > \hat{u}_3$, then this equilibrium is stable for all r. As an example of the effect of niche construction here, consider $r = 0$, $f = 1$, $\eta_1 = 0.9$, $\alpha_1 = 0.8$, $\alpha_2 = 0.5$. If $\varepsilon = 0$, $\lambda_1 = \lambda_2 > 1$ and the equilibrium is unstable. If $\varepsilon = 0.4$, however, with the same values of the other parameters, $\lambda_1 = \lambda_2 < 1$

and the equilibrium is stable. Now consider $r = 0$, $f = 1$, $\eta_1 = 1.02$, $\alpha_1 = 0.5$, $\alpha_2 = 0.6$; with $\varepsilon = 0$, the equilibrium is stable, but if $\varepsilon = 0.4$, the equilibrium is unstable. However, if $\varepsilon = 0.4$ and $r = 0.2$, say, then the equilibrium is stable. Cases in which $\varepsilon > 0$ and $\hat{u}_1 < \hat{u}_3$ or $\varepsilon < 0$ and $\hat{u}_1 > \hat{u}_3$ are those most likely to produce such complications when $\eta_1 > 1$. If $\eta_1 < 1$, then pairs of $\varepsilon > 0$, $\hat{u}_1 > \hat{u}_3$, or $\varepsilon < 0$, $\hat{u}_1 < \hat{u}_3$ may produce dependence on r. As with classical two-locus theory, if the equilibrium is stable for $r = r_0$, it is stable for $r \geq r_0$. Of course, all of these considerations have analogues at the a-fixation edge equilibrium where η_2 takes the role of η_1 in the above analysis.

(d) Interior Stability of Fixation at the E/e Locus

Suppose allele E is fixed with alleles A and a segregating. In this edge a polymorphic equilibrium may exist, and if so it is given by

$$\hat{u}_1 = \frac{\alpha_1(1-\eta_2)}{\alpha_1(2-\eta_1-\eta_2)-\varepsilon},\ \hat{u}_2 = \frac{\alpha_1(1-\eta_1)-\varepsilon}{\alpha_1(2-\eta_1-\eta_2)-\varepsilon}. \quad \text{(A1.18)}$$

The two chromosomal fixation states $\hat{u}_1 = 0$ and $\hat{u}_1 = 1$ are unstable if $\eta_2 < 1$ and $\varepsilon < \alpha_1(1 - \eta_1)$, respectively, which suggests that if these inequalities hold the polymorphism is stable in the edge. Notice that in the absence of niche construction ($\varepsilon = 0$), if $\eta_2 < 1$ and $\eta_1 > 1$ no polymorphism would be possible. However, if $\eta_1 > 1$ and $\varepsilon < 0$ there may be a polymorphism; an example is $f = 1$, $\eta_1 = 1.1$, $\eta_2 = 0.8$, $\alpha_1 = 0.9$, $\varepsilon = -0.4$. Here again, the niche construction produces a protected (in the edge) polymorphism which could not even exist under standard selection.

Now consider the stability of this edge equilibrium to invasion by allele e. From recursion A1.3, ignoring small terms of quadratic order, local linear stability analysis produces the matrix $\mathbf{T} = \|T_{ij}\|$ for changes in u_3 and u_4:

$$T_{11} = \frac{(\eta_1+\varepsilon)\alpha_1(1-\eta_2)+[\alpha_1(1-\eta_1)-\varepsilon](1-r)}{W_1^*}, \quad \text{(A1.19a)}$$

$$T_{12} = \frac{r\alpha_1(1-\eta_2)}{W_1^*}, \tag{A1.19b}$$

$$T_{21} = \frac{r[\alpha_1(1-\eta_1)-\varepsilon]}{W_1^*}, \tag{A1.19c}$$

$$T_{22} = \frac{\alpha_1(1-\eta_2)(1-r)+\eta_2[\alpha_1(1-\eta_1)-\varepsilon]}{W_1^*}, \tag{A1.19d}$$

where $W_1^* = \alpha_1^2(1 - \eta_1\eta_2) - \varepsilon\alpha_1\eta_2$. (A1.19e)

Now if $r = 0$, the two eigenvalues of $\mathbf{T}$ are

$$\lambda_1 = \frac{(\eta_1+\varepsilon)\alpha_1(1-\eta_2)+\alpha_1(1-\eta_1)-\varepsilon}{\alpha_1^2(1-\eta_1\eta_2)-\varepsilon\alpha_1\eta_2} \tag{A1.20a}$$

and

$$\lambda_2 = \frac{\alpha_1(1-\eta_2)+\eta_2[\alpha_1(1-\eta_1)-\varepsilon]}{\alpha_1^2(1-\eta_1\eta_2)-\varepsilon\alpha_1\eta_2}. \tag{A1.20b}$$

Clearly, when $\varepsilon = 0$ we have $\lambda_1 > 1$ and $\lambda_2 > 1$ if $\alpha_1 < 1$, with these inequalities reversed when $\alpha_1 > 1$. Hence with $\varepsilon = 0$ and $\alpha_1 < 1$, the equilibrium is unstable toward the interior. For $\varepsilon \neq 0$, $\lambda_1 < 1$ if

$$\alpha_1(1 - \alpha_1)(1 - \eta_1\eta_2) < \varepsilon(1 - \alpha_1) \tag{A1.21}$$

and $\lambda_2 < 1$ if

$$\alpha_1(1 - \alpha_1)(1 - \eta_1\eta_2) < \varepsilon\eta_2(1 - \alpha_1). \tag{A1.22}$$

But if $\alpha_1 < 1$, then $\lambda_1 < 1$ entails $\varepsilon > \alpha_1(1 - \eta_1\eta_2) > \alpha_1(1 - \eta_1)$ since $\eta_2 < 1$. This is impossible, since for stability of the equilibrium at the *EA*-*Ea* edge we require $\varepsilon < \alpha_1(1 - \eta_1)$. Hence $\lambda_1 > 1$ if $\alpha_1 < 1$. Similarly, since the equilibrium mean fitness $\hat{W} = W^*/[\alpha_1(2 - \eta_1 - \eta_2) - \varepsilon]$ is positive, we have $\alpha_1(1 - \eta_1\eta_2) > \varepsilon\eta_2$, so $\lambda_2 > 1$. Thus when $r = 0$ both eigenvalues are greater than unity. Since the value of the characteristic polynomial of $\mathbf{T}$ at unity is linear in r, this value cannot change sign twice. Hence at least one eigenvalue is greater than unity for all r in $0 \leq r \leq 1/2$, and, if $\alpha_1 < 1$, the edge equilibrium is therefore unstable and e invades. If $\varepsilon < 0$ under these conditions, the stability of the edge equilibrium depends on whether $\eta_1\eta_2$ is greater than unity. If $\eta_1\eta_2 < 1$ then $\varepsilon < 0$ entails that the equilibrium is unstable. But if $\eta_1\eta_2 > 1$ then the equilibrium may be stable or unstable with $\varepsilon < 0$, depending on A1.21 and A1.22.

For the case $\varepsilon > 0$, if $\alpha_1 > 1$ then $\lambda_1 < 1$ if $\alpha_1(1 - \eta_1\eta_2) > \varepsilon$ and $\lambda_2 < 1$ if $\alpha_1(1 - \eta_1\eta_2) > \varepsilon\eta_2$. Again, we have $\varepsilon < \alpha_1(1 - \eta_1)$ for the equilibrium to be stable in the edge, and since $\alpha_1(1 - \eta_1) < \alpha_1(1 - \eta_1\eta_2)$ we have $\lambda_1 < 1$. With $\eta_2 < 1$, it follows that $\lambda_2 < 1$. Hence, $\alpha_1 > 1$ entails that at $r = 0$ both eigenvalues are less than unity and equilibrium is stable. It is not difficult to see that when $\alpha_1 > 1$ the coefficient of r in the characteristic quadratic of $\mathbf{T}$ evaluated at unity is positive. Hence for no value of $r > 0$ can the eigenvalues become greater than unity, and we conclude that for $\alpha_1 > 1$ the equilibrium is stable and e cannot invade. If $\varepsilon < 0$, then the argument again depends on the sign of $(1 - \eta_1\eta_2)$, as in the preceding paragraph. A symmetric argument with α_2 replacing α_1 can be made near the equilibrium

$$\hat{u}_3 = \frac{\alpha_2(1-\eta_2)-\varepsilon}{\alpha_2(2-\eta_1-\eta_2)-\varepsilon}, \quad \hat{u}_4 = \frac{\alpha_2(1-\eta_1)}{\alpha_2(2-\eta_1-\eta_2)-\varepsilon} \qquad \text{(A1.23)}$$

at the e-fixation edge.

EXTERNAL SELECTION AT THE E LOCUS: $\eta_1 = \eta_2 = 1$

Suppose $\alpha_1 > 1 > \alpha_2$ and $\varepsilon > 0$. Then from A1.11 fixation of *EA* is locally stable, fixations on *eA* and *ea* are unstable, while *Ea* has a leading eigenvalue of 1. Consideration of the second derivative, however, shows that in the *E*-fixation edge, fixation on *Ea* is locally unstable with $\varepsilon > 0$ and locally stable when $\varepsilon < 0$. If $\alpha_1 < 1 < \alpha_2$ then $\varepsilon > 0$ produces local stability of *ea* at a geometric rate while if $\varepsilon < 0$, fixation on *eA* occurs at the slower algebraic rate. In the cases of algebraic convergence, the population moves to the *E*-fixation edge ($\alpha_1 > 1 > \alpha_2$) first and then slowly toward fixation on *Ea* ($\varepsilon < 0$), with analogous trajectories for the e-fixation edge.

EXTERNAL SELECTION AT THE A LOCUS: $\alpha_1 = \alpha_2 = 1$

In this case the edges with A fixed or a fixed and E/e segregating comprise neutral curves of equilibria with respect to perturbation along the edges themselves. Toward the interior of the tetrahedron, however, stability conditions are more complicated. With these assumptions on the parameters we have $w_{11} = w_{12} = w_{13}$, $w_{21} = w_{22} = w_{23}$, $w_{31} = w_{32} = w_{33}$, with the result that

$$Wp' = w_{11}p^2 + w_{21}pq, \tag{A1.24a}$$

$$x' = x + \frac{D(w_{11}p + w_{21}q - w_{21}p - w_{31}q)}{W}, \tag{A1.24b}$$

$$D' = DM, \tag{A1.24c}$$

where $W = p^2w_{11} + 2pqw_{21} + q^2w_{31}$ and M is a function of p, r, w_{11}, w_{21}, and w_{31} that is positive. Let us assume that $n = 1$ and $f = 1$ so that $R = x$ and $w_{11} = \eta_1 + \varepsilon x$, $w_{21} = 1 + \varepsilon\sqrt{x(1 - x)}$, $w_{31} = \eta_2 + \varepsilon(1 - x)$. Clearly, the sign of D remains unchanged over time and, if $D = 0$, the frequency of E (x) remains unchanged. Also, if the value of p takes the form

$$\hat{p} = \frac{w_{31} - w_{21}}{w_{11} + w_{31} - 2w_{21}} = \frac{1 - \eta_2 + \varepsilon\left[\sqrt{x(1-x)} - (1-x)\right]}{2 - \eta_1 - \eta_2 + \varepsilon\left[2\sqrt{x(1-x)} - 1\right]}, \tag{A1.25}$$

then x remains unchanged and p is also unchanged. In other words, equation A1.25 describes a curve of (p, x) values that constitute equilibria of the system when the values are biologically feasible.

The boundary edge $p = 1$ will be unstable to the introduction of allele a if $w_{21} > w_{11}$; that is, if

$$1 - \eta_1 + \varepsilon\sqrt{x}[\sqrt{1 - x} - \sqrt{x}] > 0. \tag{A1.26}$$

Hence, if $\eta_1 < 1$ and $x < 1/2$, then (A1.26) holds and a will increase when rare if $\varepsilon > 0$. If (A1.26) is reversed then fixation on A is locally stable. This occurs if either

$$x \le \frac{2(1-\eta_1)+\varepsilon-\sqrt{4(1-\eta_1)(\eta_1+\varepsilon-1)+\varepsilon^2}}{4\varepsilon}, \text{ or} \qquad \text{(A1.27a)}$$

$$x \ge \frac{2(1-\eta_1)+\varepsilon+\sqrt{4(1-\eta_1)(\eta_1+\varepsilon-1)+\varepsilon^2}}{4\varepsilon}. \qquad \text{(A1.27b)}$$

If the value under the radical in A1.27 is negative then fixation on A is stable for all $0 \le x \le 1$. Analogously, fixation on allele a is stable if

$$1 - \eta_2 + \varepsilon\sqrt{1-x}\,[\sqrt{x} - \sqrt{1-x}] < 0; \qquad \text{(A1.28)}$$

that is, if either

$$x \ge \frac{3\varepsilon-2(1-\eta_2)+\sqrt{4(1-\eta_2)(\eta_2+\varepsilon-1)+\varepsilon^2}}{4\varepsilon}, \text{ or} \qquad \text{(A1.29a)}$$

$$x \le \frac{3\varepsilon-2(1-\eta_2)-\sqrt{4(1-\eta_2)(\eta_2+\varepsilon-1)+\varepsilon^2}}{4\varepsilon}. \qquad \text{(A1.29b)}$$

Figure 3.4b shows that if $\eta_1 = 1.1$, $\eta_2 = 0.9$ and $\varepsilon = 0.3$, $p = 1$ (fixation on A) is stable for all x; in this case the value under the radical in A1.27 is negative and A1.26 is reversed for all x. However, in this case A1.28 fails if $x > 1/2$, so A1.29a is irrelevant, while if x is sufficiently less than 0.25, A1.28 holds and $p = 0$ is stable. With $\varepsilon = -0.3$, the value under the radical in A1.29 is negative, A1.28 cannot hold, and $p = 0$ is unstable. Also, for x large enough A1.26 holds, and, as seen in figure 3.4c, a polymorphism for A and a is possible at equilibrium. Here equilibria are given by the curve of (p, x) values specified by A1.25. Note that this curve under the parameters of figure 3.4b is unstable, but the fixation states are stable.

OVERDOMINANCE WITH NO LINKAGE DISEQUILIBRIUM

Two-locus, two-allele viability models appear to have a maximum of 15 possible equilibria; four chromosome fixation states, four allelic fixation states, and seven interior equilibria (Karlin and Feldman

1970). First we explore equilibria of the model when there is no linkage disequilibrium (which occurs when $r > r_0$ defined below). At the $D = 0$ polymorphic equilibrium, the frequency-dependent selection does not change the equilibrium frequency of E, but $\hat{p}$ becomes a function of ε. For $n = 1$, the equilibrium is given by

$$\hat{x} = \frac{(1-\alpha_2)}{(2-\alpha_1-\alpha_2)},$$

$$\hat{p} = \frac{(2-\alpha_1-\alpha_2)^{f-1}(1-\eta_2)(1-\alpha_1\alpha_2) - \varepsilon(1-\alpha_1)^{f/2}[(1-\alpha_1)^{f/2} - (1-\alpha_2)^{f/2}]}{(2-\alpha_1-\alpha_2)^{f-1}(2-\eta_1-\eta_2)(1-\alpha_1\alpha_2) - \varepsilon[(1-\alpha_1)^{f/2} - (1-\alpha_2)^{f/2})]^2}, \tag{A1.30}$$

which in the special case $f = 1$ reduces to

$$\hat{p} = \frac{(1-\eta_2)(1-\alpha_1\alpha_2) - \varepsilon\left[(1-\alpha_1) - \sqrt{(1-\alpha_1)(1-\alpha_2)}\right]}{(2-\eta_1-\eta_2)(1-\alpha_1\alpha_2) - \varepsilon\left[(2-\alpha_1-\alpha_2) - 2\sqrt{(1-\alpha_1)(1-\alpha_2)}\right]}. \tag{A1.31}$$

With $f = 1$, if $\alpha_1 = \alpha_2$ then $\hat{x} = 1/2$, and $\hat{p}$ reduces to $\hat{p} = (1 - \eta_2)/(2 - \eta_1 - \eta_2)$, which is the same equilibrium as in the standard multiplicative model (Moran 1964), so that the frequency-dependent selection has no effect on this equilibrium. We would therefore expect weak dependence of the equilibrium on niche construction for values of x close to one-half and D small enough.

OVERDOMINANCE WITH LINKAGE DISEQUILIBRIUM

(a) The Symmetric Multiplicative Viability Model

Suppose first that $\alpha_1 = \alpha_2$ and $\eta_1 = \eta_2$ (making the fixed fitness component equivalent to the symmetric multiplicative viability model). Here, we first look for symmetric solutions of the form $\hat{u}_1 = \hat{u}_4$, $\hat{u}_2 = \hat{u}_3$ or $\hat{x} = 1/2$, $\hat{p} = 1/2$ (Karlin and Feldman 1970), by using the equations

$$W(u_1 - u_4) = (u_1^2 - u_4^2)[\alpha_1\eta_1 + \varepsilon(1/2)^f]$$
$$+ (u_1u_2 - u_3u_4)[\alpha_1 + \varepsilon(1/2)^f] + (u_1u_3 - u_2u_4)[\eta_1 + \varepsilon(1/2)^f]$$

$$W(u_2 - u_3) = (u_2^2 - u_3^2)[\alpha_1\eta_1 + \varepsilon(1/2)^f]$$
$$+ (u_1u_2 - u_3u_4)[\alpha_1 + \varepsilon(1/2)^f] - (u_1u_3 - u_2u_4)[\eta_1 + \varepsilon(1/2)^f].$$

Substituting $u_1 = u_4 = 1/4 + D$, and $u_2 = u_3 = 1/4 - D$, into (A1.3), we obtain

$$W = \frac{(1+\alpha_1)(1+\eta_1)}{4} + \varepsilon(1/2)^f + 4D^2(1-\alpha_1)(1-\eta_1).$$

Using the u_1 recursion at equilibrium, this produces the cubic equation

$$D\left[(1-\alpha_1)(1-\eta_1)\left(4D^2 - \frac{1}{4}\right) + r[1+\varepsilon((1/2)^f)]\right] = 0,$$

whose solutions are $\hat{D} = 0$, and

$$\hat{D} = \pm\frac{1}{4}\sqrt{1 - \frac{4r(1+\varepsilon(1/2)^f)}{(1-\alpha_1)(1-\eta_1)}}. \tag{A1.32}$$

The right-hand side of A1.32 is real if

$$r < r_0 = \frac{(1-\alpha_1)(1-\eta_1)}{4[1+\varepsilon(1/2)^f]}.$$

Thus for values of $r < r_0$, in addition to the $\hat{D} = 0$ equilibrium, there are two symmetric equilibria, at which the frequencies of the gametes are given by $\hat{u}_1 = \hat{u}_4 = 1/4 + \hat{D}$ and $\hat{u}_2 = \hat{u}_3 = 1/4 - \hat{D}$, and $\hat{D}$ is one of the two values given by A1.32.

The three eigenvalues of the linearized recursions near the point $u_i = 1/4$, i.e., $\hat{D} = 0$, are

$$\frac{2\eta_1(1+\alpha_1) + 4\varepsilon(1/2)^f}{(1+\alpha_1)(1+\eta_1) + 4\varepsilon(1/2)^f}, \qquad \frac{2\alpha_1(1+\alpha_1) + 4\varepsilon(1/2)^f}{(1+\alpha_1)(1+\eta_1) + 4\varepsilon(1/2)^f},$$

$$\frac{2\{1+\alpha_1\eta_1+2\varepsilon(1/2)^f-2r[1+\varepsilon(1/2)^f]\}}{(1+\alpha_1)(1+\eta_1)+4\varepsilon(1/2)^f},$$

which produce the following three conditions for stability:

$$\text{(i) } (1+\alpha_1)(1-\eta_1)>0,$$

$$\text{(ii) } (1-\alpha_1)(1+\eta_1)>0,$$

$$\text{(iii) } r>\frac{(1-\alpha_1)(1-\eta_1)}{4[1+\varepsilon(1/2)^f]}=r_0. \tag{A1.33}$$

The first two conditions entail overdominance at both loci, while the last gives the critical value of r below which $\hat{D} = 0$ is unstable. Note that $\hat{D} = 0$ is stable when the $\hat{D} \neq 0$ equilibria given by (A1.32) do not exist. It is clear that in this case r_0 will be larger for negative values of ε and smaller for positive values of ε. In other words, positive values of ε support smaller values of stable linkage disequilibrium in this symmetric case. Expressions (A1.32) and (A1.33) give a clear indication of the effect of manipulating the power of the frequency dependence (f). In this case as f increases the effect of niche construction as determined by ε decreases.

(b) An Asymmetric Multiplicative Viability Model

In the case where $\alpha_1 = \alpha_2$ but $\eta_1 \neq \eta_2$ and $\eta_1, \eta_2 < 1$, the fixed fitness component of the model is asymmetric (Karlin and Feldman 1970). Here the condition for the stability of the equilibrium with $\hat{x} = 1/2$, $\hat{p} = (\eta_2 - 1)/(\eta_1 + \eta_2 - 2)$, and $\hat{D} = 0$ is given by

$$r>\frac{(1-\alpha_1)(1-\eta_1)(1-\eta_2)}{2(2-\eta_1-\eta_2)(1+\varepsilon(1/2)^f)}=r_0, \tag{A1.34}$$

which simplifies to (A1.33) in the case where $\eta_1 = \eta_2$. The effects of ε and f are the same as in the symmetric viability model. For $r < r_0$, the $\alpha_1 = \alpha_2$, $\eta_1 \neq \eta_2$ condition may generate equilibria with $D \neq 0$ and $x \neq 1/2$. If the **E** and **A** loci are loosely linked, then $\alpha_1 =$

$\alpha_2 < 1$ always gives a frequency of E of $\hat{x} = 1/2$ at the stable polymorphic equilibrium, but if they are tightly linked then $\hat{x} \neq 1/2$ (unless $\eta_1 = \eta_2$). This is of interest here, since the frequency-dependent selection has no effect on the equilibrium frequency of allele A if $x = 1/2$, but does if $x \neq 1/2$.

(c) More Complex Cases

With $\alpha_1 \neq \alpha_2$ and $\eta_1 \neq \eta_2$ the multiplicative model becomes intractable, and we have carried out a numerical analysis of this situation. This reveals that the frequency-dependent selection always increases the frequency of two gametes that increase linkage disequilibrium, and decreases the frequency of two gametes that decrease linkage disequilibrium, and hence its effect at any point in time is sensitive to the gamete frequencies. Since positive ε increases u_4 and decreases u_3, when x is small the frequency dependence should increase $+D$, and decrease $-D$. The reverse is true for negative ε. However, for x close to 1, since positive ε increases u_1 and decreases u_2, it tends to increase the $+D$, and decrease the $-D$ equilibria, with the reverse true for negative ε. Thus, whether or not positive or negative values of ε increase or decrease the amount of linkage disequilibrium depends upon the frequency of allele E, and the sign of the disequilibrium.

While it is not generally true that the absolute change in the amount of linkage disequilibrium at equilibrium due to the frequency-dependent selection is greater as x approaches 1 and 0, and moves away from 1/2, the proportional change in the amount of linkage disequilibrium (the absolute change divided by the amount of linkage disequilibrium at $\varepsilon = 0$) does seem to fit this pattern. This is because while the frequency-dependent selection acting on locus **A** becomes stronger as the frequency of E moves away from 1/2, the amount of disequilibrium typically shows the opposite pattern.

Although the above results are for $n = 1$, a numerical analysis reveals that the effect of increasing the number of generations of niche construction that affect R is primarily to change the rates at which populations converge on the above equilibria, while the equilibria themselves are little affected by n (in the range $n = 1$ to 100).

APPENDIX 2

Model 1b: A Simple Model of Niche Construction for Haplodiploids or Sex-Linked Loci

We consider an isolated population of randomly mating, haplodiploid individuals, defined at two loci, **E** (with alleles E and e) and **A** (with alleles A and a). The model also represents a diallelic two-locus system where both loci are sex-linked and the niche-constructing locus is on the X chromosome. At generation t, the frequencies of the four male haplotypes and four female gametes (EA, Ea, eA, and ea) are given by $y_1(t)$, $y_2(t)$, $y_3(t)$, $y_4(t)$ and $z_1(t)$, $z_2(t)$, $z_3(t)$, $z_4(t)$, respectively. The frequencies of alleles E and A are $x_M(t) = y_1(t) + y_2(t)$ and $p_M(t) = y_1(t) + y_3(t)$ in males, and $x_F(t) = z_1(t) + z_2(t)$ and $p_F(t) = z_1(t) + z_3(t)$ in females. At each generation, the amount of the resource created or destroyed as a consequence of niche construction is assumed to be a function of the frequency of allele E in females (x_F) at that time t. The simplest such function is linear with equal weighting of each generation. The frequency of the resource at time t is given by

$$R_t = \frac{1}{n} \sum_{i=t-n+1}^{t} x_{Fi}. \qquad \text{(A2.1)}$$

Most generally, we assume the following fitness structure for diploid female individuals, where the rows give the female's genotype at the **A** locus and the columns give the female's genotype at the **E** locus:

	EE	Ee	ee
AA	f_{11}	f_{12}	f_{13}
Aa	f_{21}	f_{22}	f_{23}
aa	f_{31}	f_{32}	f_{33}

Note that we have assumed that linkage phase does not influence fitness. In a similar manner, we will assume the following fitness structure for haploid male individuals, where rows give the male's genotype at the **A** locus and columns give the male's genotype at the **E** locus:

	E	e
A	m_{11}	m_{12}
a	m_{21}	m_{22}

The life cycle is as follows: gametes unite at random (random mating) and then, after recombination, individuals undergo viability selection. This produces the following set of gametic recursions:

$$\begin{aligned}\overline{W}_F z_1' = {} & f_{11} z_1 y_1 + f_{21}\left(\frac{z_1 y_2 + z_2 y_1}{2}\right) + f_{12}\left(\frac{z_1 y_3 + z_3 y_1}{2}\right) \\ & + f_{22}\left(\frac{z_1 y_4 + z_4 y_1}{2}\right) - f_{22}\frac{r}{2}D,\end{aligned} \tag{A2.2a}$$

$$\begin{aligned}\overline{W}_F z_2' = {} & f_{21}\left(\frac{z_1 y_2 + z_2 y_1}{2}\right) + f_{31} z_2 y_2 + f_{22}\left(\frac{z_2 y_3 + z_3 y_2}{2}\right) \\ & + f_{32}\left(\frac{z_2 y_4 + z_4 y_2}{2}\right) + f_{22}\frac{r}{2}D,\end{aligned} \tag{A2.2b}$$

$$\begin{aligned}\overline{W}_F z_3' = {} & f_{12}\left(\frac{z_1 y_3 + z_3 y_1}{2}\right) + f_{22}\left(\frac{z_2 y_3 + z_3 y_2}{2}\right) + f_{13} z_3 y_3 \\ & + f_{23}\left(\frac{z_3 y_4 + z_4 y_3}{2}\right) + f_{22}\frac{r}{2}D,\end{aligned} \tag{A2.2c}$$

$$\overline{W}_F z_4' = f_{22}\left(\frac{z_1y_4 + z_4y_1}{2}\right) + f_{32}\left(\frac{z_2y_4 + z_4y_2}{2}\right) + f_{23}\left(\frac{z_3y_4 + z_4y_3}{2}\right) + f_{33}z_4y_4 - f_{22}\frac{r}{2}D, \tag{A2.2d}$$

$$\overline{W}_M y_1' = m_{11}z_1, \tag{A2.2e}$$

$$\overline{W}_M y_2' = m_{12}z_2, \tag{A2.2f}$$

$$\overline{W}_M y_3' = m_{21}z_3, \tag{A2.2g}$$

$$\overline{W}_M y_4' = m_{22}z_4, \tag{A2.2h}$$

where $\overline{W}_F$ and $\overline{W}_M$ are the sums of the right-hand sides of A2.2a–A2.2d and A2.2e–A2.2h, respectively, r is the recombination fraction, and $D = z_1y_4 + z_4y_1 - z_2y_3 - z_3y_2$ plays the part of a linkage disequilibrium term.

Female and male genotypic fitnesses are given in tables 3.1 and 3.2, respectively. As before, the fitnesses are assumed to be functions of a fixed viability component and a frequency-dependent viability component. We assume male fitnesses correspond to those of the four female double-homozygote genotypes.

SOME GENERAL BOUNDARY RESULTS WITH SELECTION

(a) Chromosome Fixations

Let us assume that the gamete type *ea* is fixed in both eggs and sperm. Now, imagine introducing small frequencies of the other types of gametes. We denote the small frequencies of *EA*, *Ea*, and *eA* eggs by ξ_1, ξ_2, and ξ_3, respectively; and we denote the small frequencies of *EA*, *Ea*, and *eA* sperm by ψ_1, ψ_2, and ψ_3, respectively. After linearizing recursions A2.2, we have the following:

$$\begin{bmatrix} \xi_1 \\ \xi_2 \\ \xi_3 \\ \psi_1 \\ \psi_2 \\ \psi_3 \end{bmatrix}' = \begin{bmatrix} \frac{(1-r)}{2(\alpha_2\eta_2+\varepsilon)} & 0 & 0 & \frac{(1-r)}{2(\alpha_2\eta_2+\varepsilon)} & 0 & 0 \\ \frac{r}{1(\alpha_2 n_2+\varepsilon)} & \frac{\eta_2+\varepsilon}{2(\alpha_2\eta_2+\varepsilon)} & 0 & \frac{r}{2(\alpha_2\eta_2+\varepsilon)} & \frac{\eta_2+\varepsilon}{(2\alpha_2\eta_2+\varepsilon)} & 0 \\ \frac{r}{2(\alpha_2\eta_2+\varepsilon)} & 0 & \frac{\alpha_2}{2(\alpha_2\eta_2+\varepsilon)} & \frac{r}{2(\alpha_2\eta_2+\varepsilon)} & 0 & \frac{\alpha_2}{2(\alpha_2\eta_2+\varepsilon)} \\ \frac{\alpha_1\eta_1}{\alpha_2\eta_2+\varepsilon} & 0 & 0 & 0 & 0 & 0 \\ 0 & \frac{\alpha_2\eta_1}{\alpha_2\eta_2+\varepsilon} & 0 & 0 & 0 & 0 \\ 0 & 0 & \frac{\alpha_1\eta_2}{\alpha_2\eta_2+\varepsilon} & 0 & 0 & 0 \end{bmatrix} \begin{bmatrix} \xi_1 \\ \xi_2 \\ \xi_3 \\ \psi_1 \\ \psi_2 \\ \psi_3 \end{bmatrix}. \tag{A2.3}$$

The eigenvalues of the above matrix are

$$\lambda_{1,2} = \frac{(1-r) \pm \sqrt{(1-r)^2 + 8(1-r)\alpha_1\eta_1}}{4(\alpha_2\eta_2+\varepsilon)},$$

$$\lambda_{3,4} = \frac{\alpha_2 \pm \sqrt{\alpha_2{}^2 + 8\alpha_2(\alpha_1\eta_2+\varepsilon)}}{4(\alpha_2\eta_2+\varepsilon)},$$

$$\lambda_{5,6} = \frac{(\eta_2+\varepsilon) \pm \sqrt{(\eta_2+\varepsilon)^2 + 8(\eta_2+\varepsilon)\alpha_2\eta_1}}{4(\alpha_2\eta_2+\varepsilon)}.$$

APPENDIX 3

Model 2: An Extended Model of Niche Construction

In each generation, the amount of the resource at time t is given by

$$R_t = \lambda_1 R_{t-1}(1 - \gamma x_t) + \lambda_2 x_t + \lambda_3, \qquad \text{(A3.1)}$$

where λ_1 is a coefficient that determines the degree of independent depletion (if $\lambda_1 = 1$ there is no independent depletion while if $\lambda_1 < 1$ independent depletion occurs), λ_2 is a coefficient that determines the effect of positive niche construction (if $\lambda_2 = 0$ there is no positive niche construction influencing the amount of the resource), λ_3 is a coefficient that determines the degree of independent renewal (if $\lambda_3 = 0$ there is no independent renewal influencing the amount of the resource), and γ is a coefficient that determines the effect of negative niche construction (if $\gamma = 0$ there is no negative niche construction influencing the amount of the resource). We assume $0 < \lambda_1, \lambda_2, \lambda_3, \gamma < 1$, and $\lambda_1 + \lambda_2 + \lambda_3 \leq 1$. The term $\lambda_1 R_{t-1}$ represents the proportion of the resource that remains from the previous generation after independent depletion, while $(1 - \gamma x_t)$ represents any further decay in R due to negative niche construction on the part of the current generation, and $\lambda_2 x_t$ represents any increase in R due to positive niche construction by the current generation. This treatment of the interaction between the population and the resource is extremely simplified. A more realistic treatment would involve general distributions of the resource, more complex ecological dynamics of resource and population, and ecological models that would take the density of niche constructors into account. Our assumptions concerning resource dynamics are made largely on the basis of analytical convenience.

With only positive niche construction ($R_t = \lambda_1 R_{t-1} + \lambda_2 x_t + \lambda_3$, $\gamma = 0$), from equation A3.1 when $n = 1$ the equilibrium value of R as a function of an equilibrium value of the E-allele frequency, $\hat{x}$, is given by

$$\hat{R} = \frac{\lambda_3 + \hat{x}\lambda_2}{1 - \lambda_1}. \qquad \text{(A3.2)}$$

With only negative niche construction [$R_t = \lambda_1 R_{t-1}(1 - \gamma x_t) + \lambda_3$, $\lambda_2 = 0$], when $n = 1$ the equilibrium value of R for any given value of $\hat{x}$ is given by

$$\hat{R} = \frac{\lambda_3}{1 - \lambda_1(1 - \gamma\hat{x})}. \qquad \text{(A3.3)}$$

In the numerical analyses, for the case $n > 1$, R_t was computed by iterated substitution of the value of R in the previous generation into equation A3.1 for n steps. For instance, with only positive niche construction ($R_t = \lambda_1 R_{t-1} + \lambda_2 x_t + \lambda_3$, $\gamma = 0$), R_t is given by

$$R_t = \lambda_1{}^n R_{t-n} + \lambda_2 \sum_{i=t-n+1}^{t} x_i \lambda_1^{t-i} + \left(\frac{1 - \lambda_1^n}{1 - \lambda_1}\right)\lambda_3. \qquad \text{(A3.4)}$$

When there is external selection at the **A** locus a curve of equilibria analogous to that described for model 1 by A1.25 is given here by

$$\hat{p} = \frac{w_{31} - w_{21}}{w_{11} + w_{31} - 2w_{21}} = \frac{1 - \eta_2 + \varepsilon[\sqrt{\hat{R}(1 - \hat{R})} - (1 - \hat{R})]}{2 - \eta_1 - \eta_2 + \varepsilon[2\sqrt{\hat{R}(1 - \hat{R})} - 1]}. \qquad \text{(A3.5)}$$

As with model 1 (A1.27), fixation on A is locally stable if either

$$\hat{R} \le \frac{2(1 - \eta_1) + \varepsilon - \sqrt{4(1 - \eta_1)(\eta_1 + \varepsilon - 1) + \varepsilon^2}}{4\varepsilon}, \text{ or} \qquad \text{(A3.6a)}$$

$$\hat{R} \ge \frac{2(1 - \eta_1) + \varepsilon + \sqrt{4(1 - \eta_1)(\eta_1 + \varepsilon - 1) + \varepsilon^2}}{4\varepsilon}. \qquad \text{(A3.6b)}$$

Similarly, as with model 1 (A1.29), fixation on allele *a* is stable if either

$$\hat{R} \geq \frac{3\varepsilon - 2(1-\eta_2) + \sqrt{4(1-\eta_2)(\eta_2 + \varepsilon - 1) + \varepsilon^2}}{4\varepsilon}, \text{ or} \quad \text{(A3.7a)}$$

$$\hat{R} \leq \frac{3\varepsilon - 2(1-\eta_2) - \sqrt{4(1-\eta_2)(\eta_2 + \varepsilon - 1) + \varepsilon^2}}{4\varepsilon}. \quad \text{(A3.7b)}$$

With overdominance at both loci, there may be a polymorphic equilibrium with $\hat{D} = 0$ at which the frequencies of E and A are given by

$$\hat{x} = \frac{(1-\alpha_2)}{(2-\alpha_1-\alpha_2)},$$

$$\hat{p} = \frac{(1-\eta_2)(1-\alpha_1\alpha_2) - \varepsilon(2-\alpha_1-\alpha_2)[1-\hat{R}-\sqrt{\hat{R}(1-\hat{R})}]}{(2-\eta_1-\eta_2)(1-\alpha_1\alpha_2) - \varepsilon(2-\alpha_1-\alpha_2)[1-2\sqrt{\hat{R}(1-\hat{R})}]}. \quad \text{(A3.8)}$$

APPENDIX 4

Models 3 and 4: Cultural Niche Construction

Consider a population of humans, or human ancestors, capable of the cultural transmission of information from one generation to the next. Individuals express culturally transmitted information, which we call **E**, with two states E and e, representing the presence or absence (or greater or lesser impact) of a socially learned niche-constructing behavior. This cultural niche-constructing behavior affects the amount R of a resource **R** (where $0 < R < 1$) in the environment. Genotypes at locus **A** (with alleles A and a) have fitnesses that are functions of the frequency with which the resource **R** is encountered by organisms in their environment. The three genotypes and two cultural states can be found in six possible phenogenotype combinations, namely, AAE, AAe, AaE, Aae, aaE, and aae, which have frequencies z_1–z_6 respectively, and fitnesses w_{ij} (given in table 6.2). We shall use p to denote the frequency of allele A, and x that of trait E, where $p = z_1 + z_2 + (z_3 + z_4)/2$ and $x = z_1 + z_3 + z_5$. We assume that the population is isolated from other populations, and that individuals mate randomly (that is, independently of their genotype or cultural state).

Model 3 describes the simple case in which the amount of the resource is proportional to the amount of cultural niche construction over the previous n generations, where R is given by equation A1.1 in appendix 1. Model 4 includes the more complex R function, for which, in each generation, the amount of the resource at time t is given by equation A3.1 in appendix 3. In each case, however, x_t is the frequency of the cultural state E in generation t. Positive niche construction corresponds to cases where $\lambda_2 > 0$ and $\gamma = 0$ for equation A3.1 in appendix 3, while negative niche construction refers to cases where $\lambda_2 = 0$ and $\gamma > 0$ for equation A3.1 in appendix 3. Vertical cultural transmission occurs according to standard rules (Cavalli-Sforza and Feldman 1981), given in table 6.3.

Following reproduction, cultural transmission, and selection, we derive the six phenogenotype recursions that describe the frequency of each combination of cultural state and genotype in terms of their frequencies in the previous generation. These can be reduced to a more tractable system of four allelophenotype recursions, which give the frequency of alleles A and a among E and e individuals. The allelophenotypes AE, aE, Ae, and ae have frequencies u_1–u_4 in the present generation, with u'_1–u'_4 the corresponding frequencies in the next generation, and

$$\begin{aligned}
Wu'_1 &= [u_1^2 b_3 + u_1 u_3 (b_1 + b_2) + u_3^2 b_0] w_{11} \\
&\quad + [u_1 u_2 b_3 + (u_1 u_4 + u_2 u_3)(b_1 + b_2)/2 + u_3 u_4 b_0] w_{21}, \\
Wu'_2 &= [u_1 u_2 b_3 + (u_1 u_4 + u_2 u_3)(b_1 + b_2)/2 + u_3 u_4 b_0] w_{21} \\
&\quad + [u_2^2 b_3 + u_2 u_4 (b_1 + b_2) + u_4^2 b_0] w_{31}, \\
Wu'_3 &= [u_1^2 (1 - b_3) + u_1 u_3 (2 - b_1 - b_2) \\
&\quad + u_3^2 (1 - b_0)] w_{12} + [u_1 u_2 (1 - b_3) \\
&\quad + (u_1 u_4 + u_2 u_3)(2 - b_1 - b_2)/2 + u_3 u_4 (1 - b_0)] w_{22}, \\
Wu'_4 &= [u_1 u_2 (1 - b_3) + (u_1 u_4 + u_2 u_3)(2 - b_1 - b_2)/2 \\
&\quad + u_3 u_4 (1 - b_0)] w_{22} + [u_2^2 (1 - b_3) + u_2 u_4 (2 - b_1 - b_2) \\
&\quad + u_4^2 (1 - b_0)] w_{32},
\end{aligned} \tag{A4.1}$$

where W is the sum of the right-hand sides of equations A4.1 (Feldman and Zhivotovsky 1992). The frequency of the cultural state E is now given by $x = u_1 + u_2$, the frequency of allele A is given by $p = u_1 + u_3$, and the interaction between the gene and the cultural state can be specified by the quantity $D = u_1 u_4 - u_2 u_3$. In the special cases where $w_{11} = w_{12}$, $w_{21} = w_{22}$, $w_{31} = w_{32}$, and with unbiased transmission ($b_3 = 1$, $b_2 = b_1 = 1/2$, $b_0 = 0$) we can write useful recursions for x, p, and D:

$$Wx' = Wx + D[pw_{11} + qw_{21} - pw_{21} - qw_{31}], \tag{A4.2a}$$

$$Wp' = p[pw_{11} + qw_{21}], \tag{A4.2b}$$

and

$$W^2D' = D\left[\frac{p^2w_{11}w_{21}}{2} + pqw_{11}w_{31} + \frac{q^2w_{21}w_{31}}{2}\right], \quad \text{(A4.2c)}$$

where

$$W = p^2w_{11} + 2pqw_{21} + q^2w_{31}.$$

With no external selection ($\alpha_1 = \eta_1 = \alpha_2 = \eta_2 = 1$) and unbiased cultural transmission ($b_3 = 1$, $b_2 = b_1 = 1/2$, $b_0 = 0$), if there is no statistical association between the cultural trait and allele frequencies, i.e., $D = 0$, then from equation A4.2a, x is constant, and given by the frequency of E. Under such circumstances, R is now unaffected by n, and equation A1.1 in appendix 1 simplifies to the $n = 1$ case. Numerical analysis reveals that if the gene-culture association D is initially non-zero, it breaks down after a small number of generations ($D \to 0$), and is rarely large enough to drive meaningful change in the frequency of the cultural state.

For model 3 in the absence of selection with $D = 0$, if $\delta > 0$ and $b \neq 1/2$, all populations remain polymorphic for the cultural trait (Cavalli-Sforza and Feldman 1981), and there is convergence on a line of gene-frequency equilibria with the frequency of E given by

$$\hat{x} = \frac{1 + 2\delta - 2b - \sqrt{1 - 4b + 4\delta^2 + 4b^2}}{2(1 - 2b)}. \quad \text{(A4.3)}$$

When there is external selection at the A locus only ($\alpha_1 = \alpha_2 = 1$, $\eta_1 \neq 1$, $\eta_2 \neq 1$) and unbiased cultural transmission ($b_3 = 1$, $b_2 = b_1 = 0.5$, $b_0 = 0$), by summing equations A4.1a and A4.1c, we derive a recursion for the frequency p of A, whose equilibria are given by $\hat{p} = 0$, $\hat{p} = 1$, and the polymorphism given by equation A1.25 in appendix 1. The $p = 1$ boundary is stable for $w_{1i} > w_{2i}$, and we may rewrite this inequality in terms of R, and derive the range or ranges of R values compatible with the stability of $\hat{p} = 1$, which for model 3 is given by A1.27 of appendix 1, and for model 4 by A3.6 of appendix 3. A similar analysis for $\hat{p} = 0$, when $w_{3i} > w_{2i}$, yields equations A1.29 of appendix 1 and A3.7 of appendix 3. Thus these inequalities are identical to those in the genetic system, namely, inequalities (5a–5d) in Laland et al. (1999). (Full details can be found in Laland et al. [1996, 1999, 2001].)

With selection on the cultural trait ($\alpha_1 \neq 1$, $\alpha_2 \neq 1$, $\eta_1 = \eta_2 = 1$) and incomplete transmission ($b_3 = 1 - \delta$, $b_2 = b_1 = b$, $b_0 = \delta$, with $\delta > 0$), the frequency of the cultural trait at gene-fixation equilibria is given by a solution to the cubic equation

$$\begin{aligned} x^3[(w_{i1} - w_{i2})(1 - 2b)] + x^2[2(w_{i1} - w_{i2})(b - \delta) \\ - w_{i1}(1 - 2b)] + x[(w_{i2} - w_{i1})(1 - \delta) \\ + w_{i1}(1 - 2b + 2\delta)] - \delta w_{i1} = 0, \end{aligned} \tag{A4.4}$$

where $i = 1$ at $\hat{p} = 0$ and $i = 3$ at $\hat{p} = 1$. It can be seen that the frequency of the cultural trait typically differs at the two genetic fixation states, although this difference is small for realistic parameter values. The difference occurs because $w_{11}/w_{12} \neq w_{31}/w_{32}$ when $R \neq 0.5$.

APPENDIX 5

Model 5: Niche Construction and the Evolution of the Sex Ratio in Hymenoptera

Niche-constructing activities such as nest building and maintenance can be regarded as altruistic, in the sense that they are energetically costly to the constructor and beneficial to other members of the colony. As nest-construction and maintenance activities are generally carried out by females, the differing costs and benefits of niche construction in different species may result in differential patterns of investment in males and females. Hence a niche-construction perspective may help to predict sex ratios among species that exhibit such differences, for example, the Hymenoptera.

Here we consider a single-locus model and ask whether an allele for altruistic niche-constructing behavior, E, can spread in a haplo-diploid population. We use this to derive an expression for the expected sex ratio of a niche-constructing population. The model is based on Uyenoyama and Feldman's (1981) sex-linked model of altruism from sisters to siblings, but relaxes their assumption of an equal sex ratio. Here the proportion of males in all broods at birth is given by χ. We assume that workers fail to discriminate between brothers and sisters with probability ε. Hence ε is a measure of the proportion of niche construction carried out by workers that affects all members of the colony, including, for instance, nest construction, defense, and maintenance. This parameter may also represent the degree to which workers are unable to invest in sisters without also providing incidental benefits for males. When $\varepsilon = 0$, females can discriminate between brothers and sisters accurately and direct their altruism exclusively to females, while when $\varepsilon = 1$, females behave equally to both sexes. If γ is the cost of altruism and β is the benefit to the recipient, then the benefit to males is $\beta\varepsilon\chi$ and to females is $\beta(1 - \varepsilon\chi)$. Genotypes *EE*, *Ee*, and *ee* have frequencies u, v, and w in females, where $p = 1 - q = u + v/2$, and E and e have frequencies x and y in males. The level of dominance of E over e is given by h. That is, *EE* is always altruistic, *Ee* is altruistic with probability h, while *ee* is never altruistic.

Recursions for the frequency of each genotype in each sex in the offspring generation are computed by adding the values of the contributions to fitness of each genotype in the broods producing that genotype with weights corresponding to the product of the frequency of each mating and the probability that the mating produces that genotype. This gives

$$
\begin{aligned}
W_F u' &= x[p(1 - \gamma) + \beta(1 - \varepsilon\chi)(u + v/4 + vh/4)],\\
W_F v' &= (xq + yp)(1 - h\gamma)\\
&\quad + \beta(1 - \varepsilon\chi)[vx/4 + (uy + wx + v/4)h],\\
W_F w' &= qy + \beta(1 - \varepsilon\chi)yvh/4,\\
W_M x' &= p + \beta\varepsilon\chi[(u + v/4)x + (uy + v/4)h],\\
W_M y' &= q + \beta\varepsilon\chi[vx/4 + (v/4 + wx)h],
\end{aligned}
$$

where

$$
\begin{aligned}
W_F &= 1 + [\beta(1 - \varepsilon\chi) - \gamma][px + (xq + yp)h],\\
W_M &= 1 + \beta\varepsilon\chi[px + (py + qx)h]. \qquad \text{(A5.1)}
\end{aligned}
$$

A local stability analysis near the $w = y = 1$ equilibrium, where e is fixed, gives the following condition for the invasion of allele E:

$$
\gamma < \frac{\beta(1 + 2(1 - \varepsilon\chi) + \beta h\varepsilon\chi(1 - \varepsilon\chi)}{4 + \beta\varepsilon h\chi}. \qquad \text{(A5.2)}
$$

When $\varepsilon = 0$, sisters only help sisters, and A5.2 reduces to $\gamma < 3\beta/4$. When sisters niche construct in a manner that benefits all sibs ($\varepsilon = 1$), then for $\chi = 1/2$ equation A5.2 reduces to $\gamma < \beta/2$. These values can be seen to correspond to Hamilton's rule. However, as the sex ratio becomes more female-biased, the probability that the altruistic niche-constructing behavior will invade increases.

On the assumption that investment in the two sexes is equal, Trivers and Hare (1976) proposed that

$$\frac{W_M \chi}{W_F(1-\chi)} = \frac{r_M}{r_F}, \tag{A5.3}$$

where r_M and r_F are the degrees of relatedness of the altruist to male and female recipients. Assuming that no workers lay eggs, r_M and r_F are 1/4 and 3/4, respectively. At the equilibrium with altruism fixed ($p = x = 1$), from A5.1 it can be seen that

$$W_F = 1 + [\beta(1 - \varepsilon\chi) - \gamma],$$
$$W_M = 1 + \beta\varepsilon\chi. \tag{A5.4}$$

Substituting these values into the Trivers and Hare formulation allows us to derive an expression for χ that enables A5.3 to hold. At $\varepsilon = 0$, the proportion of males is given by

$$\chi = \frac{1+\beta-\gamma}{4+\beta-\gamma}. \tag{A5.5a}$$

Thus, assuming that the benefits of altruism to the recipient are greater than the costs to the altruist, or $\beta - \gamma > 0$, when sisters help only sisters, the sex ratio is expected to exhibit a higher frequency of males than the 1/4 fraction or 1:3 ratio predicted by Trivers and Hare, and will be a function of the costs and benefits of altruism. This may help to explain why the average sex-investment ratio across 40 species of monogynous ants is less female-biased than the 25% male ratio that Trivers and Hare predicted, being around $\chi = 0.37$ (Bourke 1997). Trivers and Hare's formulation may be inaccurate because it does not take into account the costs and benefits of altruism.

If $\varepsilon > 0$, the proportion of males is given by

$$\chi = \frac{-4-\beta+\gamma-\beta\varepsilon+\sqrt{\beta^2\varepsilon^2+16\beta\varepsilon+10\beta^2\varepsilon-10\beta\varepsilon\gamma+16+8\beta-8\gamma+\beta^2-2\beta\gamma+\gamma^2}}{4\beta\varepsilon}. \tag{A5.5b}$$

Although expression A5.5b is difficult to interpret intuitively, numerical analyses reveal that χ ranges from 0.2 to 0.4 for realistic values of ε, β, and γ. In general, as ε increases, thus increasing niche-constructing behavior that benefits the entire colony, χ shows a corresponding decrease, and the sex ratio becomes more female-biased. Thus, niche construction for the general good of the colony selects for a more female-biased sex ratio. This niche-construction perspective, as expressed in equations A5.5a and A5.5b, may help to explain some of the variability in sex ratios observed among social insects.

Glossary of New Terms

Ecological inheritance — The inheritance, via an external environment, of one or more natural selection pressures previously modified by niche-constructing organisms.

EMGAs (environmentally mediated genotypic associations) — Indirect but specific connections between distinct genotypes mediated either by biotic or abiotic environmental components in the external environment. EMGAs may either associate different genes in a single population or associate different genes in different populations. Where EMGAs arise, the expression of semantic information by niche-constructing organisms in one population may affect the acquisition of semantic information in the same or a second population through the modification of natural selection pressures generated via one or more intermediate components, with evolutionary or ecological consequences (see chapter 5).

Extended evolutionary theory — Evolutionary theory that includes niche construction and ecological inheritance as evolutionary processes, in addition to natural selection, and genetic inheritance.

Niche — The sum of all the natural selection pressures to which the population is exposed (see chapter 2). A population O's niche is specified at time t by a "niche function" $\mathcal{N}(t)$ where

$$\mathcal{N}(t) = h(O,E).$$

O is the population of organisms, and E is O's environment, both specified at time t. The temporal dynamics of $\mathcal{N}(t)$, equivalent to niche evolution, are driven by both O's niche-constructing acts, and selection from sources in E that have previously been modified by O's niche-constructing acts, as well as by the dynamics of E that are independent of O's niche construction.

Niche construction — The process whereby organisms, through their metabolism, their activities, and their choices, modify their own and/or each other's niches. Niche construction may result in changes in one or more natural selection pressures in the external environment of populations. Niche-constructing organisms may alter the natural selection pressures of their own population, of other populations, or of both.

Perturbational niche construction – occurs when organisms physically change one or more components of their external environments.

Relocational niche construction – occurs when organisms actively move in space, as well as choose or bias the direction and the distance in space through which they travel, and the time when they travel, thereby modifying natural selection pressures.

Inceptive niche construction – occurs when organisms either perturb components of their environments, or relocate in their environments in such a way as to introduce a change in one or more natural selection pressures.

Counteractive niche construction – occurs when organisms either perturb components of their environments or relocate in their environments in such a way as to wholly or partly reverse or neutralize some prior change in one or more natural selection pressures in their environments. These prior changes in natural selection could be caused either by independent processes in an environment, or by the prior niche-constructing activities of other organisms, or by the prior niche-constructing activities of the organisms themselves, and/or their ancestors.

Positive niche construction – refers to niche-constructing acts that, on average, increase the fitness of the niche-constructing organisms. In the short run most niche construction by individual organisms is expected to be positive.

Negative niche construction – refers to niche-constructing acts that, on average, decrease the fitness of the niche-constructing organisms. In the long run we expect some niche-constructing activities of organisms to become negative for their populations.

Semantic information — Information, typically encoded in DNA, that specifies the adaptations of organisms. It pertains to the life requirements and functioning of organisms in their local environments. Semantic information is accrued primarily through population genetic processes, although it can also be acquired through other processes, for instance, learning. Semantic information is therefore "meaningful" information for organisms in their niches. Unlike configurational information, semantic information cannot be represented by binary digits, or bits (see chapter 4).

Phenogenotype — A property of individuals in a human population defined by a specified combination of a genotype and a variant of a cultural trait. Phenogenotypes are useful abstractions in mathematical analyses of human evolution, and are used to model human cultural processes and human genetic processes simultaneously. Phenogenotypes are the basic evolutionary units when biological and cultural transmission and inheritance processes are operating in association with each other (see chapters 6 and 9).

Bibliography (indexed)

Note: Numbers in brackets following an entry are the page numbers of citations in the text.

Abrams, P. A. 1988. Resource productivity-consumer species diversity: Simple models of competition in spatially heterogeneous environments. Ecology 69:1418–1433. [24, 119]

———. 1996. Evolution and the consequences of species introductions and deletions. Ecology 77:1321–1328. [24, 124]

———. 2000. Character shifts of prey species that share predators. American Naturalist 156:S45–S61. [119, 125, 219]

———. 2001. The effect of density-independent mortality on the coexistence of exploitative competitors for renewing resources. American Naturalist 158:459–470. [205]

Adenzato, M. 2000. Gene-culture coevolution does not replace standard evolutionary theory. Behavioral and Brain Sciences 23:146. [248]

Aiello, L. C., and P. Wheeler. 1995. The expensive-tissue hypothesis. Current Anthropology 36:199–221. [242, 343, 344, 353]

Alcock, J. 1972. The evolution of the use of tools by feeding animals. Evolution 26:464–473. [21, 258]

Alexander, R. D. 1979. *Darwinism and Human Affairs*. Seattle: University of Washington Press. [243]

Allen, T.F.H., and T. B. Starr. 1982. *Hierarchy: Perspectives for Ecological Complexity*. Chicago: University of Chicago Press. [59]

Alper, J. 1998. Ecosystem "engineers" shape habitats for other species. Science 280:1195–1196. [7]

Ambrose, S. H. 2001. Paleolithic technology and human evolution. Science 291:1748–1753. [346]

Andrewartha, H. G., and L. C. Birch. 1954. *The Distribution and Abundance of Animals*. Chicago: University of Chicago Press. [310]

Aoki, K., and M. W. Feldman. 1987. Toward a theory for the evolution of cultural communication: Coevolution of signal transmission and reception. Proceedings of the National Academy of Sciences, USA 84:7164–7168. [247, 355]

———. 1989. Pleiotropy and preadaptation in the human language capacity. Theoretical Population Biology 35:181–194. [247]

———. 1991. Recessive hereditary deafness, assortative mating and persistence of a sign language. Theoretical Population Biology 39:358–372. [247]

———. 1997. A gene-culture coevolutionary model for brother-sister mating. Proceedings of the National Academy of Sciences, USA 94: 13046–13050. [247]

Arbib, M. A. 1969. Self reproducing automata–some implications for theoretical biology. In *Towards a Theoretical Biology. 2 Sketches*, edited by C. H. Waddington. Edinburgh, Scotland: Edinburgh University Press. [172]

Armstrong, R. A. 1979. Prey species replacement along a gradient of nutrient enrichment: A graphical approach. Ecology 60:76–84. [119, 219]

Attenborough, D. 1990. *The Trials of Life*. London: BBC Books. [75, 76, 80, 83, 90, 104, 108, 112]

———. 1998. *The Life of Birds*. London: BBC Books. [64]

Aunger, R. 2000a. *Darwinizing Culture: The Status of Memetics as a Science*. Oxford: Oxford University Press. [249, 252, 259, 261, 263, 358]

———. 2000b. Conclusion. Pages 205–232 in *Darwinizing Culture: The Status of Memetics as a Science*, edited by R. Aunger. Oxford: Oxford University Press. [281, 358]

———. 2000c. The life history of culture learning in a face-to face society. Ethos 28:1–38. [356]

———. 2002. *The Electric Meme: A New Theory of How We Think and Communicate*. New York: The Free Press. [261, 281, 358]

Axelrod, R. 1984. *The Evolution of Cooperation*. New York: Basic Books. [189, 298]

Axelrod R., and W. D. Hamilton. 1981. The evolution of cooperation. Science 211:1390–1396. [122]

Balser, T. C., A. P. Kinzig, and M. K. Firestone. 2001. Linking soil microbial communities and ecosystem functioning. In *The Functional Consequences of Biodiversity*, edited by A. P. Kinzig, S. W. Pacala, and D. Tilman. Princeton, NJ: Princeton University Press. [322, 329]

Barkow, J. H. 2000. Our shared species-typical evolutionary psychology. Behavioral and Brain Sciences 23:148. [281]

Barkow, J., L. Cosmides, and J. Tooby. 1992. *The Adapted Mind: Evolutionary Psychology and the Generation of Culture*. New York: Oxford University Press. [243, 246, 263]

Barnes, R.S.K., and R. N. Hughes. 1999. *An Introduction to Marine Ecology*. 3rd ed. Oxford: Blackwell. [59]

Barton, N. H., and S. Rouhani. 1987. The frequency of shifts between alternative equilibria. Journal of Theoretical Biology 125:397–418. [301]

Basnet, K., G. E. Likens, F. N. Scatena, and A. E. Lugo. 1992. Hurricane Hugo: Damage to a tropical rain forest in Puerto Rico. Journal of Tropical Ecology 8:47–55. [217]

Bateson, P.P.G. 1988. The active role of behaviour in evolution. In *Evolutionary Processes and Metaphors*, edited by M.W. Ho and S. Fox. New York: Wiley. [30, 113, 166, 353, 354]

Begon, M., J. L. Harper, and C. R.Townsend. 1996. *Ecology: Individuals, Populations and Communities*. 3rd ed. Oxford: Blackwell. [6, 56, 118]

Bennett, C. H. 1987. Demons, engines and the second law. Scientific American 275(5):88–96. [171, 172, 177, 186n]

Benton, M. J. 1988. Burrowing by vertebrates. Nature 331:17–18. [62]

Berg, L. R. 1997. *Introductory Botany: Plants, People, and the Environment*. Orlando, FL: Saunders. [110, 111]

Bergman, A., and M. W. Feldman. 1995. On the evolution of learning: Representation of a stochastic environment. Theoretical Population Biology 48:251–276. [355]

Bernado, J. 1996. Maternal effects in animal ecology. American Zoologist 36:83–105. [20, 125]

Bertness, M. D., and G. H. Leonard. 1997. The role of positive interactions in communities: Lessons from intertidal habitats. Ecology 78:1976–1989. [216]

Bock, W. J. 1980. The definition and recognition of biological adaptation. American Zoologist 20:217–227. [41]

Bodmer, W. F., and L. L. Cavalli-Sforza. 1976. *Genetics, Evolution and Man*. San Francisco: Freeman. [241]

Bodmer, W. F., and J. Felsenstein. 1967. Linkage and selection: Theoretical analysis of the two-locus random mating model. Genetics 57:237–265. [127, 134, 143, 144]

Bohannan, B.J.M., and R. E. Lenski. 2000. Linking genetic change to community evolution: Insights from studies of bacteria and bacteriophage. Ecology Letters 3:362–377. [119]

Bolhuis, J. J., and E. M. MacPhail. 2001. A critique of the neuroecology of learning and memory. Trends in Cognitive Sciences 5:426–433. [288]

Bolles, R. C. 1970. Species-specific defense reactions and avoidance learning. Psychological Review 77:32–48. [257]

Bond, W. J., and J. J. Midgley. 1995. Kill thy neighbour: An individualistic argument for the evolution of flammability. Oikos 73:79–85. [59]

Bond, W. J., and B. W. van Wilgen. 1996. *Fire and Plants*. London: Chapman and Hall. [124]

Boots, M., and A. Sasaki. 1999. Small worlds and the evolution of virulence: Infection occurs locally and at a distance. Proceedings of the Royal Society of London, Series B 266:1933–1938. [124]

Borgerhoff Mulder, M. 1998. The demographic transition: Are we any closer to an evolutionary explanation? Trends in Ecology and Evolution 13:266–270. [362, 363]

Bourke, A.F.G. 1997. Sociality and kin selection in insects. Pages 203–227 in *Behavioural Ecology: An Evolutionary Approach*, edited by J. R. Krebs and N. B. Davies. 4th ed. Oxford: Blackwell. [287, 300, 417]

Bowles, S. 2000. Economic institutions as ecological niches. Behavioral and Brain Sciences 23:148–149. [249, 281, 361]

Boyd, R., and P. J. Richerson. 1985. *Culture and the Evolutionary Process*.

Chicago: University of Chicago Press. [126, 128, 129, 242, 246–50, 254, 259, 260, 353, 354, 355, 359, 360, 361, 366, 367]

———. 1988. The evolution of reciprocity in sizeable groups. Journal of Theoretical Biology 132:337–356. [359]

Brace, C. L. 1999. Comments on R. W. Wrangham, J. H. Jones, G. Laden, D. Pilbeam, and N. Conklin Brittain. The raw and the stolen. Cooking and the ecology of human origins. Current Anthropology 40:577–579. [344]

Bradshaw, W. E., and C. M. Holzapfel. 1992. Resource limitation, habitat segregation, and species interactions of British tree-hold mosquitoes in nature. Oecologia 90:227–237. [56]

Brandon, R., and J. Antonovics. 1996. The coevolution of organism and environment. Pages 161–178 in *Concepts and Methods in Evolutionary Biology*, edited by R. Brandon. Cambridge, England: Cambridge University Press. [30]

Bristowe, W. S. 1958. *The World of Spiders*. London: Collins. [78, 81]

Bronstein, J. L. 2001. Mutualisms. In *Evolutionary Ecology: Concepts and Case Studies*, edited by C. W. Fox, D. A. Roff, and D. J. Fairbairn. New York: Oxford University Press. [317]

Brown, J. H. 1995. Organisms and species as complex adaptive systems: Linking the biology of populations with the physics of ecosystems. In *Linking Species and Ecosystems*, edited by C. G. Jones and J. H. Lawton. New York: Chapman and Hall. [195, 201]

Brown, J. S. 1990. Habitat selection as an evolutionary game. Evolution 44:732–746. [124]

———. 1998. Game theory and habitat selection. Pages 188–220 in *Game Theory and Animal Behavior*, edited by L. A. Dugatkin and H. K. Reeve. Oxford: Oxford University Press. [123, 124]

Brown, J. S., and N. B. Pavlovic. 1992. Evolution in heterogeneous environments—Effects of migration on habitat specialization. Evolutionary Ecology 6:360–382. [124]

Bubier, J. L., and T. R. Moore. 1994. An ecological perspective on methane emissions from northern wetlands. Trends in Ecology and Evolution 9:460–464. [55]

Byrne, R., and A. Whiten. 1988. *Machiavellian Intelligence: Social Expertise and the Evolution of Intellect in Monkeys, Apes and Humans*. Oxford: Clarendon Press. [242, 352]

Campbell, D. T. 1960. Blind variation and selective retention in creative thought as in other knowledge processes. Psychological Review 67:380–400. [254]

———. 1974. Evolutionary epistemology. Pages 413–463 in *The Philosophy of Karl R. Popper*, edited by P. A. Schilpp. La Salle, IL: Open Court. [180, 254]

Canfeld, D. E., and A. Teske. 1996. Late Proterozoic rise in atmospheric oxygen concentration inferred from phylogenetic and sulphur-isotope studies. Nature 382:127–132. [54]

Capone, D. G., J. P. Zehr, H. W. Paerl, B. Bergman, and E. J. Carpenter. 1997. *Trichodesmium*: A globally significant marine cynobacterium. Science 276:1221–1229. [55]

Carpenter, S. R., and D. M. Lodge. 1986. Effects of submerged macrophytes on ecosystem processes. Aquatic Botany 26:341–370. [214]

Carvalho, G., A. C. Barros, P. Moutinho, and D. Nepstad. 2001. Sensitive development could protect Amazonia instead of destroying it. Nature 409:131. [58]

Cavalli-Sforza, L. L., and M. W. Feldman. 1973a. Cultural versus biological inheritance: Phenotypic transmission from parent to children (a theory of the effect of parental phenotypes on children's phenotypes). American Journal of Human Genetics 25:618–637. [246, 364]

———. 1973b. Models for cultural inheritance: I. Group mean and within-group variation. Theoretical Population Biology 4:42–55. [246]

———. 1981. *Cultural Transmission and Evolution: A Quantitative Approach*. Princeton, NJ: Princeton University Press. [34, 126, 246, 247, 248, 259, 260, 267, 270, 339, 364–69 passim, 412, 413]

———. 1983. Cultural versus genetic adaptation. Proceedings of the National Academy of Sciences, USA 79:1331–1335. [355]

Cavalli-Sforza, L. L., M. W. Feldman, K. H. Chen, and S. M. Dornbusch. 1982. Theory and observation in cultural transmission. Science 218: 19–27. [356]

Chao, L., and B. R. Levin. 1981. Structured habitats and the evolution of anticompetitor toxins in bacteria. Proceedings of the National Academy of Sciences, USA 78:6324–6328. [53]

Chapin, F. S., B. H. Walker, R. J. Hobbs, D. U. Hooper, J. H. Lawton, O. E. Sala, and D. Tilman. 1997. Biotic control over the functioning of ecosystems. Science 277:500–504. [213]

Chapin, F. S., E. S. Zavaleta, V. T. Eviners, R. L. Naylor, P. M. Vitousek, H. L. Reynolds, D. U. Hooper, S. Lavorel, O. E. Sala, S. E. Hobbie, M. C. Mack, and S. Diaz. 2000. Consequences of changing biodiversity. Nature 405:234–242. [383]

Charlson, R. J., J. E. Lovelock, M. O. Andreae, and S. J. Warren. 1987. Ocean phytoplankton, atmospheric sulphur, cloud albedo and climate. Nature 326:655–661. [55]

Chesson, P. 2000. Mechanisms of maintenance of species diversity. Annual Review of Ecology and Systematics 31: 343–366. [118]

Cheverud, J. M. 1984. Evolution by kin selection: A quantitative genetic model illustrated by maternal performance in mice. Evolution 38:766–777. [299, 300n]

Christiansen, F. B. 1990. The generalized multiplicative model for viability selection at multiple loci. Journal of Mathematical Biology 29:99–129. [134]

———. 2000. *Population Genetics of Multiple Loci*. New York: Wiley. [127, 134]

Christiansen, F. B., and V. Loeschcke. 1980a. Evolution and intraspecific exploitative competition I. One-locus theory for small additive genetic effects. Theoretical Population Biology 19:378–419. [121]

———. 1980b. Intraspecific competition and evolution. Pages 151–170 in *Vito Volterra Symposium on Mathematical Models in Biology*, edited by C. Barigozzi. Berlin: Springer-Verlag. [121]

Clark, A. G. and M. W. Feldman. 1981. Density-dependent fertility selection in experimental populations of *Drosophila melanogaster*. Genetics 98: 849–69. [121]

Clarke, B. 1972. Density-dependent selection. American Naturalist 106:1–13. [120]

Clements, F. E. 1916. *Plant Succession: An Analysis of the Development of Vegetation*. Publication 242. Washington, DC: Carnegie Institute. [334]

Cockerham, C. C., P. M. Burrows, S. S. Young, and T. Prout. 1972. Frequency-dependent selection in randomly mating populations. American Naturalist 106:493–515. [121]

Cody, M. L. 1985. *Habitat Selection in Birds*. Orlando, FL: Academic. [123]

Cogger, H. G. 1998. *Encyclopedia of Reptiles and Amphibians*. 2nd ed. San Diego, CA: Academic. [63, 64, 72–82 passim, 89–99 passim, 103, 105, 108, 112]

Cohen, D. 1991. The equilibrium distribution of optimal search and sampling effort of foraging animals in patchy environments. Pages 173–191 in *Adaptation in Stochastic Environments*, edited by J. Yoshimura and C. W. Clark. Berlin: Springer-Verlag. [124]

Cohen, J. E. 1978. *Food Webs and Niche Space*. Princeton, NJ: Princeton University Press. [43]

Cohen, M. N. 1998. The emergence of health and social inequalities in the archaeological record. *Society for the Study of Human Biology, Symposium 39* edited by S. S. Strickland and P. S. Shetty. Cambridge, England: Cambridge University Press. [361]

Combes, C. 1995. *Interactions Durables. Ecologie et Evolution du Parasitisme*. Paris: Masson. [251]

Connor, R. C. 1995. The benefits of mutualism: A conceptual framework. Biological Reviews 70:427– 457. [299]

Cowley, D. E., and W. R. Atchley. 1992. Quantitative genetic models for development, epigenetic selection, and phenotypic evolution. Evolution 46:495–518. [10, 125, 126]

Cronk, L. 1995. Comment on K. N. Laland, J. Kumm, and M. W. Feldman's "Gene-culture coevolutionary theory." Current Anthropology 36:147–148. [265]

Cronk, L., N. Chagnon, and W. Irons, eds. 2000. *Adaptation and Human Behavior: An Anthropological Perspective*. New York: Aldine de Gruyter. [242, 243, 246]

Crook, J. H. 1963. A comparative analysis of nest structure in the weaver birds. Proceedings of the Zoological Society of London 142:217–255. [85]

Daborn, G. R., C. L. Amos, M. Brylinsky, H. Christian, G. Drapeau, R. W. Fass, J. Grant, B. Long, D. M. Paterson, G.M.E. Perillo, and M. C. Piccolo. 1993. An ecological cascade effect: Migratory birds affect stability of intertidal sediments. Limnology and Oceanography 38: 225–231. [326]

Daly, M., and M. Wilson. 1983. *Sex, Evolution, and Behavior*. 2nd ed. Boston: Wadsworth. [246, 263]

Darwin, C. 1881. *The Formation of Vegetable Mold through the Action of Worms, with Observations on their Habits*. London: Murray. [11]

Daufresne, T. and M. Loreau. 2001. Ecological stoichiometry, primary producer-decomposer interactions, and ecosystem persistence. Ecology 82:3069–3082. [315, 329]

Davies, N. B., R. M. Kilner, and D. G. Noble. 1998. Nestling cuckoos, *Cuculus canorus*, exploit hosts with begging calls that mimic a brood. Philosophical Transactions of the Royal Society of London, Series B 265:673–678. [10, 17]

Dawkins, R. 1982. *The Extended Phenotype*. Oxford: Freeman. [30, 131, 184, 191, 213, 382]

———. 1989. *The Selfish Gene*. 2nd ed. Oxford: Oxford University Press. [21, 243, 259, 365]

———. 1996. *Climbing Mount Improbable*. London: Penguin. [17]

Deacon, T. W. 1997. *The Symbolic Species: The Co-evolution of Language and the Brain*. New York: Norton. [246]

DeAngelis, D. L. 1992. *Dynamics of Nutrient Cycling and Food Webs*. London: Chapman and Hall. [6, 24, 43, 118, 119, 121, 149, 199–205 passim, 219, 222, 313, 316]

———. 1995. The nature and significance of feedback in ecosystems. Pages 450–467 in *Complex Ecology: The Part-Whole Relation in Ecosystems*, edited by B. C. Patten and S. E. Jorgensen. Englewood Cliffs, NJ: Prentice Hall. [330]

Dennett D. 1995. *Darwin's Dangerous Idea: Evolution and the Meanings of Life*. London: Penguin. [180, 254]

Diamond, J. 1998. *Guns, Germs and Steel*. Croydon, England: Vintage. [55, 242, 261, 347, 348]

Diaz, S., and M. Cabido. 2001. Vive la difference: Plant functional diversity matters to ecosystem processes. Trends in Ecology and Evolution 16:646–655. [320, 329]

Dobzhansky, T. 1955. *Evolution, Genetics and Man*. New York: Wiley. [31, 283]

Doney, S. C. 1997. The ocean's productive deserts. Nature 389:905–906. [55]

Dugatkin, L. A. 1992. The evolution of the con artist. Ethology and Sociobiology 13:3–18. [361]

Dugatkin, L. A., and H. K. Reeve, eds. 1998. *Game Theory and Animal Behavior*. New York: Oxford University Press. [121, 122]

Dunbar, R.I.M. 1993. Coevolution of neocortical size, group size and language in humans. Behavioral and Brain Sciences 16:681–735. [242, 352]

Durham, W. H. 1991. *Coevolution: Genes, Culture and Human Diversity.* Stanford, CA: Stanford University Press. [27, 242, 246–54 passim, 260, 263, 340, 341, 342, 343, 352, 369]

Durrett, R., and S. A. Levin. 1994. The importance of being discrete (and spatial). Theoretical Population Biology 46:363–394. [311]

———. 1997. Allelopathy in spatially distributed populations. Journal of Theoretical Biology 185:165–171. [54]

Dwyer, G., S. A. Levin, and L. Buttel. 1990. A simulation of the population dynamics and evolution of myxomatosis. Ecological Monographs 60:423–447. [233, 248]

Dyer, F. C. 1998. Cognitive ecology of navigation. In *Cognitive Ecology: The Evolutionary Ecology of Information Processing and Decision Making,* edited by R. Dukas. Chicago: University of Chicago Press. [101]

Edmunds, M. 1974. *Defense in Animals.* New York: Longmans. [9, 81, 263]

Ehrlich, P. R. 2000. *Human Natures. Genes, Cultures, and the Human Prospect.* Washington, DC: Island Press. [249, 347]

Ehrlich, P. R., and J. Roughgarden. 1987. *The Science of Ecology.* New York: Macmillan. [40]

Ehrman, L. 1968. Frequency dependence of mating success in *Drosophila pseudoobscura.* Genetical Research 11:135–140. [121]

Eigen, M. 1992. *Steps Towards Life: A Perspective on Evolution.* Oxford: Oxford University Press. [15, 168, 183]

Eldredge, N., and S. J. Gould. 1972. Punctuated equilibria: An alternative to phyletic gradualism. Pages 82–115 in *Models in Palaeobiology,* edited by T.J.M. Schopf. San Francisco: Freeman and Cooper. [304]

Ellis, S., and A. Mellor. 1995. *Soils and Environment.* London: Routledge. [11]

Elton, C. 1927. *Animal Ecology.* London: Sidgwick and Jackson. [37, 38, 43, 334]

———. 1958. *The Ecology of Invasions by Animals and Plants.* London: Chapman and Hall. [328]

Emerson, S., P. Quay, D. Karl, C. Winn, L. Tupas, and M. Landry. 1997. Experimental determination of the organic carbon flux from open-ocean surface waters. Nature 389:951–954. [53, 55]

Enattah, N.S., T. Sahi, E. Savilahti, J. D. Terwilliger, L. Peltonen, and I. Jarvela. 2002. Identification of a variant associated with adult-type hypolactasia. Nature Genetics 30:233–237. [342]

Endler, J. A. 1986. *Natural Selection in the Wild.* Princeton, NJ: Princeton University Press. [8, 184, 233, 292, 350]

Engelberg, J., and L. L. Boyarsky. 1979. The noncybernetic nature of ecosystems. American Naturalist 114:317–324. [330, 331]

Eshel, I., and M. W. Feldman. 1982. On evolutionary genetic stability of the sex ratio. Theoretical Population Biology 21:430–439. [120, 122]

———. 1984. Initial increase of new mutants and some continuity properties of ESS in two-locus systems. American Naturalist 124:631–640. [121, 122]

———. 2001. Optimality and evolutionary stability under short-term and long-term selection. Pages 161–190 in *Adaptations and Optimality*, edited by S. H. Orzack and E. Sober. Cambridge, England: Cambridge University Press. [122]

Eshel, I., M. W. Feldman, and A. Bergman. 1998. Long-term evolution, short-term evolution and population genetic theory. Journal of Theoretical Biology 191:391–396. [122]

Evans, H. E., and M. J. West-Eberhard. 1970. *The Wasps*. Ann Arbor, MI: University of Michigan Press. [70]

Ewald, P. W. 1983. Host-parasite relations, vectors and the evolution of disease severity. Annual Review of Ecology and Systematics 14:465–485. [124]

———. 1994. *Evolution of Infectious Disease*. New York: Oxford University Press. [55, 124, 233, 369]

Falkowski, P. G. 1997. Evolution of the nitrogen cycle and its influence on the biological sequestration of CO_2 in the ocean. Nature 387:272–275. [53]

Feldman, M. W., K. Aoki, and J. Kumm. 1996. Individual versus social learning: Evolutionary analysis in a fluctuating environment. Anthropological Science 104:209–232. [355]

Feldman, M. W., and L. L. Cavalli-Sforza. 1976. Cultural and biological evolutionary processes, selection for a trait under complex transmission. Theoretical Population Biology 9:238–259. [21, 68, 126, 128, 129, 161, 241, 242, 246, 247, 259, 284, 353, 354]

———. 1979. Aspects of variance and covariance analysis with cultural inheritance. Theoretical Population Biology 15:276–307. [246]

———. 1984. Cultural and biological evolutionary processes: Gene-culture disequilibrium. Proceedings of the National Academy of Sciences, USA 81:1604–1607. [246, 366]

———. 1989. On the theory of evolution under genetic and cultural transmission with application to the lactose absorption problem. In *Mathematical Evolutionary Theory*, edited by M. W. Feldman. Princeton, NJ: Princeton University Press. [27, 248, 250, 263, 342, 365]

Feldman, M. W., and K. N. Laland. 1996. Gene-culture coevolutionary theory. Trends in Ecology and Evolution 11:453–457. [128, 242, 247, 249, 254, 354, 355, 367]

Feldman, M. W., R. C. Lewontin, I. R. Franklin, and F. B. Christiansen. 1975. Selection in complex genetic systems: III. An effect of allele multiplicity with two loci. Genetics 79:333–347. [134]

Feldman, M. W., and L. A. Zhivotovsky. 1992. Gene-culture coevolution: Toward a general theory of vertical transmission. Proceedings of the National Academy of Sciences, USA 89:11935–11938. [366, 412]

Fenchel, T. M., and F. B. Christiansen. 1977. Selection and interspecific competition. Pages 477–798 in *Measuring Selection in Natural Populations*, edited by F. B. Christiansen and T. M. Fenchel, Lecture Notes in Biomathematics, vol. 19. New York: Springer-Verlag. [121]

Fenster, C. B., L. F. Galloway, and L. Chao. 1997. Epistasis and its consequences for the evolution of natural populations. Trends in Ecology and Evolution 12:282–286. [127]

Feynman, R. 1965. *The Character of Physical Law*. London: BBC Books. [172]

Fisher, J., and R. A. Hinde. 1949. The opening of milk bottles by birds. British Birds 42:347–357. [22]

Flatz, G. 1987. Genetics of lactose digestion in humans. Advances in Human Genetics 16:1–77. [342]

Fodor, J. A. 2000. *The Mind Doesn't Work That Way: The Scope and Limits of Computational Psychology*. Cambridge, MA: MIT Press. [281]

Foley, R. 1988. Hominids, humans, and hunter-gatherers: An evolutionary perspective. In *Hunters and Gatherers: History, Evolution and Social Change*, edited by T. Ingold, D. Riches, and J. Woodburn. New York: Berg. [349]

———. 1995. The adaptive legacy of human evolution: A search for the environment of evolutionary adaptedness. Evolutionary Anthropology 4:194–203. [352]

Forshaw, J. 1998. *Encyclopedia of birds*. 2nd ed. San Diego: Academic. [61, 63, 64, 92, 100]

Fox, C. W., D. A. Roff, and D. J. Fairbairn, eds. 2001. *Evolutionary Ecology: Concepts and Case Studies*. New York: Oxford University Press. [306, 307]

Fox, D. L. 1979. *Biochromy: Natural Coloration of Living Things*. Berkeley, CA: University of California Press. [114]

Fragaszy, D., and S. Perry, eds. In press. *The Biology of Traditions: Models and Evidence*. Chicago: Chicago University Press. [281]

Frautschi, S. 1988. Entropy in an expanding universe. In *Entropy, Information, and Evolution: New Perspectives on Physical and Biological Evolution*, edited by B. W. Weber, D. J. Depew, and J. D. Smith. Cambridge, MA: MIT Press. [171]

Frisch, K. v. 1975. *Animal Architecture*. London: Hutchinson. [62, 63, 70, 76–84 passim, 88–98 passim]

Futuyma, D. J. 1986. *Evolutionary Biology*. 2nd ed. Sunderland, MA: Sinauer. [121]

———. 1998. *Evolutionary Biology*. 3rd ed. Sunderland, MA: Sinauer. [12, 30, 113, 120, 123, 124]

Futuyma, D. J., and M. Slatkin. 1983. *Coevolution*. Sunderland, MA: Sinauer. [24, 44, 124, 125, 165, 207]

Galef, B. G., Jr. 1988. Imitation in animals: History, definition, and interpretation of data from the psychological laboratory. In *Social Learning: Psychological and Biological Perspectives*, edited by T. R. Zentall and B. G. Galef, Jr. Hillsdale, NJ: Erlbaum. [250, 356]

Garcia, J., F. R. Ervin, and R. A. Koelling. 1966. Learning with prolonged delay of reinforcement. Psychonomic Science 5:121–122. [257]

Garcia-Dorado, A. 1986. The effect of niche preference on polymorphism protection in a heterogeneous environment. Evolution 40:936–945. [123]

Gaston, K. J. 2000. Global patterns in biodiversity. Nature 405:220–227. [383]

Gaston, K. J., T. M. Blackburn, and J. I. Spicer. 1998. Rapoport's rule: Time for an epitaph? Trends in Ecology and Evolution 13:70–74. [288, 348]

Gavrilets, S. 1997. Coevolutionary chase in exploiter-victim systems with polygenic characters. Journal of Theoretical Biology 186:527–534. [124]

Getz, W. M. 1999. Population and evolutionary dynamics of consumer-resource systems. In *Advanced Ecological Theory: Principles and Applications*, edited by J. McGlade. Oxford: Blackwell. [306]

Gilbert, L. E. 1983. Coevolution and mimicry. In *Coevolution*, edited by D. J. Futuyma and M. Slatkin. Sunderland, MA: Sinauer. [102, 107, 112]

Gillespie, J. H. 1973. Natural selection with varying selection coefficients—a haploid model. Genetical Research 21:115–120. [130]

Gingerich, P. D. 1983. Rates of evolution: Effects of time and temporal scaling. Science 222:159–161. [234]

Gintis, H., E. Alden Smith, and S. Bowles. 2001. Costly signaling and cooperation. Journal of Theoretical Biology 213:103–119. [189]

Ginzburg, L. R. 1998. Inertial growth: Population dynamics based on maternal effects. Pages 42–53 in *Maternal Effects as Adaptations*, edited by T. A. Mousseau and C. W. Fox. New York: Oxford University Press. [125, 126]

Ginzburg, L., and D. Taneyhill. 1995. Higher growth rate implies shorter cycle, whatever the cause. A reply to Berryman. Journal of Animal Ecology 64:294–295. [126]

Giraldeau, L. A. 1997. The ecology of information use. Pages 42–68 in *Behavioural Ecology: An Evolutionary Approach*, edited by J. R. Krebs and N. B. Davies. 4th ed. Oxford: Blackwell. [303]

Giraldeau, L. A., and T. Caraco. 2001. *Social Foraging Theory*. Princeton, NJ: Princeton University Press. [122, 124]

Gleason, H. A. 1917. The structure and development of plant association. Bulletin of the Torrey Botanical Club 43:463–481. [334]

———. 1926. The individualist concept of the plant association. Bulletin of the Torrey Botanical Club 53:7–26. [334]

Godfrey-Smith, P. 1996. *Complexity and the Function of Mind in Nature*. Cambridge, England: Cambridge University Press. [371, 373]

Golenser, J., J. Miller, D. T. Spira, T. Navok, and M. Chevion. 1983. The effect of a favism inducing agent on the in vitro development of *Plasmodium falciparum* in normal and glucose-6-phosphate dehydrogenase deficient erythrocytes. Blood 61:507–510. [346]

Golubic, S. 1992a. Microbial mats of Abu Dhabi. In *Environmental Evolution: Effects of the Origin and Evolution of Life on Planet Earth*, edited by L. Margulis and L. Olendzenski. Cambridge, MA: MIT Press. [55]

———. 1992b. Stromatolites of Shark Bay. In *Environmental Evolution: Effects of the Origin and Evolution of Life on Planet Earth*, edited by L. Margulis and L. Olendzenski. Cambridge, MA: MIT Press. [55]

Gomulkiewicz, R. S. 1998. Game theory, optimization, and quantitative genetics. Pages 283–303 in *Game Theory and Animal Behavior*, edited by L. A. Dugatkin and H. K. Reeve. Oxford: Oxford University Press. [122]

Gomulkiewicz, R. S., and A. Hastings. 1990. Ploidy and evolution by sexual selection: A comparison of haploid and diploid female choice models near fixation equilibria. Evolution 44:757–770. [138]

Gomulkiewicz, R. S., J. N. Thompson, R. D. Holt, S. L. Nuismer, and M. E. Hochberg. 2000. Hot spots, cold spots, and the geographic mosaic theory of coevolution. American Naturalist. 156:156–174. [129]

Gould, S. J., and R. C. Lewontin. 1979. The spandrels of San Marco and the Panglossian paradigm. A critique of the adaptationist programme. Proceedings of the Royal Society of London, Series B 205:581–598. [29]

Gould, S. J., and E. S. Vrba. 1982. Exaptation—A missing terms in the science of form. Paleobiology 8:4–15. [46, 290]

Grant, P. R. 1986. *Ecology and Evolution of Darwin's finches*. Princeton, NJ: Princeton University Press. [263]

Grant, P. R., and B. R. Grant. 1995a. The founding of a new population of Darwin's finches. Evolution 49:229–240. [233, 248]

———. 1995b. Predicting microevolutionary responses to directional selection on heritable variation. Evolution 49:241–251. [233, 248]

Greenland, D. J., and J.M.L. Kowal. 1960. Nutrient content of a moist tropical forest of Ghana. Plant Soil 12:154–174. [318, 319]

Griffiths, P. E., and R. D. Gray. 2001. Darwinism and developmental systems. In *Cycles of Contingency: Developmental Systems and Evolution*, edited by S. Oyama, P. E. Griffiths, and R. D. Gray. Cambridge, MA: MIT Press. [30, 281]

Grimm, N. B. 1995. Why link species and ecosystems? A perspective from ecosystem ecology. In *Linking Species and Ecosystems*, edited by C. G. Jones and J. H. Lawton. New York: Chapman and Hall. [201]

Grinnell, J. 1917. The niche-relationship of the California Thrasher. Auk 34:427–433. [37, 38, 43]

———. 1924. Geography and evolution. Ecology 5:225–229. [37–38, 43]

———. 1928. Presence and absence of animals. University of California Chronicle 30:429–450. [37]

Grover, J. D. 1997. *Resource Competition*. New York: Chapman and Hall. [39, 118]

Grun, P. 1976. *Cytoplasmic Inheritance and Evolution*. New York: Columbia University Press. [126]

Guglielmino, C. R., C. Viganotti, B. Hewlett, and L. L. Cavall-Sforza. 1995. Cultural variation in Africa: Role of mechanisms of transmission and adaptation. Proceedings of the National Academy of Sciences, USA 92:7585–7589. [357, 368]

Gullan, P. J., and P. S. Cranston. 1994. *The Insects: An Outline of Entomology*. London: Chapman and Hall. [20, 61, 65, 69, 70, 81, 85, 88, 93, 94, 99, 101–9 passim]

Gurney, W.S.C., and J. H. Lawton. 1996. The population dynamics of ecosystem engineers. Oikos 76:273–283. [25, 165, 213, 236, 306–7, 310, 313]

Hacker, S. D., and M. D. Bertness. 1995. Morphological and physiological consequences of a positive plant interaction. Ecology 76:2165–2175. [216]

Hacker, S. D., and S. D. Gaines. 1997. Some implications of direct positive interactions for community species diversity. Ecology 78:1990–2003. [216]

Hagen, J. B. 1989. Research perspectives and the anomalous status of modern ecology. Biology and Philosophy 4:433–455. [201]

———. 1992. *An Entangled Bank: The Origins of Ecosystem Ecology*. New Brunswick, NJ: Rutgers University Press. [201]

Haldane, J.B.S., and S. D. Jayakar. 1963. Polymorphism due to selection of varying direction. Journal of Genetics 58:237–242. [130]

Hamilton, W. D. 1964. The genetical evolution of social behaviour I. Journal of Theoretical Biology 7:1–16. [298]

———. 1967. Extraordinary sex ratios. Science 156:477–478. [122]

———. 1971. Geometry for the selfish herd. Journal of Theoretical Biology 31:295–311. [297]

Hammerstein, P. 1996. Darwinian adaptation, population genetics and the streetcar theory of evolution. Journal of Mathematical Biology 34:511–532. [122]

Hanney, P. W. 1975. *Rodents: Their Lives and Habits*. London: David and Charles. [77, 109]

Hansell, M. H. 1984. *Animal Architecture and Building Behaviour*. New York: Longmans. [61–67 passim, 75–101 passim, 105, 106, 108, 112, 215, 283]

———. 1987. Nest building as a facilitating and limiting factor in the evolution of eusociality in the Hymenoptera. Pages 155–181 in *Oxford Surveys in Evolutionary Biology*, vol. 4, edited by P. H. Harvey and L. Partridge. Oxford: Oxford University Press. [287]

———. 1993. The ecological impact of animal nests and burrows. Functional Ecology 7:5–12. [109, 215, 223, 237, 287]

———. 1996. Wasps make nests: Nests make conditions. In *Natural History and Evolution of Paper-Wasps*, edited by S. Turillazzi and M. J. West-Eberhard. New York: Oxford University Press. [287]

Hanski, I. 1999. *Metapopulation Ecology*. New York: Oxford University Press. [121, 310]

Hanski, I., and M. C. Singer. 2001. Extinction-colonization and host-plant choice in butterfly metapopulations. American Naturalist 158:341–353. [123]

Hartl, D. L., and A. G. Clark. 1989. *Principles of Population Genetics*. 2nd ed. Sunderland, MA: Sinauer. [131]

Hartl, D. L., and R. D. Cook. 1973. Balanced polymorphisms of quasineutral alleles. Theoretical Population Biology 4:163–172. [130]

Harvey, P. H., and M. D. Pagel. 1991. *The Comparative Method in Evolutionary Biology*. Oxford: Oxford University Press. [114, 286, 287, 345]

Hastings, A. 1981. Simultaneous stability of $D = 0$ and $D \neq 0$ for multiplicative viabilities at two loci: An analytical study. Journal of Theoretical Biology 89:69–81. [34]

Hayes, M.H.B. 1983. Darwin's "vegetable mold" and some modern concepts of humus structure and soil aggregation. In *Earthworm Ecology: From Darwin to Vermiculture*, edited by J. E. Satchell. London: Chapman and Hall. [11, 291]

Healy, S. D., and J. R. Krebs. 1992. Food storing and the hippocampus in corvids: Amount and volume are correlated. Proceedings of the Royal Society of London, Series B 248:241–245. [288]

———. 1996. Food storing and the hippocampus Paridae. Brain, Behavior and Evolution 47:195–199. [288]

Heesterbeek, J.A.P., and M. G. Roberts. 1995. Mathematical models for microparasites of wildlife. Pages 90–122 in *Ecology of Infectious Diseases in Natural Populations*, edited by B. T. Grenfell and A. P. Dobson. Cambridge, England: Cambridge University Press. [24, 124]

Henschel, J. R. 1995. Tool use by spiders: Stone selection and placement by corolla spiders *Ariadna* (Segestriidae) of the Namib desert. Ethology 101:187–199. [81, 85]

Hewlett, B. S., and L. L. Cavalli-Sforza. 1986. Cultural transmission among Aka pygmies. American Anthropologist 88:922–934. [356, 357]

Heyes, C., and L. Huber, eds. 2000. *The Evolution of Cognition*. Cambridge, MA: MIT Press. [281]

Hibbett, D. S., L. B. Gilbert, and M. J. Donoghue. 2000. Evolutionary instability of ectomycorrhizal symbioses in basidiomycetes. Nature 407: 506–508. [57]

Hill, W. G., and H. Caballero. 1990. Artificial selection experiments. Annual Review of Ecology and Systematics 23:287–310. [233]

Hinde, R. A., and J. Fisher. 1951. Further observations on the opening of milk bottles by birds. British Birds 44:393–396. [22]

Hinde, R. A., and J. Stevenson-Hinde. 1973. *Constraints on Learning*. New York: Academic. [257]

Hoekstra, R. F., R. Bijlsma, and A. J. Dolman. 1985. Polymorphism from environmental heterogeneity: Models are only robust if the heterozygote is close in fitness to the favoured homozygote in each environment. Genetical Research 45:299–314. [121, 123]

Hofbauer, J., and K. Sigmund. 1998. *Evolutionary Games and Replicator Dynamics*. Cambridge, England: Cambridge University Press. [122]

Holden, C., and R. Mace. 1997. Phylogenetic analysis of the evolution of lactose digestion in adults. Human Biology 69:605–628. [27, 342, 343]

Holland, H. D. 1995. Atmospheric oxygen in the biosphere. In *Linking Species and Ecosystems*, edited by C. G. Jones and J. H. Lawton. New York: Chapman and Hall. [54, 58]

Holland, J. H. 1992. *Adaptation in Natural and Artificial Systems*. Cambridge, MA: MIT Press. [15, 178, 183, 253]

———. 1995. *Hidden Order: How Adaptation Builds Complexity*. New York: Addison-Wesley. [15, 178, 183, 332, 333]

Holland, J. H., K. J. Holyoak, R. E. Nisbett, and P. R. Thagard. 1986. *Induction: Processes of Inference Learning and Discovery*. Cambridge, MA: MIT Press.

Hölldobler, B., and E. O. Wilson. 1994. *Journey to the Ants: A Story of Scientific Exploration*. Cambridge, MA: Belknap Press of Harvard University Press. [3, 4, 61, 62, 64, 81, 86, 87, 88, 92, 94, 105, 106, 107, 108, 190]

Holling, C. S. 1992. Cross-scale morphology, geometry and dynamics of ecosystems. Ecological Monographs 62:447–502. [58, 216]

Holloway, R. L. 1981. Culture, symbols and human brain evolution: A synthesis. Dialectical Anthropologist 5:287–303. [242]

Hollox, E. J., M. Poulter, M. Zvarik, V. Ferak, A. Krause, T. Jenkins, N. Saha, A. I. Kozolov, and D. M. Swallow. 2001. Lactase haplotype diversity in the Old World. American Journal of Human Genetics 68:160–172. [342]

Holt, R. D. 1977. Predation, apparent competition, and the structure of prey communities. Theoretical Population Biology 12:197–229. [125]

———. 1985. Density-independent mortality, nonlinear competitive interactions, and species coexistence. Journal of Theoretical Biology 116:479–493. [24, 119]

———. 1987. Population dynamics and evolutionary processes: The manifold roles of habitat selection. Evolutionary Ecology 1:331–347. [124]

———. 1995. Linking species and ecosystems: Where's Darwin? In *Linking Species and Ecosystems*, edited by C. G. Jones and J. H. Lawton. New York: Chapman and Hall. [195, 305, 379]

———. 1996. Demographic constraints in evolution: Towards unifying the evolutionary theories of senescence and niche conservatism. Evolutionary Ecology 10:1–11. [129, 130, 197]

Holt, R. D., and M. S. Gaines. 1992. Analysis of adaptation in heterogeneous landscapes: Implications for the evolution of fundamental niches. Evolutionary Ecology 6:433–447. [30, 124, 129, 130]

Holt, R. D., J. Grover, and D. Tilman. 1994. Simple rules for interspecific dominance in systems with exploitative and apparent competition. American Naturalist 144:741–771. [24, 119, 219]

Houghton, R. A., D. L. Skole, C. A. Nobre, J. L. Hackler, K. T. Lawrence, and W. H. Chomentowski. 2000. Annual fluxes of carbon from deforestation and regrowth in the Brazilian Amazon. Nature 403:301–304. [58]

Hrdy, S. B. 1999. *Mother Nature: Natural Selection and the Female of the Species*. London: Chatto and Windus. [347]

Hull, D. L., R. E. Langman, and S. S. Glenn. 2001. A general account of selection: Biology, immunology, and behavior. Behavioral and Brain Sciences 24:511–573. [180, 183, 186, 254]

Hutchinson, G. E. 1944. Limnological studies in Connecticut. VII. A critical examination of the supposed relationship between phytoplankton periodicity and chemical changes in later waters. Ecology 25:3–26. [37, 38, 39, 40]

———. 1948. Circular causal systems in ecology. Annals of the New York Academy of Sciences 50:221–246. [202]

———. 1957. Concluding remarks. Cold Spring Harbor Symposia on Quantitative Biology 22:415–427. [37, 38]

———. 1978. *An Introduction to Population Ecology*. New Haven, CT: Yale University Press. [37]

Huxley, J. S. 1942. *Evolution, the Modern Synthesis*. London: Allen and Unwin. [29]

Iwasa, Y., M. Nakamuru, and S. A. Levin. 1998. Allelopathy of bacteria in a lattice population: Competition between colicin-sensitive and colicin-producing strains. Evolutionary Ecology 12:785–802. [54]

Jablonka, E. 2001. The systems of inheritance. In *Cycles of Contingency: Developmental Systems and Evolution*, edited by S. Oyama, P. E. Griffiths, and R. D Gray. Cambridge, MA: MIT Press. [30]

Jackson, F. 1996. The coevolutionary relationship of humans and domesticated plants. Physical Anthropology 39:161–176. [251, 345, 346]

Jaenike, J. 1982. Environmental modification of oviposition behaviour in *Drosophila*. American Naturalist 119:784–802. [123, 124]

Jaenike, J., and R. D. Holt. 1991. Genetic variation for habitat preference: Evidence and explanations. American Naturalist 137 Supplement: S67–S90. [304]

James, R., C. Kleanhous, and G. R. Moore. 1996. The biology of E colicins: Paradigms and paradoxes. Microbiology 142:1569–1580. [53]

Jaynes, E. T. 1996. The Gibbs paradox. In *Maximum Entropy and Bayesian Methods*, edited by C. R. Smith, G. J. Erickson, and P. O. Neudorfer. Dordrecht, Holland: Kluwer Academic. [172, 173, 182, 187]

Jennings, D. H., and G. Lysek. 1999. *Fungal Biology*. 2nd ed. Oxford: Bios. [56]

Johnson, L. 1995. The far-from-equilibrium ecological hinterlands. In *Complex Ecology: The Part-Whole Relation in Ecosystems*, edited by B. C. Patten and S. E. Jorgensen. Englewood Cliffs, NJ: Prentice Hall. [168]

Johnstone, R. A. 1995. Sexual selection, honest advertisement and the handicap principle: Reviewing the evidence. Biological Reviews 70:1–65. [122]

———. 1998. Game theory and communication. Pages 94–117 in *Game Theory and Animal Behavior*, edited by L. A. Dugatkin and H. K. Reeve. Oxford: Oxford University Press. [122]

Jones, C. G., and J. H. Lawton, eds. 1995. *Linking Species and Ecosystems*. New York: Chapman and Hall. [6, 43, 195, 201, 224, 229]

Jones, C. G., J. H. Lawton, and M. Shachak. 1994. Organisms as ecosystem engineers. Oikos 69:373–386. [6, 25, 55, 58, 64, 67, 82, 83, 84, 90, 91, 105, 106, 107, 111, 112, 194, 213–216, 225, 236, 311, 326, 379]

———. 1997. Positive and negative effects of organisms as physical ecosystem engineers. Ecology 78:1946–1957. [6–7, 8, 25, 56, 64, 114, 194, 213–16, 225, 226, 236, 283, 286, 304, 317, 326, 379, 384]

Jones, C. G., and M. Shachak. 1990. Fertilisation of the desert soil by rock-eating snails. Nature 346:839–841. [214, 216]

Jones, J. S. 1980. Can genes choose habitats? Nature 286:757–758. [123]

Jones, J. S., and R. F. Probert. 1980. Habitat selection maintains a deleterious allele in a heterogeneous environment. Nature 287:632–633. [297]

Karlin, S. 1975. General two-locus selection models: Some objectives, results and interpretations. Theoretical Population Biology 7:364–398. [143]

Karlin, S., and M. W. Feldman. 1970. Linkage and selection: Two-locus symmetric viability model. Theoretical Population Biology 1:39–71. [127, 134, 143, 144, 399, 400, 402]

———. 1978. Simultaneous stability of $D = 0$ and $D \neq 0$ for multiplicative viabilities at two loci. Genetics 90:813–825. [143, 144, 163]

Karlin, S., and S. Lessard. 1986. *Theoretical Studies on Sex Ratio Evolution*. Princeton, NJ: Princeton University Press. [120]

Karlin, S., and U. Liberman. 1974. Random temporal variation in selection intensities: Case of large population size. Theoretical Population Biology 6:355–382. [130]

———. 1979. Central equilibria in multilocus systems. I. Generalized non-epistatic regimes. Genetics 91:777–798. [134]

Kauffman, S. A. 1993. *The Origins of Order: Self-Organisation and Selection in Evolution*. Oxford: Oxford University Press. [178]

Kaufman, K. 1996. *Lives of North American Birds*. New York: Houghton Mifflin. [75, 79]

Kawecki, T. J. 1995. Demography of source-sink populations and the evolution of ecological niches. Evolutionary Ecology 9:38–44. [124]

Kawecki, T. J., N. H. Barton, and J. D. Fry. 1997. Mutational collapse of fitness in marginal habitats and the evolution of ecological specialisation. Journal of Evolutionary Biology 10:407–429. [129, 130]

Keller, L. 1995. Social life: The paradox of multiple-queen colonies. Trends in Ecology and Evolution 10:355–360. [300]

Kendrick, W. B., and A. Burges. 1962. Biological aspects of the decay of *Pinus sylvestris* leaf litter. Nova Hedwiga 4:313–342. [56]

Kerr, B., R. A. Riley, M. W. Feldman, and B.J.M. Bohannan. 2002. Local dispersal promotes biodiversity in a real-life game of rock-paper-scissors. Nature 418:171–174. [54, 119, 311, 324]

Kerr, B., D. W. Schwilk, A. Bergman, and M. W. Feldman. 1999. Rekindling an old flame: A haploid model for the evolution and impact of flammability in resprouting plants. Evolutionary Ecology Research 1:807–833. [56, 60, 234, 311]

Kershaw, A. P. 1986. Climatic change and Aboriginal burning in north-east Australia during the last two glacial/interglacial cycles. Nature 322: 47–49. [347]

Kiene, R. P. 1999. Sulphur in the mix. Nature 402:363–364. [55]

Kimura, M. 1954. Process leading to quasi-fixation of genes in natural populations due to random fluctuations of selection intensities. Genetics 39:280–295. [130]

King, A. W., and S. L. Pimm. 1983. Complexity and stability: A reconciliation of theoretical and experimental results. American Naturalist 122: 229–239. [328]

Kingsolver, J. G., H. E. Hoekstra, J. M. Hoekstra, D. Berrigan, S. N. Vignieri, C. E. Hill, A. Hoang, P. Gilbert, and P. Beerli. 2001. The strength of phenotypic selection in natural populations. American Naturalist 157:245–261. [233, 234, 249]

Kinzig, A. P., S. W. Pacala, and D. Tilman, eds. 2001. *The Functional Consequences of Biodiversity*. Princeton, NJ: Princeton University Press. [320, 328]

Kirkpatrick, M. 1982. Sexual selection and the evolution of female choice. Evolution 36:1–12. [138]

Kirkpatrick, M., and R. Lande. 1989. The evolution of maternal characters. Evolution 43:485–503. [10, 21, 68, 125, 126, 128, 161, 284]

Kitching, R. L. 1971. An ecological study of water-filled tree-holes and their position in the woodland ecosystem. Journal of Animal Ecology 40: 281–302. [56]

Klein, R. G. 1999. *The Human Career: Human Biological and Cultural Origins*. 2nd ed. Chicago: University of Chicago Press. [242, 249, 251, 340, 346, 347, 348, 350]

Krebs, J. R. and N. B. Davies. 1993. *An Introduction to Behavioural Ecology*. 3rd ed. Oxford: Blackwell. [11, 109]

Krebs, J. R., A. Kacelnick, and P. Taylor. 1978. Test of optimal sampling by foraging great tits. Nature 275:27–31. [124]

Kretzschmar, A. 1983. Soil transport as a homeostatic mechanism for stabilizing the earthworm environment. In *Earthworm Ecology: From Darwin to Vermiculture*, edited by J. E. Satchell. London: Chapman and Hall. [11]

Kruger, O., and N. B. Davies. 2001. The evolution of cuckoo parasitism: A comparative analysis. Proceedings of the Royal Society of London, Series B 269:375–381. [45]

Kumm, J., K. N. Laland, and M. W. Feldman.1994. Gene-culture coevolution and sex ratios: The effects of infanticide, sex-selective abortion, sex selection, and sex-biased parental investment on the evolution of sex ratios. Theoretical Population Biology 46:249–278. [247, 248, 249]

Kuper, A. 1988. *The Invention of Primitive Society*. London: Routledge. [241]

Lack, D. 1954. *The Natural Regulation of Animal Numbers*. Oxford: Oxford University Press. [310]

Laing, R. 1989. Artificial organisms: History, problems, directions. In *Artificial Life*, edited by C. G. Langton. Redwood City, CA: Addison-Wesley. [172]

Laland, K. N. 1994. Sexual selection with a culturally transmitted mating preference. Theoretical Population Biology 45:1–15. [247, 249]

Laland, K. N., and G. Brown. 2002. *Sense and Nonsense: Evolutionary Perspectives on Human Behaviour*. Oxford: Oxford University Press. [239, 241]

Laland, K. N., J. Kumm, J. D. Van Horn, and M. W. Feldman. 1995. A gene-culture model of handedness. Behavior Genetics 25:433–445. [247, 248]

Laland, K. N., F. J. Odling-Smee, and M. W. Feldman. 1996. On the evolutionary consequences of niche construction. Journal of Evolutionary Biology 9:293–316. [21, 68, 113, 114, 117, 127, 128, 138, 141, 142, 145, 146, 278, 413]

———. 1999. Evolutionary consequences of niche construction and their implications for ecology. Proceedings of the National Academy of Sciences, USA 96:10242–10247. [26, 68, 114, 117, 128, 155, 158, 276, 278, 413]

———. 2000a. Group selection: A niche construction perspective. Journal of Consciousness Studies 7:221–224. [359, 362]

———. 2000b. Niche construction, biological evolution, and cultural change. Behavioral and Brain Sciences 23:131–175. [14, 113, 242, 245, 251, 253, 257, 264, 298, 301, 359]

———. 2001. Cultural niche construction and human evolution. Journal of Evolutionary Biology 14:22–33. [265, 383, 413]

Laland, K. N., P. J. Richerson, and R. Boyd. 1993. Animal social learning: Towards a new theoretical approach. Pages 249–277 in *Perspectives in Ethology 10: Behavior and Evolution*. New York: Plenum. [356]

———. 1996. Developing a theory of animal social learning. In *Social Learning in Animals: The Roots of Culture*, edited by C. M. Heyes and B. G. Galef, Jr. New York: Academic. [355]

Lansing, J. S., J. N. Kremer, and B. B. Smuts. 1998. System-dependent selection, ecological feedback and the emergence of functional structure in ecosystems. Journal of Theoretical Biology 192:377–391. [351]

Law, R. 1985. Evolution in a mutualistic environment. In *The Biology of Mutualism*, edited by D. H. Boucher. London: Croom Helm. [57, 110]

———. 1999. Theoretical aspects of community assembly. In *Advanced Ecological Theory*, edited by J. McGlade. Oxford: Blackwell. [320]

Lawton, J. H., and C. G. Jones. 1995. Linking species and ecosystems: Organisms as ecosystem engineers. In *Linking Species and Ecosystems*, edited by C. G. Jones and J. H. Lawton. New York: Chapman and Hall. [202]

Lawton, J. H., and R. M. May. 1995. *Extinction Rates*. Oxford: Oxford University Press. [369]

Layzer, D. 1988. Growth of order in the universe. In *Entropy, Information, and Evolution: New Perspectives on Physical and Biological Evolution*, edited by B. W. Weber, D. J. Depew, and J. D. Smith. Cambridge, MA: MIT Press. [71, 177]

Lee, K. E. 1985. *Earthworms: Their Ecology and Relation with Soil and Land Use*. London: Academic. [11, 111, 291]

Leibold, M. A. 1995. The niche concept revisited: Mechanistic models and community context. Ecology 76:1371–1382. [37, 38, 197n]

Leith, H. 1973. Primary production: Terrestrial ecosystem. Human Ecology 1:303–332. [320]

———. 1975. Primary productivity of the major vegetal units of the world. Pages 201–215 in *Primary Productivity of the Biosphere*, edited by H. Leith and R. H. Whittaker. Berlin: Springer-Verlag. [320]

Lenton, T. M. 1998. Gaia and natural selection. Nature 394:439–447. [189]

Levene, H. 1953. Genetic equilibrium when more than one niche is available. American Naturalist 87:331–333. [121, 123]

Levin, B. R. 1988. Frequency-dependent selection in bacterial populations. Philosophical Transactions of the Royal Society of London, Series B 319:459–472. [53, 54]

Levin, S. A. 1989. Challenges in the development of a theory of community and ecosystem structure and function. In *Perspectives in Ecological Theory*, edited by J. Roughgarden, R. M. May, and S. A. Levin. Princeton, NJ: Princeton University Press. [201]

Levine, S. H. 1976. Competitive interactions in ecosystems. American Naturalist 110:903–910. [119]

Levins, R. and R. C. Lewontin. 1985. *The Dialectical Biologist*. Cambridge, MA: Harvard University Press. [30]

Lewin, R. 1998. *Principles of Human Evolution: A Core Textbook*. Oxford: Blackwell. [350]

Lewontin, R. C. 1982. Organism and environment. In *Learning, Development and Culture*, edited by H. C. Plotkin. New York: Wiley. [375]

———. 1983. Gene, organism, and environment. In *Evolution from Molecules to Men*, edited by D. S. Bendall. Cambridge, England: Cambridge University Press. [16, 17, 18, 19, 42, 117, 162, 164, 168, 186, 372, 373, 374, 375, 385]

———. 2000. *The Triple Helix. Gene, Organism, and Environment.* Cambridge, MA: Harvard University Press. [17, 162, 283, 375]

Lewontin, R. C., and D. Cohen. 1969. On population growth in a randomly varying environment. Proceedings of the National Academy of Sciences, USA 62:1056–1066. [130]

Lewontin, R. C., and K. Kojima 1960. The evolutionary dynamics of complex polymorphisms. Evolution 14:458–472. [144]

Lloyd, S. 1999. Rolf Landauer (1927–99): Head and heart of the physics of information. Nature 400:720. [186]

Loeschcke, V., and F. B. Christiansen. 1984. Evolution and intraspecific exploitative competition II. A two-locus model for additive gene effects. Theoretical Population Biology 26:228–264. [121]

Lovelock, J. E. 1979. *Gaia: A New Look at Life on Earth.* Oxford: Oxford University Press. [54, 334]

———. 1988. *The Ages of Gaia. A Biography of Our Living Earth.* Oxford: Oxford University Press. [334]

Lumsden, C. J., and E. O. Wilson. 1981. *Genes, Mind and Culture.* Cambridge, MA: Harvard University Press. [245]

Lyam, C. P., J. S. Willis, A. Malan, and L.H.C. Wang. 1982. *Hibernation and Torpor in Animals.* London: Academic. [63]

MacArthur, R. H., and R. Levins. 1967. The limiting similarity, convergence and divergence of coexisting species. American Naturalist 101:377–385. [39, 135]

MacNally, R. C. 1995. *Ecological Versatility and Community Ecology.* Cambridge, England: Cambridge University Press. [149]

McCann, K. S. 2000. The diversity-stability debate. Nature 405:228–233. [383]

McFarland, D. 1987. *The Oxford Companion to Animal Behaviour.* Oxford: Oxford University Press. [67, 104]

McGlade. J. M., ed. 1999a. *Advanced Ecological Theory.* Oxford: Blackwell. [327]

———. 1999b. Ecosystem analysis and the governance of natural resources. In *Advanced Ecological Theory*, edited by J. McGlade. Oxford: Blackwell. [383]

McLennan, D. A., D. R. Brooks, and J. D. McPhail. 1988. The benefits of communication between comparative ethology and phylogenetic systematics. A case study using gasterosteid fishes. Canadian Journal of Zoology 66:2177–2190. [99, 114, 287]

McNaughton, S. J. 1985. Ecology of a grazing ecosystem: The Serengeti. Ecological Monographs 55:259–294. [328]

McNaughton, S. J., and M. B. Coughenour. 1981. The cybernetic nature of ecosystems. American Naturalist 117:985–990. [330]

Margalef, R. 1963. On certain unifying principles in ecology. American Naturalist 97:357–374. [330]

———. 1995. Information theory and complex ecology. Pages 40–50 in *Complex Ecology: The Part-Whole Relation in Ecosystems*, edited by B. C. Patten and S. E. Jorgensen. Englewood Cliffs, NJ: Prentice Hall. [330]

Margules, C. R., and R. L. Pressey. 2000. Systematic conservation planning. Nature 405:243–253. [383]

Marrow, P., U. Dieckmann, and R. Law. 1996. Evolutionary dynamics of predator-prey systems: An ecological perspective. Journal of Mathematical Biology 34:556–578. [194]

Matessi, C., and S. D. Jayakar. 1976. Models of density-frequency dependent selection for exploitation of resources. Pages 707–21 in *Population Genetics and Ecology*, edited by S. Karlin and E. Nevo. New York: Academic. [121]

Mathews, R. W., and J. R. Mathews. 1978. *Insect Behavior*. New York: Wiley. [62, 89]

May, R. M. 1973. *Stability and Complexity in Model Ecosystems*. Princeton, NJ: Princeton University Press. [39, 118]

———. 1974. *Stability and Complexity in Model Ecosystems*. 2nd ed. Princeton, NJ: Princeton University Press. [39, 328]

———, ed. 1981. *Theoretical Ecology: Principles and Applications*. 2nd ed. Oxford: Blackwell. [118, 306]

May, R. M., and R. M. Anderson. 1983. Parasite-host coevolution. Pages 186–206 in *Coevolution*, edited by D. J. Futuyma and M. Slatkin. Sunderland, MA: Sinauer. [124]

———. 1990. Parasite-host co-evolution. Parasitology 100:S89–S101. [124]

Maynard Smith, J. 1962. Disruptive selection, polymorphism, and sympatric speciation. Nature 195:60–62. [124]

———. 1966. Sympatric speciation. American Naturalist 107:171–198. [123]

———. 1974. *Models in Ecology*. Cambridge, England: Cambridge University Press. [306]

———. 1982. *Evolution and the Theory of Games*. Cambridge, England: Cambridge University Press. [122]

———. 1989. *Evolutionary Genetics*. Oxford: Oxford University Press. [121]

Maynard Smith, J., and G. R. Price. 1973. The logic of animal conflict. Nature 246:15–18. [122]

Maynard Smith, J., and E. Szathmary. 2000. *The Origins of Life: From the Birth of Life to the Origins of Language*. Oxford: Oxford University Press. [182, 200]

Mayr, E. 1963. *Animal Species and Evolution.* Cambridge, MA: Harvard University Press. [28–29]

———. 1982. *The Growth of Biological Thought: Diversity, Evolution, and Inheritance.* Cambridge, MA: Harvard University Press. [243]

Mazancourt, C. de, M. Loreau, and U. Dieckmann. 2001. Can the evolution of plant defense lead to plant-herbivore mutualism? American Naturalist 158:109–123. [329]

Michel, G. F., and C. L. Moore. 1995. *Developmental Psychobiology: An Interdisciplinary Science.* Cambridge, MA: MIT Press. [30]

Michod, R. E. 1999. *Darwinian Dynamics: Evolutionary Transitions in Fitness and Individuality.* Princeton, NJ: Princeton University Press. [184]

Milinski, M., and G. A. Parker. 1991. Competition for resources. In *Behavioural Ecology: An Evolutionary Approach,* edited by J. R. Krebs and N. B. Davies. 3rd ed. Oxford: Blackwell. [169]

Milton, K. 1991. Comparative aspects of diet in Amazonian forest-dwellers. Philosophical Transactions of the Royal Society of London, Series B 334:253–263. [350]

Montagu, A. 1980. *Sociobiology Examined.* Oxford: Oxford University Press. [346, 367]

Moore, A. J., E. D. Brodie III, and J. B. Wolf. 1997. Interacting phenotypes and the evolutionary process: I. Direct and indirect genetic effects of social interactions. Evolution 51:1352–1362. [30, 128, 284]

Moore, A. J., J. B. Wolf, and E. D. Brodie III. 1998. The influence of direct and indirect genetic effects on the evolution of behaviour: Social and sexual selection meet maternal effects. Pages 22–41 in *Maternal Effects as Adaptations,* edited by T. A. Mousseau and C. W. Fox. New York: Oxford University Press. [128]

Mora, C. I., S. G. Driese, and L. A. Colarusso. 1996. Middle to late Paleozoic atmospheric CO_2 levels from soil carbonate and organic matter. Science 271:1105–1107. [58]

Moran, P.A.P. 1964. On the nonexistence of adaptive topographies. Annals of Human Genetics 27:383–393. [143, 400]

Morris, D. W. 1994. Habitat matching: Alternatives and implications to populations and communities. Evolutionary Ecology 4:387–406. [123]

Mount, A. B. 1964. The interdependence of the eucalypts and forest fires in southern Australia. Australian Forestry 28:166–172. [59]

Mousseau, T. A., and H. Dingle. 1991. Maternal effects in infant life histories. Annual Review of Entomology 35:511–534. [20, 125]

Mousseau, T. A., and C. W. Fox. 1998a. The adaptive significance of maternal effects. Trends in Ecology and Evolution 13:403–407. [10, 11, 20, 21, 125, 126, 284]

———, eds. 1998b. *Maternal Effects as Adaptations.* New York: Oxford University Press. [10, 21, 125, 126, 283, 284]

Mutch, R. W. 1970. Wildland fires and ecosystems—a hypothesis. Ecology 51:1046–1051. [59]

Naeem, S. 2001. Autotrophic-heterotrophic interactions and their impacts on biodiversity and ecosystem functioning. In *The Functional Consequences of Biodiversity*, edited by A. P. Kinzig, S. W. Pacala, and D. Tilman. Princeton, NJ: Princeton University Press. [329]

———. 2002. Biodiversity equals instability? Nature 416:23–24. [329]

Naeem, S., and S. Li. 1997. Biodiversity enhances ecosystem reliability. Nature 390:507–509. [324]

Naiman, R. J. 1988. Animal influences on ecosystem dynamics. BioScience 38:750–752. [213]

Naiman, R. J., C. A. Johnston, and J. C. Kelley. 1988. Alteration of North American streams by beaver. BioScience 38:753–762. [213]

Noirot, C. 1970. The nests of termites. In *Biology of Termites*, edited by K. Krishma and F. M. Weesner. New York: Academic. [87]

Nowak, R. M. 1991. *Walker's Mammals of the World*. 5th ed. Baltimore, MD: John Hopkins University Press. [62, 63, 64, 67, 77, 83, 84, 87, 91, 92, 93, 109, 110]

O'Brien, W. J. 1974. The dynamics of nutrient limitation of phytoplankton: A model reconsidered. Ecology 55:135–141. [119]

Odling-Smee, F. J. 1983. Multiple levels in evolution: An approach to the nature-nurture issue via "applied epistemology." Pages 135–158 in *Animal Models of Human Behaviour*, edited by G.C.L. Davey. Chichester, England: John Wiley. [180, 254]

———. 1988. Niche constructing phenotypes. Pages 73–132 in *The Role of Behavior in Evolution*, edited by H. C. Plotkin. Cambridge, MA: MIT Press. [1, 13, 168, 178, 180]

Odling-Smee, F. J., K. N. Laland, and M. W. Feldman. 1996. Niche construction. American Naturalist 147:641–648. [13, 113–14]

Odum, E. P. 1971. *Fundamentals of Ecology*. Philadelphia: Saunders. [202]

———. 1989. *Ecology and Our Endangered Life-Support Systems*. Sunderland, MA: Sinauer. [6, 40, 54, 202]

Odum, H. T. 1988. Self-organization, transformity and information. Science 242:1132–1139. [330]

O'Neill, R. V., D. L. DeAngelis, J. B. Waide, and T.F.H. Allen. 1986. *A Hierarchical Concept of Ecosystems*. Princeton, NJ: Princeton University Press. [6, 26, 43, 44, 168, 189, 196–205 passim, 327, 379]

Otto, S. P., F. B. Christiansen, and M. W. Feldman. 1995. Genetics and cultural inheritance of continuous traits. Morrison Institute for Population and Resource Studies Working Paper 64, Stanford University, Stanford, CA. [248]

Ovington, J. D. 1962. Quantitative ecology and the woodland ecosystem concept. Advances in Ecological Research 1:103–192. [318, 319]

Oyama, S., P. E. Griffiths, and R. D. Gray. 2001. *Cycles of Contingency:*

Developmental Systems and Evolution. Cambridge, MA: MIT Press. [30, 126]

Pacala, S., and D. Tilman. 2001. The transition from sampling to complementarity. In *The Functional Consequences of Biodiversity,* edited by A. P. Kinzig, S. W. Pacala, and D. Tilman. Princeton, NJ: Princeton University Press. [328, 329]

Pagel, M. D. 1992. A method for the analysis of comparative data. Journal of Theoretical Biology 156:531–442. [343]

———. 1999. Inferring the historical patterns of biological evolution. Nature 401:877–884. [346]

Palumbi, S. R. 2001a. Humans as the world's greatest evolutionary force. Science 293:1786–1790. [233, 247, 369]

———. 2001b. *The Evolution Explosion.* New York: Norton. [233, 247, 369]

Parrish, J. K. 1998. Fish behavior. In *Encyclopedia of Fishes,* edited by J. R. Paxton and W. N. Eschmeyer. 2nd ed. New York: Academic. [99]

Passera, L., S. Aron, E. L. Vargo, and L. Keller, 2001. Queen control of sex ratios in fire ants. Science 293:1308–1310. [300]

Patten, B. C., and S. E. Jorgensen. 1995. *Complex Ecology: The Part-Whole Relation in Ecosystems.* Englewood Cliffs, NJ: Prentice-Hall. [6, 189, 202]

Patten, B. C., and E. P. Odum. 1981. The cybernetic nature of ecosystems. American Naturalist 118:886–895. [330]

Paxton, J. R., and W. N. Eschmeyer, eds. 1998. *Encyclopedia of Fishes.* 2nd ed. New York: Academic. [61, 65, 66, 67, 72, 95–104 passim, 108, 112]

Pearce, M. J. 1997. *Termites, Biology and Pest Management.* New York: CAB International. [61, 87, 92, 94]

Pfisterer, A. B., and B. Schmid. 2002. Diversity-dependent production can decrease the stability of ecosystem functioning. Nature 416:84–86. [328, 329]

Phillips, J. 1931. The biotic community. Journal of Ecology 19:1–24. [334]

———. 1934. Succession, development, the climax, and the complex organism: An analysis of concepts. Part I. Journal of Ecology 22:554–571. [334]

———. 1935. Succession, development, the climax, and the complex organism: An analysis of concepts. Part II. Development and the climax. Journal of Ecology 23:210–246. [334]

Phillips, P. C., S. P. Otto, and M. C. Whitlock. 2000. Beyond the average: The evolutionary importance of gene interactions and variability of epistatic effects. In *Epistasis and the Evolutionary Process,* edited by J. B. Wolf, E. D. Brodie III, and M. J. Wade. Oxford: Oxford University Press. [127]

Pimm, S. L. 1999. The dynamics of the flows of matter and energy. In *Advanced Ecological Theory,* edited by J. McGlade. Oxford: Blackwell. [320, 328]

Pinker, S. 1994. *The Language Instinct: The New Science of Language and Mind*. St. Ives, England: Allen Lane Penguin. [246]

Plotkin, H. C. 1996. Non-genetic transmission of information: Candidate cognitive processes and the evolution of culture. Behavioral Processes 35:207–213. [259]

———, ed. 1988. *The Role of Behavior in Evolution*. Cambridge, MA: MIT Press. [29, 30, 113, 166, 353, 354]

Plotkin, H. C., and F. J. Odling-Smee. 1981. A multiple-level model of evolution and its implications for sociobiology. Behavioral and Brain Sciences 4:225–268. [254, 257]

Polis, G. A., and D. R. Strong. 1996. Food web complexity and community dynamics. American Naturalist 147:813–846. [149]

Preston-Mafham, K., and R. Preston-Mafham. 1996. *The Natural History of Spiders*. Ramsbury, England: Crowood. [9, 61, 69, 70, 71, 78, 81]

Prigogine, I., and I. Stengers. 1984. *Order out of Chaos: Man's New Dialogue with Nature*. London: Bantam. [168, 177]

Purvis, A., and A. Hector. 2000. Getting the measure of biodiversity. Nature 405:212–219. [383]

Qvamstrom, A., and E. Forsgren. 1998. Should females prefer dominant males? Trends in Ecology and Evolution 13:498–501. [100]

Rasmussen, J. A., and E. L. Rice. 1971. Allelopathic effects of *Sporobolus pyramidatus* on vegetational patterning. American Midland Naturalist 86:309–326. [59]

Rausher, M. D. 1984. The evolution of habitat preference in subdivided populations. Evolution 38:596–608. [123]

Reader, S. M., and K. N. Laland. 2002. Social intelligence, innovation and enhanced brain size in primates. Proceedings of the National Academy of Sciences, USA 99:4436–4441. [353]

Reid, R. P., P. T. Visscher, A. W. Decho, J. F. Stolz, B. M. Bebout, C. Du-Praz, I. G. MacIntyre, H. W. Paerl, J. L. Pickney, L. Prufert-Bebout, T. F. Steppe, and D. J. Desmarais. 2000. The role of microbes in accretion, lamination and early lithification of modern marine stromatolites. Nature 406:989–992. [55]

Reiners, W. R. 1986. Complementary models for ecosystems. American Naturalist 127:59–73. [7, 189, 195, 202, 236, 237, 238, 323]

Reznick, D. N., F. H. Shaw, H. Rodd, and R. G. Shaw. 1997. Evaluation of the rate of evolution in natural populations of guppies (*Poecilia reticulata*). Science 275:1934–1936. [233, 248]

Rice, E. L. 1984. *Allelopathy*. 2nd ed. Orlando, FL: Academic. [59, 110]

Rice, S. H. 1998. The evolution of canalization and the breaking of Von Baer's laws: Modeling the evolution of development with epistasis. Evolution 52:647–656. [127]

Richards, B. N. 1974. *Introduction to the Soil Ecosystem*. Harlow, England: Longmans. [56]

Ricklefs, R. E., and G. L. Miller. 2000. *Ecology*. 4th ed. New York: Freeman. [309, 317, 318, 319, 320]

Riebesell, J. 1974. Paradox of enrichment: Destabilization of exploitation ecosystems in ecological time. Science 171:385–387. [119]

Roach, D., and R. Wulf. 1987. Maternal effects in plants. Annual Review of Ecology and Systematics 18:209–236. [20, 125]

Robertson, D. S. 1991. Feedback theory and Darwinian evolution. Journal of Theoretical Biology 152:469–484. [12, 21, 68, 131, 376]

Robson Brown, K. 2000. The meaning of hominid species—culture as process and product? Behavioral and Brain Sciences 23:157. [281]

Rogers, E. M. 1995. *Diffusion of Innovations*. 4th ed. New York: Free Press. [355, 369]

Ronce, O., and M. Kirkpatrick. 2001. When sources become sinks: Migrational meltdown in heterogeneous habitats. Evolution 55:1520–1531. [124]

Rose, M. R., and G. V. Lauder. 1996. *Adaptation*. San Diego, CA: Academic. [41, 170]

Rosenzweig, M. L. 1981. A theory of habitat selection. Ecology 62:327–335. [123, 124]

———. 1987. Habitat selection as a source of biological diversity. Evolutionary Ecology 1:315–330. [123, 124]

———. 1991. Habitat selection and population interactions. American Naturalist 137 (Supplement):S5–S28. [123]

———. 1995. *Species Diversity in Space and Time*. Cambridge, England: Cambridge University Press. [6]

Rossiter, M. C. 1996. Incidence and consequences of inherited environmental effects. Annual Review of Ecology and Systematics 27:451–476. [60]

Roughgarden, J. 1971. Density-dependent natural selection. Ecology 52:453–468. [120]

———. 1976. Resource partitioning among competing species—a coevolutionary approach. Theoretical Population Biology 9:388–424. [121]

———. 1979. *Theory of Population Genetics and Evolutionary Ecology: An Introduction*. New York: Macmillan. [118, 306, 307, 328, 329]

———. 1998. *Primer of Ecological Theory*. Upper Saddle River, NJ: Prentice Hall. [120, 316, 317]

Roughgarden, J., R. M. May, and S. A. Levin. 1989. *Perspectives in Ecological Theory*. Princeton, NJ: Princeton University Press. [201]

Rowlett, R. M. 1999. Comments on R. W. Wrangham, J. H. Jones, G. Laden, D. Pilbeam, and N. Conklin Brittain, "The raw and the stolen: Cooking and the ecology of human origins." Current Anthropology 40:584–585. [344]

Rush, R. T., and J. A. McKenzie. 1987. Ecological genetics of insecticide and acaricide resistance. Annual Review of Entomology 32:361–380. [233]

Sahlins, M. D. 1976. *The Use and Abuse of Biology: An Anthropological Critique of Sociobiology*. Ann Arbor, MI: University of Michigan Press. [241, 246, 367]

Schlichting, C. D., and M. Pigliucci. 1998. *Phenotypic Evolution: A Reaction Norm Perspective*. Sunderland, MA: Sinauer. [170, 191, 381, 382, 383]

Schluter, D. 2001. Ecology and the origin of species. Trends in Ecology and Evolution 16:372–380. [302]

Schluter, D., and L. Gustafsson. 1993. Maternal inheritance of condition and clutch size in the Collared Flycatcher. Evolution 47:658–667. [10, 126]

Schmid, B., J. Joshi, and F. Schlapfer. 2001. Empirical evidence for biodiversity–ecosystem functioning relationships. In *The Functional Consequences of Biodiversity*, edited by A. P. Kinzig, S. W. Pacala, and D. Tilman. Princeton, NJ: Princeton University Press. [329]

Schmidt, R. S. 1964. *Apicotermes* nests. American Zoologist 4:221–225. [86]

Schnell, G. D. 1973. A reanalysis of nest structure in the weavers (Ploceinae) using numerical taxonomic techniques. Ibis 115:93–106. [85]

Schoener, T. 1974. Resource partitioning in ecological communities. Science 185:27–39. [135]

———. 1982. The controversy over interspecific competition. American Scientist 70:586–595. [39]

———. 1986. Mechanistic approaches to community ecology: A new reductionism? American Zoologist 49:667–685. [201]

———. 1989. The ecological niche. In *Ecological Concepts: The Contribution of Ecology to an Understanding of the Natural World*, edited by J. M. Cherrett. Oxford: Blackwell. [37, 38, 39]

Schrödinger, E. 1944. "What Is Life?" Lecture. [1992. *"What Is Life?" with "Mind and Matter" and "Autobiographical Sketches."* Cambridge, England: Cambridge University Press]. [28, 168, 171, 192]

Schwilk, D. W. 2002. Plant evolution in fire-prone environments. Pages 63–76 in doctoral dissertation, Stanford University, Stanford, CA. [342]

Scrimshaw, N. S., and E. B. Murray. 1988. The acceptability of milk and milk products in populations with a high prevalence of lactose intolerance. American Journal of Clinical Nutrition 48:1083–1159. [342]

Seger, J. 1985. Intraspecific resource competition as a cause of sympatric speciation. Pages 43–53 in *Evolution*, edited by P. J. Greenwood, P. H. Harvey, and M. Slatkin. Cambridge, England: Cambridge University Press. [303]

———. 1992. Evolution of exploiter-victim relationships. Pages 3–25 in *Natural Enemies: The Population Biology of Predators, Parasites, and Disease*, edited by M. J. Crawley. Oxford: Blackwell. [125, 303]

Segerstrale, U. 2000. *Defenders of the Truth: The Sociobiology Debate*. Oxford: Oxford University Press. [241, 367]

Seligman, M.E.P. 1970. On the generality of the laws of learning. Psychological Review 77:406–418. [257]

Shachak, M., and C. G. Jones. 1995. Ecological flow chains and ecological systems: Concepts for linking species and ecosystem perspectives. In *Linking Species and Ecosystems*, edited by C. G. Jones and J. H. Lawton. New York: Chapman and Hall. [8, 227–234 passim]

Shachak, M., C. G. Jones, and S. Brand. 1995. The role of animals in an arid ecosystem: Snails and isopods as controllers of soil formation, erosion and desalinization. Advances in GeoEcology 28:37–50. [216]

Shachak, M., C. G. Jones, and Y. Granot. 1987. Herbivory in rocks and weathering of a desert. Science 236:1098–1099. [7, 8, 214, 216]

Shannon, C. 1948. The mathematical theory of communication. Bell Systems Technical Journal 27:379–423. [181]

Shannon, C., and W. Weaver. 1949. *The Mathematical Theory of Communication*. Urbana, IL: Illinois University Press. [181]

Sharples, J. M., A. A. Meharg, S. M. Chambers, and J.W.G. Cairney. 2000. Symbiotic solution to arsenic contamination. Nature 404:951–952. [57, 58]

Sherry, D. F., and B. G. Galef. 1984. Cultural transmission without imitation: Milk bottle opening by birds. Animal Behaviour 32:937–938. [22]

Sherry, D., A. L. Vaccarino, K. Buckenham, and R. S. Herz. 1989. The hippocampal complex of food-storing birds. Brain, Behavior and Evolution 34:308–317. [288]

Shukla, J., C. Nobre, and P. Sellers. 1990. Amazon deforestation and climate change. Science 247:1322–1325. [58]

Sigmund, K. 1993. *Games of Life: Explorations in Ecology, Evolution, and Behaviour*. Oxford: Oxford University Press. [189]

Simo, R. 2001. Production of atmospheric sulphur by oceanic plankton: Biogeochemical, ecological and evolutionary links. Trends in Ecology and Evolution 16:287–294. [55]

Simonton, D. K. 2000. Human creativity, cultural evolution, and niche construction. Behavioral and Brain Sciences 23:159–160. [281]

Simpson, G. G. 1949. *The Meaning of Evolution*. New Haven, CT: Yale University Press. [31, 283]

Sinervo, B. 1998. Adaptation of maternal effects in the wild: Path analysis of natural variation and experimental tests of causation. Pages 288–306 in *Maternal Effects as Adaptations*, edited by T. A. Mousseau and C. W. Fox. New York: Oxford University Press. [126]

Skutch, A. F. 1987. *Helpers at Bird's Nests: A Worldwide Survey of Cooperative Breeding and Related Behavior*. Iowa City, IA: University of Iowa Press. [62]

Slatkin, M. 1979a. The evolutionary response to frequency- and density-dependent interactions. American Naturalist 114:384–398. [120]

———. 1979b. Frequency- and density-dependent selection in a quantitative character. Genetics 93:755–771. [121]

———. 1980. Ecological character displacement. Ecology 61:163–177. [125]

Slobodkin, L. B., and A. Rapoport. 1974. An optimal strategy of evolution. Quarterly Review of Biology 49:181–200. [180]

Smith, E. 2000. Three styles in the evolutionary analysis of human behavior. In *Adaptation and Human Behavior: An Anthropological Perspective*, edited by L. Cronk, N. Chagnon, and W. Irons. New York: Aldine de Gruyter. [242, 243, 246, 248, 252]

Smouse, P. E. 1976. The implications of density-dependent population growth for frequency- and density-dependent selection. American Naturalist 110:849–860. [121]

Smuts, J. C. 1926. *Holism and Evolution*. New York: Macmillan. [334]

Snyder, J. R. 1984. The role of fire: Mutch ado about nothing? Oikos 43:404–405. [59]

Sober, E., and D. S. Wilson. 1998. *Unto Others: The Evolution and Psychology of Unselfish Behavior*. Cambridge, MA: Harvard University Press. [360]

Sol, D., and L. Lefebvre. 2000. Behavioural flexibility predicts invasion success in birds introduced to New Zealand. Oikos 90:599–605. [325, 349]

Sperber, D. 2000. An objection to the memetic approach to culture. In *Darwinizing Culture: The Status of Memetics as a Science*, edited by R. Aunger. Oxford: Oxford University Press. [259]

Spradbery, J. P. 1973. *Wasps*. London: Sidgwick and Jackson. [62, 89, 92]

Stanley, S. M. 1989. *Earth and Life through Time*. 2nd ed. New York: Freeman. [54]

Stephens, D. W., and J. R. Krebs. 1986. *Foraging Theory*. Princeton, NJ: Princeton University Press. [124]

Sterelny, K. 2001. Niche construction, developmental systems, and the extended replicator. In *Cycles of Contingency: Developmental Systems and Evolution*, edited by S. Oyama, P. E. Griffiths, and R. D. Gray. Cambridge, MA: MIT Press. [30]

———. In press. *Thought in a Hostile World*. New York: Blackwell. [281]

Sterner, R. W. 1995. Elemental stoichiometry of species in ecosystems. In *Linking Species and Ecosystems*, edited by C. G. Jones and J. H. Lawton. New York: Chapman and Hall. [195]

Stiner, M. C., N. D. Munro, and T. A. Surovell. 2000. The tortoise and the hare: Small-game use, the broad-spectrum revolution, and Paleolithic demography. Current Anthropology 41:39–73. [347]

Stout, J. D. 1983. Organic matter turnover by earthworms. In *Earthworm Ecology: From Darwin to Vermiculture*, edited by J. E. Satchell. London: Chapman and Hall. [11]

Straskraba, M. 1995. Cybernetic theory of complex ecosystems. Pages 104–129 in *Complex Ecology: The Part-Whole Relation in Ecosystems*, edited by B. C. Patten and S. E. Jorgensen. Englewood Cliffs, NJ: Prentice Hall. [330]

Stuart, C.I.J.M. 1985. Bio-information equivalence. Journal of Theoretical Biology 113:611–636. [183, 187, 200, 330]

Sundstrom, L., M. Chapuisat, and L. Keller. 1996. Conditional manipulation of sex ratios by ant workers: A test of kin selection theory. Science 274:993–995. [300]

Szathmary, E., and J. Maynard Smith. 1995. The major evolutionary transitions. Nature 374:227–231. [253]

Szilard, L. 1929. Uber die Entropieverminderung in einem thermodynamischen System bei Eingriffen intelligenter Wesen. Zeitschrift für Physik 53:840–856. [1964. On the decrease of entropy in a thermodynamic system by the intervention of intelligent beings. Behavioural Science 9:301–310.] [173]

Tan, Y., and M. A. Riley. 1997. Positive selection and recombination: Major molecular mechanisms in colicin diversification. Trends in Ecology and Evolution 12:348–352. [53, 54]

Tansley, A. G. 1949. *Britain's Green Mantle*. London: George Allen and Unwin. [58]

Tebbich, S., M. Taborsky, B. Febl, and D. Blomqvist. 2001. Do woodpecker finches acquire tool-use by social learning? Proceedings of the Royal Society of London, Series B 268:2189–2193. [21, 258, 263, 296, 377]

Thompson, J. N. 1994. *The Coevolutionary Process*. Chicago: University of Chicago Press. [24, 74, 104, 124, 125, 201, 207]

———. 1998. Rapid evolution as an ecological process. Trends in Ecology and Evolution 13:329–332. [233, 234, 249]

Thompson, N. S. 2000. Niche construction and group selection. Behavioral and Brain Sciences 23:161–162. [281]

Tilman, D. 1982. *Resource Competition and Community Structure*. Princeton, NJ: Princeton University Press. [24, 39, 118, 119, 219]

———. 1986. Resources, competition and the dynamics of plant communities. Pages 51–74 in *Plant Ecology*, edited by M. J. Crawley. Oxford: Blackwell. [118]

———. 1989. Discussion: Population dynamics and species interactions. Pages 89–100 in *Perspectives in Ecological Theory*, edited by J. Roughgarden, R. M. May, and S. A. Levin. Princeton, NJ: Princeton University Press. [201, 202]

———. 1990. Mechanisms of plant competition for nutrients: The elements of a predictive theory of competition. Pages 117–141 in *Perspectives on Plant Competition*, edited by J. B. Grace and D. Tilman. New York: Academic. [106, 118]

———. 1997. Biotic control over the functioning of ecosystems. Science 277:500–504. [60]

———. 2000. Causes, consequences and ethics of biodiversity. Nature 405:208–211. [383]

Tilman, D., and J. A. Downing. 1994. Biodiversity and stability in grasslands. Nature 367:363–365. [328]

Tilman, D., and C. Lehman. 2001. Biodiversity, composition, and ecosystem processes: theory and concepts. In *The Functional Consequences of*

Biodiversity, edited by A. P. Kinzig, S. W. Pacala, and D. Tilman. Princeton, NJ: Princeton University Press. [329]

Tooby, J., and L. Cosmides. 1990. The past explains the present: Emotional adaptations and the structure of ancestral environments. Ethology and Sociobiology 11:375–424. [243, 245]

Trivers, R. L. 1985. *Social Evolution*. Menlo Park, CA: Benjamin/Cummings. [243, 298]

Trivers, R. L., and H. Hare. 1976. Haplodiploidy and the evolution of the social insects. Science 191:249–263. [300, 417]

Troumbis, A. S., and L. Trabaud. 1989. Some questions about flammability in fire ecology. Acta Oecologica—Oecologia Plantarum 10:167–449. [59]

Turing, A. M. 1936. On computable numbers, with an application to the Entscheidungsproblem. Proceedings of the London Mathematical Society, Series 2 42:230–265. [172]

———. 1937. A correction. Proceedings of the London Mathematical Society, Series 2 43:544–546. [172]

Turnbull, A. L. 1964. The search for prey by a web-building spider *Achaearanea tepidariorum* (C. L. Koch) (Araneae Theridiidae). Canadian Entomologist 95:568–1579. [64]

Turner, A., and C. Rose. 1989. *A Handbook to the Swallows and Martins of the World*. London: Christopher Helm. [104]

Turner, J. S. 2000. *The Extended Organism: The Physiology of Animal-Built Structures*. Cambridge, MA: Harvard University Press. [12, 55, 67, 168, 169, 210, 283, 291, 297, 374, 375, 376]

Tylor, E. B. 1871. *Primitive Culture*. London: Murray. [247, 263]

Ulijaszek, S. J., and S. S. Strickland. 1993. *Nutritional Anthropology: Prospects and Perspectives*. London: Smith-Gordon. [341, 342]

Uyenoyama, M., and M. W. Feldman. 1979. Evolutionary effects of contagious and familial transmission. Proceedings of the National Academy of Sciences, USA 76:420–424. [125]

———. 1980. Theories of kin and group selection: A population genetics perspective. Theoretical Population Biology 17:380–414. [300, 360]

———. 1981. On relatedness and adaptive topography in kin selection. Theoretical Population Biology 19:87–123. [415]

Uyenoyama, M., M. W. Feldman, and L. D. Mueller. 1981. Population genetic theory of kin selection: Multiple alleles at one locus. Proceedings of the National Academy of Sciences, USA 78:5036–5040. [360]

Van Valen, L. 1973. A new evolutionary law. Evolutionary Theory 1:1–30. [131]

Verner, J., and G. H. Englesen. 1970. Territories, multiple nesting, and polygyny in the long-billed marsh wren. Auk 87:557–567. [98]

Via, S. 2001. Sympatric speciation in animals: The ugly duckling grows up. Trends in Ecology and Evolution 16:381–390. [302, 303]

Vilà, C., P. Savolainen, J. E. Maldonado, I. R. Amorim, J. E. Rice, R. L.

Honeycutt, K. A. Crandall, J. Lundeberg, and R. K. Wayne. 1997. Multiple and ancient origins of the domestic dog. Science 276:1687–1689. [346]

Vincent, T.L.S., D. Scheel, J. S. Brown, and T. L. Vincent. 1996. Trade-offs and coexistence in consumer-resource models: It all depends on what and where you eat. American Naturalist 148:1038–1058. [119]

Vitousek, P. M. 1986. Biological invasions and ecosystem properties: Can species make a difference? In *Ecology of Biological Invasions of North America and Hawaii*, edited by H. A. Mooney and J. A. Drake. New York: Springer-Verlag. [200, 236, 320]

Vitousek, P. M., H. A. Mooney, J. Lubchenco, and J. M. Melillo. 1997. Human domination of earth's ecosystems. Science 277:494–499. [202]

Vollrath, F. 1988. Untangling the spider's web. Trends in Ecology and Evolution 3:331–335. [17]

———. 1992. Analysis and interpretation of orb spider exploration and web building behavior. Advances in the Study of Behavior 21:147–199. [17]

von Neumann, J. 1956. Probabilistic logics and the synthesis of reliable organisms from unreliable components. Pages 43–98 in *Automata Studies*. Princeton, NJ: Princeton University Press. [172n]

———. 1966. In *The Theory of Self-Reproducing Automata*, edited by A. W. Burks. Urbana, IL: University of Illinois Press. [172n]

Vrba, E. S. 1992. Mammals as a key to evolutionary theory. Journal of Mammalogy 73:1–28. [288]

Waddington, C. H. 1959. Evolutionary adaptation. Pages 381–402 in *Evolution after Darwin*, edited by S. Jax. Chicago: University of Chicago Press. [29, 383]

———. 1969. Paradigm for an evolutionary process. Pages 106–128 in *Towards a Theoretical Biology*, edited by C. H. Waddington. Edinburgh, Scotland: Edinburgh University Press. [29]

Wade, M. 1998. The evolutionary genetics of maternal effects. Pages 5–21 in *Maternal Effects as Adaptations*, edited by T. A. Mousseau and C. W. Fox. New York: Oxford University Press. [11, 125, 126]

Warner, R. R. 1988. Traditionality of mating site preferences in a coral reef fish. Nature 335:719–721. [103]

Wcislo, W. T. 1989. Behavioral environments and evolutionary change. Annual Review of Ecology and Systematics 20:137–169. [30, 60]

Weatherby, A., P. H. Warren, and R. Law. 1998. Coexistence and collapse: An experimental investigation of the persistent communities of a protist species pool. Journal of Animal Ecology 67:554–566. [324]

Werner, E. E. 1992. Individual behavior and higher-order species interactions. American Naturalist 140 (supplement):S5–S32. [101]

West, M. J., A. P. King, and A. A. Arberg. 1988. The inheritance of niches: The role of ecological legacies in ontogeny. Pages 41–62 in *Handbook of Behavioral Neurobiology*, edited by E. Blass. New York: Plenum. [30]

West, N. E. 1990. Structure and function of microphytic soil crusts in wildland ecosystems of arid to semi-arid regions. Advances in Ecological Research 20:180–223. [7]

West-Eberhard, M. J. 1987. Flexible strategy and social evolution. In *Animal Societies: Theories and Facts*, edited by Y. Brown and J. Kikkawa. Tokyo: Japan Scientific Societies Press. [30, 353]

Whitlock, M. C. 1995. Variance-induced peak shifts. Evolution 46:1674–1697. [301]

———. 1997. Founder effects and peak shifts without genetic drift: Adaptive peak shifts occur easily when environments fluctuate slightly. Evolution 51:1044–1048. [301]

Williams, G. C. 1966. *Adaptation and Natural Selection: A Critique of Some Current Evolutionary Thought*. Princeton, NJ: Princeton University Press. [300, 360]

———. 1992. Gaia, nature worship, and biocentric fallacies. Quarterly Review of Biology 67:479–486. [289]

Williamson, G. B. 1990. Allelopathy. Koch's postulates and the neck riddle. In *Perspectives on Plant Competition*, edited by J. B. Grace and D. Tilman. San Diego, CA: Academic. [59, 60]

Williamson, G. B., and E. M. Black. 1981. High temperature of forest fires under pines as a selective advantage over oaks. Nature 293:643–644. [60]

Wilson, A. C. 1985. The molecular basis of evolution. Scientific American 253(3):148–157. [30, 353 354]

Wilson, D. S., and E. Sober. 1994. Reintroducing group selection to the human behavioral sciences. Behavioral and Brain Sciences 17:585–654. [360]

Wilson, E. O. 1975. *Sociobiology: The New Synthesis*. Cambridge, MA: Harvard University Press. [27, 243, 245]

———. 1994. *Naturalist*. New York: Warner. [246]

Wilson, J. B., and A.D.Q. Agnew. 1992. Positive-feedback switches in plant communities. Advances in Ecological Research 23:263–336. [58]

Wolf, J. B. 2000. Indirect genetic effects and gene interactions. In *Epistasis and the Evolutionary Process*, edited by J. B. Wolf, E. D. Brodie III, and M. J. Wade. Oxford: Oxford University Press. [128]

Wolf, J. B., E. D. Brodie III, J. M. Cheverud, A. J. Moore, and M. J. Wade. 1998. Evolutionary consequences of indirect genetic effects. Trends in Ecology and Evolution 13:64–69. [23, 30, 127, 128, 221, 284, 302]

Wolf, J. B., E. D. Brodie III, and M. J. Wade, eds. 2000. *Epistasis and the Evolutionary Process*. Oxford: Oxford University Press. [21, 127, 284]

Wood, B., and A. Brooks. 1999. We are what we ate. Nature 400:219–220. [346]

Wrangham, R. W., J. H. Jones, G. Laden, D. Pilbeam, and N. Conklin Brittain. 1999. The raw and the stolen: Cooking and the ecology of human origins. Current Anthropology 40:567–594. [242, 344, 346, 353]

Wright, S. 1984. *Evolution and the Genetics of Populations*. Vol. 2, *The Theory of Gene Frequencies*. Chicago: University of Chicago Press. [120]

Zaady, E., and M. Shachak. 1994. Microphytic soil crust and ecosystem leakage in the Negev desert. American Journal of Botany 81:109. [7]

Zedler, P. H. 1995. Are some plants born to burn? Trends in Ecology and Evolution 10:393–395. [59]

Index

MONOGRAPHS IN POPULATION BIOLOGY

EDITED BY SIMON A. LEVIN AND HENRY S. HORN

1. *The Theory of Island Biogeography*, by Robert H. MacArthur and Edward O. Wilson
2. *Evolution in Changing Environments: Some Theoretical Explorations*, by Richard Levins
3. *Adaptive Geometry of Trees*, by Henry S. Horn
4. *Theoretical Aspects of Population Genetics*, by Motoo Kimura and Tomoko Ohta
5. *Populations in a Seasonal Environment*, by Steven D. Fretwell
6. *Stability and Complexity in Model Ecosystems*, by Robert M. May
7. *Competition and the Structure of Bird Communities*, by Martin L. Cody
8. *Sex and Evolution*, by George C. Williams
9. *Group Selection in Predator-Prey Communities*, by Michael E. Gilpin
10. *Georgraphic Variation, Speciation, and Clines*, by John A. Endler
11. *Food Webs and Niche Space*, by Joel E. Cohen
12. *Caste and Ecology in the Social Insects*, by George F. Oster and Edward O. Wilson
13. *The Dynamics of Arthropod Predator-Prey Systems*, by Michael P. Hassel
14. *Some Adaptations of Marsh-Nesting Blackbirds*, by Gordon H. Orians
15. *Evolutionary Biology of Parasites*, by Peter W. Price
16. *Cultural Transmission and Evolution: A Quantitative Approach*, by L. L. Cavalli-Sforza and M. W. Feldman
17. *Resource Competition and Community Structure*, by David Tilman
18. *The Theory of Sex Allocation*, by Eric L. Charnov
19. *Mate Choice in Plants: Tactics, Mechanisms, and Consequences*, by Nancy Burley and Mary F. Wilson
20. *The Florida Scrub Jay: Demography of a Cooperative-Breeding Bird*, by Glen E. Woolfenden and John W. Fitzpatrick

21. *Natural Selection in the Wild*, by John A. Endler
22. *Theoretical Studies on Sex Ratio Evolution*, by Samuel Carlin and Sabin Lessard
23. *A Hierarchial Concept of Ecosystems*, by R. V. O'Neill, D. L. DeAngelis, J. B. Waide, and T.F.H. Allen
24. *Population Ecology of the Cooperatively Breeding Acorn Woodpecker*, by Walter D. Koening and Ronald L. Mumme
25. *Population Ecology of Individuals*, by Adam Lomnicki
26. *Plant Strategies and the Dynamics and Structure of Plant Communities*, by David Tilman
27. *Population Harvesting: Demographic Models of Fish, Forest, and Animal Resources*, by Wayne M. Getz and Robert G. Haight
28. *The Ecological Detective: Confronting Models with Data*, by Ray Hilborn and Marc Mangel
29. *Evolutionary Ecology across Three Trophic Levels: Goldenrods, Gallmakers, and Natural Enemies*, by Warren G. Abrahamson and Arthur E. Weis
30. *Spatial Ecology: The Role of Space in Population Dynamics and Interspecific Interactions*, edited by David Tilman and Peter Kareiva
31. *Stability in Model Populations*, by Laurence D. Mueller and Amitabh Joshi
32. *The Unified Neutral Theory of Biodiversity and Biogeography*, by Stephen P. Hubbell
33. *The Functional Consequences of Biodiversity: Emperical Progress and Theoretical Extensions*, edited by Ann P. Kinzig, Stephen J. Pacala, and David Tilma
34. *Communities and Ecosystems: Linking the Aboveground and Belowground Components*, by David A. Wardle
35. *Complex Population Dynamics: A Theoretical/Empirical Synthesis*, by Peter Turchin
36. *Consumer-Resource Dynamics*, by William Murdoch, Cherie Briggs, and Roger Nisbet
37. *Niche Construction: The Neglected Process in Evolution*, by F. John Odling-Smee, Kevin N. Laland, and Marcus W. Feldman